D1595902

Monographs in Computer Science

Editors

David Gries
Fred B. Schneider

Monographs in Computer Science

Abadi and Cardelli, **A Theory of Objects**

Alagic, **Rational Database Technology**

Bauer and Wössner, **Algorithmic Language and Program Development**

Benosman and Kang [editors], **Panoramic Vision: Sensors, Theory, and Applications**

Bhanu, Lin, and Krawiec, **Evolutionary Synthesis of Pattern Recognition Systems**

Bronstein, Bronstein, and Kimmel, **Numerical Geometry of Non-Rigid Shapes**

Broy and Stolen, **Specification and Development of Interactive Systems**

Brzozowki and Seger, **Asynchronous Circuits**

Burgin, **Super-Recursive Algorithms**

Cantone, Omodeo, Policriti, and Schwartz, **Set Theory for Computing: From Decision Procedures to Declarative Programming with Sets**

Downey and Fellows: **Parameterized Complexity**

Feijen, Gasteren, Gries, and Misra [editors], **Beauty is our Business: A Birthday Salute to Edsger W. Dijkstra**

Feijen and Gasteren, **On a Method of Multiprogramming**

Gries, **The Science of Programming**

Gries and Schneider, **A Logical Approach to Discrete Math**

Grune and Jacobs, **Parsing Technique: A Practical Guide**

Hehner, **A Practical Theory of Programming**

Herbert and Spärck Jones [editors], **Computer Systems: Theory, Technology, and Applications**

Heydon, Levin, Mann, and Yu, **Software Configuration Management Using Vesta**

Kozen, **The Design and Analysis of Algorithms**

McIver and Morgan, **Abstraction, Refinement, and Proof for Probabilistic Systems**

McIver and Morgan [editors], **Programming Methodology**

Mishra, **Algorithmic Algebra**

Misra, **A Discipline of Multiprogramming: Programming Theory for Distributed Applications**

Partsch, **Specification and Transformation of Programs: A Formal Approach to Software Development**

Paton [editor], **Active Runes in Database Systems**

Poernomo, Crossley, and Wirsing, **Adapting Proofs-as-Programs: The Curry-Howard Protocol**

Preparata and Shamos, **Computational Geometry: An Introduction**

Randell [editor], **The Origins of Digital Computers: Selected Papers**

Reps and Teitelbaum, **The Synthesizer Generator Reference Manual**

Selig, **Geometric Fundamentals of Robotics**

Sasha and Zhu, **High Performance Discovery in Time Series: Techniques and Case Studies**

Tonella and Potrich, **Reverse Engineering of Object Oriented Code**

Vitányi, **Algorithmic Statistics**

Alexander M. Bronstein
Michael M. Bronstein
Ron Kimmel

Numerical Geometry of Non-Rigid Shapes

With 10 Color Figures

Alexander M. Bronstein
Technion-Israel Institute of Technology
Haifa, Israel
bron@cs.technion.ac.il

Michael M. Bronstein
Technion-Israel Institute of Technology
Haifa, Israel
mbron@cs.technion.ac.il

Ron Kimmel
Technion-Israel Institute of Technology
Haifa, Israel
ron@cs.technion.ac.il

ISSN: 0172-603x
ISBN: 978-0-387-73300-5 e-ISBN: 978-0-387-73301-2
DOI 10.1007/978-0-387-73301-2

Library of Congress Control Number: 2008934481

Mathematics Subject Classification (2000): 53AØ5, 52C25, 49M37

© 2008 Springer Science+Business Media, LLC
All rights reserved. This work may not be translated or copied in whole or in part without the written permission of the publisher (Springer Science+Business Media, LLC, 233 Spring Street, New York, NY 10013, USA), except for brief excerpts in connection with reviews or scholarly analysis. Use in connection with any form of information storage and retrieval, electronic adaptation, computer software, or by similar or dissimilar methodology now known or hereafter developedis forbidden.
The use in this publication of trade names, trademarks, service marks and similar terms, even if they are not identified as such, is not to be taken as an expression of opinion as to whether or not they are subject to proprietary rights.

Printed on acid-free paper

springer.com

The mediocre teacher tells.
The good teacher explains.
The great teacher inspires.

W. A. WARD

Dedicated to our teachers.

Foreword

Once upon a time, a child imagined a fierce boa constrictor that swallowed an elephant, giving rise to a most peculiar shape. The child made a drawing of it and showed it to grown-ups around him, expecting them to be appalled by the scene. Alas, nobody saw in his drawing more than a hat. The child then proceeded to draw an explanatory drawing showing the elephant inside the snake's expansible stomach...

Non-rigid objects are all around us, from snakes to octopuses, from ropes to the pages of this book, from the surface of the sea to the pudding in our plates, and we have no particular problems in dealing with them in our daily lives. Yet the mathematical tools we have for their description and analysis are few, and only relatively recently researchers in graphics and computer vision have started paying due attention to them. This book offers a rare opportunity to encounter the fascinating world of the flexible, elastic, plastic, and amorphous form and shape through the looking glass of their mathematical representation and numerical treatment by our miraculous and powerful computing machines.

I hope that after reading this book, you, too, just like the Little Prince, will start to see in Saint-Exupery's childhood drawing the elephant inside the boa constrictor, rather than the obvious, boring and rigid hat.

Alfred M. Bruckstein
Ollendorff Professor of Science
Technion, Haifa, Israel

About the Authors

Ron Kimmel is a researcher in the areas of computer vision, image processing, and numerical geometry. He received his Ph.D. from the Department of Electrical Engineering, Technion – Israel Institute of Technology in 1995 and spent the post-doctoral years (1995–1998) at the University of California in Berkeley. In 2003–2004, he was a visiting professor the Stanford University. Currently, he is a professor in the Department of Computer Science, Technion.

Ron has authored more than one hundred twenty articles in scientific journals and conferences and the book *Numerical Geometry of Images*. He is a member of the editorial board of *International Journal of Computer Vision* and of *IEEE Transactions on Image Processing* and was the co-chair of the *International Conference on Scale-Space Methods in Computer Vision* in 2001 and 2005 and of the *SIAM Conference on Imaging* in 2006. Prof. Kimmel is the recipient of the Hershel Rich Technion Innovation Award (twice), the Taub Prize for Excellence in Research, the Gensler Counter-Terrorism Prize, and the Alon Fellowship. He was a consultant to HP Labs and to Net2Wireless/Jigami research and currently serves on the advisory board of MediGuide.

Alexander and Michael Bronstein met Ron in 2002, taking his class on *Numerical Geometry of Images* when still undergraduate students. According to Michael's recollection, the first phrase Prof. Kimmel referred to him was "usually I throw such students out of the class" when his cellular phone accidentally rang during the lecture. Shortly afterwards, Alex and Michael became Ron's graduate students, working together on analysis of non-rigid objects, computer vision, and image processing problems. They each received a Ph.D. from the Technion in 2007.

The Bronstein twins are the alumni of the Technion Excellence Program and the Academy of Achievement. They shared numerous awards, including the Technion Humanities and Arts Department prize, Kasher prize, Thomas Schwartz award, Hershel Rich Technion Innovation award, Gensler Counter-Terrorism Prize, and the Best Paper award at the *Copper Mountain Conference on Multigrid Methods*. In 2003, their research on three-dimensional face recognition was a hit item in CNN news and other international media.

Michael and Alex were awarded the Adams Fellowship by the Israel Academy of Science in 2005 and 2006, respectively.

Besides this book, Alex, Michael, and Ron co-authored more than a dozen journal and conference papers, dedicated mostly to the analysis of three-dimensional non-rigid shapes, and co-founded Novafora Inc. Among colleagues and friends, the harmonic trio is frequently referred to as "B^2K."

Preface

Looking around, we notice that our world is full of objects that, due to their physical properties, are non-rigid and therefore can be deformed and bent. Non-rigid shapes appear at all scales in Nature – from the human body, its organs and tissues, to tiny bacteria and microscopic protein molecules. Being so ubiquitous, such shapes are often encountered in pattern recognition and computer vision applications. The richness of the possible deformations of non-rigid shapes appears to be a nightmare for a pattern recognition researcher, who faces a vast number of degrees of freedom when trying to analyze them. For this reason, explicit analysis of non-rigid objects has been avoided for a long period in computer vision, and as it often happens in applied sciences, research has focused on simplified problems that are easier to solve.

Today, there is a gradually penetrating comprehension that in many applications the necessity to model and understand non-rigid objects is unavoidable. At the same time, recent research results have shown that problems related to non-rigid objects are not necessarily untractable. We decided to write this book because we believed that a critical mass of research has accumulated, making the field of non-rigid shape analysis sufficiently profound on one hand and having significant open research questions on the other.

In some sense, the book can be considered as a sequel to *Numerical Geometry of Images*, which focused mainly on geometric methods in image processing and analysis. In *Numerical Geometry of Non-rigid Shapes*, as the title suggests, our main theme is two- and three-dimensional non-rigid objects. We invite the reader to join us for a fascinating journey to the non-rigid world, a rapidly developing field at the crossroad of computer vision, pattern recognition, and geometry, where the last word has not yet been spoken.

Intended use

This book was initially written as lecture notes for a one-semester monographic graduate course taught at the Technion – Israel Institute of Technology.

The course was based on a corpus of recent research results, and its leitmotif was intrinsic geometric invariants and embedding methods. However, because the field of non-rigid object analysis is multi-disciplinary, bringing together theoretical and numerical geometry, optimization, graph theory, machine learning, and computer graphics, it is almost impossible to expect students to have all the necessary background. Therefore, we had to extend and elaborate the text, including a gradual introduction of the material required for understanding the mathematical machinery we use throughout the book. The reader is assumed to have basic knowledge of calculus and algebra and can acquire the required background through reading the introductory chapters.

The book is intended as a graduate-level textbook for engineers, computer scientists, and applied mathematicians. For teachers, it can be the main reference for a monographic course on non-rigid shape analysis or a supplementary material for various courses in computer vision and pattern recognition, geometry processing, computational and numerical geometry, and computer graphics. For students, the book can be both course material and a self-study reference. For mature experts and specialists in the field, the book offers frontline methods and most recent results, as well as less traditional points of view on old problems and approaches. In the book and supporting online material, the reader can find numerical recipes, ready to use codes and references to public domain and commercial software.

Features

When working on the book, we set ourselves to three main goals. First, we tried to make the book as self-contained as possible. Given the breadth of the material covered, it is obvious that in-depth study of all the fields could easily spread over ten other books. One of the most difficult tasks was to cherry-pick only the relevant material and present it consistently. Second, we were convinced that any material, however complicated, could be simply presented. We therefore tried to maintain simplicity of explanation throughout the book, often resorting to illustrative descriptions, drawing inspiration in realms ranging from fairy tales to cartography. We believe this made the reading even of the driest technical material somewhat more entertaining. Finally, a special emphasis was made on practical applications, as our largest audience is engineers and applied scientists who would like to see things work at the end of the day. We tried to keep the right balance between the simplicity of explanation and the amount of details necessary for implementation of the discussed numerical algorithms. For those willing to dive deeper into additional details, we provide references to related books and research papers. In order to increase the practical value of the book, we also provide references to public domain and commercial software.

Each chapter includes references to implementation of some of the discussed algorithms and a list of suggested literature for those interested in wider and deeper understanding. We believe this will be especially appreciated by readers using the book for self-education. In order to make the discussion as focused as possible, we omit certain details, concentrating them into notes at the end of each chapter. Proofs of some results are formulated as problems, on which the reader may test his or her understanding of the material. Solutions of selected problems appear at the end of the book; other problems are left as an exercise to the reader, which may help in using the book as supplementary class material. The level of problems varies significantly, from exercises involving basic calculus to open research questions. To leverage the background and terminology and facilitate the reading, we provide a glossary with the definitions of the most frequently used terms. At the end of the book, the reader can find a subject and an author index.

Synopsis

The first chapters are dedicated to the mathematical background necessary to create a common language and terminology. In many books, the theoretical introduction is often apparently unrelated to the problems discussed afterwards. We tried to avoid this impression, targeting our examples and illustrations at the problems we solve, thereby motivating the background chapters.

Our model of non-rigid world stands on three pillars: metric geometry, discrete geometry, and numerical optimization. We begin our mathematical introduction (Chapter 2) with geometry. Geometry of surfaces is typically presented either from the point of view of topology or by resorting to the heavy apparatus of calculus and differential geometry. This creates an apparent gap: whereas topology is too "crude" and usually does not satisfy our needs in studying surfaces, differential geometry is often cumbersome and poorly accessible by the average reader. We preferred to follow a somewhat less orthodox path and present the *metric* point of view, using notions as basic as distance and length. This, following Dmitry Burago's expression, brings geometry back "down-to-earth" where it traditionally, and literally, began. In our introduction to geometry, we explore the difference between *intrinsic* and *extrinsic* geometry, the notion of *isometry*, and invariant description of shapes. Chapters 3 and 4 present the geometric foundations through the glasses of a computer scientist, who has to convert the continuous geometric world into discrete objects tractable by a computer. Much attention in Chapter 4 is dedicated to *fast marching*, a class of efficient numerical algorithms for geodesic distance measurement. The third pillar, numerical optimization, is presented in Chapter 5.

Starting from Chapter 6, applications begin to appear. To ease the entrance into the non-rigid world, we first study a simple problem of rigid object analysis, which has a smaller number of degrees of freedom. Our emphasis in this chapter is on *iterative closest point* algorithms, a class of numerical methods for rigid object comparison and matching. In Chapters 7, 8, and 9, we discuss ways of creating deformation-invariant representations of non-rigid shapes. The main focus is on *multidimensional scaling* methods, first developed in psychology in the 1960s for multidimensional data analysis, and more recently adapted to non-rigid surface matching. Chapter 10 is devoted to problems of shape similarity, which is addressed by constructing an ideal deformation-invariant distance. We encounter the Gromov-Hausdorff distance, introduced in the early 1980s by Mikhail Gromov and first used in pattern recognition applications by Facundo Mémoli and Guillermo Sapiro in 2005, and study its theoretical properties and approaches for approximate computation. In Chapter 11, we discuss *partial similarity*, the problem of comparing shapes having similar parts. Chapter 12 deals with the problem of correspondence between non-rigid shapes. This problem is presented following the same line of shape similarity. In Chapter 13, we study the problem of face recognition, one of the most challenging computer vision applications. Face recognition appears as a playground to demonstrate and summarize the tools discussed in the book. Finally, Chapter 14 gives a short retrospective and concludes the book.

How to use this book

The book is largely structured in such a way that new material builds up progressively and thus it can be read continuously. Some sections not essential for the understanding of the entire material or those containing challenging mathematics are denoted by ⋆. These sections can be skipped without sacrificing future understanding.

Readers new to the field and those using the book for self-education are recommended to start with the introductory chapters. Experts in the field may be interested in focused topics offering new insight and state-of-the-art methods. Those include parallel fast marching (Chapter 4), multigrid and vector extrapolation accelerated multidimensional scaling (Chapter 7), generalized multidimensional scaling (Chapter 9), Gromov-Hausdorff distance (Chapter 10), Pareto similarity (Chapter 11), and expression-invariant face recognition (Chapter 13).

Teachers may use the book for a monographic course on analysis and synthesis of non-rigid shapes, using the flow of the chapters. Detailed study of examples, some technical descriptions, and solutions to selected problems (marked with ✓) can be a basis for tutorials. Problems without solutions are a potential source for home assignments. Alternatively, Chapters 4, 9, and 12 may be used as supplementary material for computer graphics related courses and Chapters 7, 8, 10, and 11 for enrichment of computer vision and pattern recognition curricula.

Acknowledgment

The publication of this book would not have been possible without the help, advice, and support of many people to whom we are deeply indebted. Particularly, we would like to mention Ronen Basri, Alexander Brook, Alfred Bruckstein, Dmitry Burago, Laurent Cohen, Mathieu Desbrun, David Donoho, Yohai Dvir, Ilya Eckstein, Michael Elad, Gene Golub, Hugues Hoppe, Martin Kilian, Nahum Kiryati, Vladimir Lin, Nathan Linial, Facundo Mémoli, Gérard Medioni, Davide Migliore, Gabriel Peyré, Boaz Porat, Helmut Pottmann, Dan Raviv, Guy Rosman, Guillermo Sapiro, Michael Saunders, Peter Schröder, James Sethian, Avraham Sidi, Nir Sochen, Alla Sheffer, Alon Spira, Pierluigi Taddei, Ofir Weber, Irad Yavneh, and Michael Zibulevsky, whose valuable comments helped us shape the manuscript at different stages of its life. Special thanks to Constantine Hadavas, Alex Hubris, Idan Shatz, Ayellet Tal, and Malene Thyssen for the permission to reproduce some of the illustrations used in the book. We are also grateful to the Springer team, Vaishali Damle and Ann Kostant, for helping us create the book you are currently reading.

Feedback and support

Because non-rigid shape analysis is an active and fast evolving research field, *Numerical Geometry of Non-rigid Shapes* is intended as a living text. We will do our best to keep the book up to date. Updates, new materials, data, and software will be published on the website `tosca.cs.technion.ac.il/book`. Comments, corrections, suggestions, and interesting solutions to problems would always be highly appreciated.

Haifa and Santa Clara,   *Alexander M. Bronstein*
April 2008   *Michael M. Bronstein*
                                                                                           Ron Kimmel

Contents

Foreword VII

About the Authors IX

Preface XI

1   **Introduction** 1
    1.1   Similarity of non-rigid shapes 3
    1.2   Correspondence problems 6
    1.3   A landscape of problems 7
    Notes 9

2   **A Taste of Geometry** 11
    2.1   Basic terms in metric geometry and topology 11
    2.2   Isometries 13
    2.3   Length spaces 17
    2.4   Manifolds 20
    2.5   Embedded surfaces 21
    2.6   Curvature and the second fundamental form 26
    2.7   Intrinsic view on geometry of surfaces 29
    2.8   Bending and rigidity 31
    2.9   Intrinsic invariants 34
    Suggested reading 37
    Problems 37
    Notes 39

3 Discrete Geometry 41
- 3.1 Point clouds and sampling 41
- 3.2 Farthest point sampling 43
- 3.3 Voronoi tessellation 46
- 3.4 Centroidal Voronoi sampling and the Lloyd-Max algorithm ... 48
- 3.5 Connectivity 52
- 3.6 Delaunay tessellation 53
- 3.7 Triangular meshes 54
- 3.8* Local feature size and curvature-dependent sampling 57
- 3.9* Approximation quality 61
- Suggested reading 63
- Software 63
- Problems 63
- Notes 64

4 Shortest Paths and Fast Marching Methods 67
- 4.1 The shortest path problem 67
- 4.2 Dijkstra's shortest path algorithm 69
- 4.3 Fast marching methods 71
- 4.4 Fast marching on parametric surfaces 81
- 4.5 Marching even faster 83
- 4.6 Parallel distance computation 85
- 4.7* Minimal geodesics 87
- Suggested reading 89
- Software 90
- Problems 90
- Notes 91

5 Numerical Optimization 93
- 5.1 Local versus global optimization 93
- 5.2 Optimality conditions 94
- 5.3 Unconstrained optimization algorithms 97
- 5.4 The quest for a descent direction 100
- 5.5 Preconditioning 104
- 5.6 Let Newton be! 105
- 5.7 Truncated Newton 106
- 5.8 Quasi-Newton algorithms 107
- 5.9 Non-convex optimization 108
- 5.10 Constrained optimization 110
- 5.11 Penalty and barrier methods 112
- 5.12* Augmented Lagrangian method 114
- Suggested reading 116
- Software 116
- Problems 116
- Notes 118

6 In the Rigid Kingdom 119
6.1 Moments of joy, moments of sorrow 120
6.2 Iterative closest point algorithms 125
6.3 Enter numerical optimization 128
6.4 Rigid correspondence 131
Suggested reading 133
Software 133
Problems 133
Notes 134

7 Multidimensional Scaling 137
7.1 Isometric embedding problem 138
7.2 Multidimensional scaling 142
7.3 SMACOF algorithm 143
7.4* Second-order methods 146
7.5 Variations on the stress theme 148
7.6 Multiresolution methods 153
7.7* Multigrid MDS 156
7.8* Vector extrapolation 160
7.9 A trouble with topology 164
Suggested reading 165
Software 166
Problems 167
Notes 167

8 Spectral Embedding 169
8.1 Classic MDS 170
8.2 Local methods 173
8.3 The Laplace-Beltrami operator 176
8.4 To hear the shape of the drum 178
8.5* Discrete Laplace-Beltrami operator 180
Suggested reading 184
Software 184
Problems 184
Notes 185

9 Non-Euclidean Embedding 187
9.1 Spherical embedding 187
9.2 Generalized multidimensional scaling 192
9.3 Representation issues 194
9.4 Geodesic distance computation 197
9.5 Minimization of the generalized stress 198
9.6 Multiresolution encore 202
Suggested reading 203
Software 203

XX   Contents

    Problems 204
    Notes. 204

10 Isometry-Invariant Similarity 205
    10.1 Equivalence, similarity, and distance 205
    10.2 Embedding distance 207
    10.3 Gromov-Hausdorff distance 208
    10.4 Intrinsic symmetry 211
    Suggested reading 214
    Problems 214

11 Partial Similarity 217
    11.1 Recognition by parts 218
    11.2 Paretian approach to partial similarity 221
    11.3 Scalar partial similarity 224
    11.4 Fuzzy approximation. 226
    11.5 Extrinsic partial similarity 229
    11.6 Intrinsic partial similarity 230
    11.7 Not only size matters 232
    Suggested reading 236
    Problems 236
    Notes. 238

12 Non-rigid Correspondence and Calculus of Shapes 239
    12.1 Intrinsic parameterization 240
    12.2 An image processing approach 241
    12.3 Minimum distortion correspondence 244
    12.4 Texture mapping and transfer 246
    12.5 Morphing 249
    12.6*Guaranteed self-intersection free morph 254
    12.7 Calculus of shapes 255
    Suggested reading 258
    Software 258
    Problems 259
    Notes. 259

13 Three-dimensional Face Recognition 261
    13.1 Some terminology 263
    13.2 A retrospective. 264
    13.3 Isometric model of facial expressions 268
    13.4 Expression-invariant face recognition 269
    13.5 Comparison of photometric properties. 273
    Suggested reading 275
    Notes. 275

14 Epilogue 277

Solutions of Selected Problems 279

Software 293

Notation 297

Acronyms 299

Glossary 301

References 307

Subject Index 327

Author Index 335

1
Introduction

> - Do you see that cloud, that's almost in shape like a camel?
> - By the mass, and 't is like a camel, indeed.
> - Methinks, it is like a weasel.
> - It is backed like a weasel.
> - Or, like a whale?
> - Very like a whale.
>
> W. SHAKESPEARE, *Hamlet*

Analysis and understanding of shapes is one of the most fundamental tasks in our interaction with the surrounding world. When we see a picture, we *understand* it by recognizing the depicted objects and relating them to concepts we have learned throughout our lives. An evidence to such an understanding is the fact that we can translate a picture into a higher-level semantic description and communicate this description to another person. Hearing about a "beautiful house made of rosy brick, with geraniums in the windows and doves on the roof" [131], most of us would create a vivid mental image of it in our imagination, that is, in some sense perform the inverse process of synthesis of the picture from its semantic description.

In our everyday experience, we are often unaware of how extremely complex the shape analysis performed by the brain is, because it is done mostly subconsciously, without involving the higher level of cognition. Just imagine how different and difficult our lives would be if this ability were gone. Waking up in the morning, we would meet a lot of new people – their faces would seem unfamiliar, because we recognize humans by the visual features of their faces. Going into the bathroom would become an adventure of epic proportions, as we would be unable to decide whether the object we pick up is a razor or a toothbrush, because such a decision requires the analysis of the geometric shape of these objects. Even continuing reading this book would become impossible, because the shapes of the letters would lose their meaning, and as a result, we would lose our ability to read. Of course, this apocalyptic scenario is exaggerated, as we do not rely only on visual information in our lives, and other senses could somehow compensate for its loss. Yet, for most of us, the most significant information about the surrounding world comes from vision, such that, rephrasing an old proverb, a picture is worth a thousand odors or touches.

In the era of computers, attempts to imitate the ability of the human visual system to understand shapes gave birth to the fields of computer vision and pattern recognition. When we say "shape" in this context, we usually imply a visual representation of the object rather than the object itself. Because

Figure 1.1. Cave paintings are the one of the earliest evidences of the interest of our prehistoric fathers in the understanding of shapes and their use as a method of communicating information. Shown here are animal shapes from Salle des Taureaux (Grotte de Lascaux, France), dating back to around 15,000 BC.

what we see is actually a two-dimensional picture perceived by our eyes, a common way to think of a shape is of a two-dimensional projection of the three-dimensional object. However, computers "see" the world differently from us humans. A computer's "eye" can be an ultrasonic sensor introduced into the human body by an intravascular catheter, or a hyperspectral camera mounted on a Martian rover crawling through the canyons of the red planet. Such sensors may provide information normally imperceivable by humans, such as sub-millimeter accurate measurement of object dimensions, or colors in spectral ranges unseen by our eyes. Accordingly, in some computer vision applications, besides the common two-dimensional shapes, we can encounter shapes represented as three-dimensional point clouds, triangular meshes, or parametric or implicit surfaces.

Usually, some model that relates the shapes to the underlying objects is assumed. We call the model of shapes used throughout this book the *non-rigid world*. In this world, objects have a certain degree of flexibility by virtue of their natural properties. Consequently, we may find a great variety of shapes produced as a result of deformations of a non-rigid object. Being able to analyze the properties of such shapes and describe their behavior is the key to understanding the non-rigid world.

Figure 1.2. The legendary Russian ballerina Maya Plisetskaya and a stretching Siberian tiger are living examples of non-rigid objects we will encounter in this book (tiger photo reproduced by courtesy of Malene Thyssen).

1.1 Similarity of non-rigid shapes

In the epigraph we chose to open the book, we quote a renowned scene from the third act of Shakespeare's *Hamlet*, in which the Prince of Denmark and Polonius argue about the shape of clouds they see from the windows of the Elsinor Castle.[1] Speaking in modern language, the topic of the two noblemen's dialogue is *non-rigid shape similarity*, or how to compare shapes that are susceptible to deformations. Clouds are only one example of such shapes; we see plenty of other non-rigid shapes in the world surrounding us at all scales.

Generally speaking, in the problem of shape similarity we are looking for a quantitative measure of "distance" between two shapes: if this distance is small, we conclude that the shapes are similar. The dispute between Hamlet and Polonius teaches us that the definition of similarity may be subjective: depending on the criterion used for similarity, one can recognize the same shape of a cloud as a camel, a weasel, or a whale.

In the non-rigid world, this problem is especially acute due to the fact that an object can assume many forms as a result of its deformations. The same object deformed in different ways may result in shapes that are apparently dissimilar. As an illustration, we resort to the example of a non-rigid object

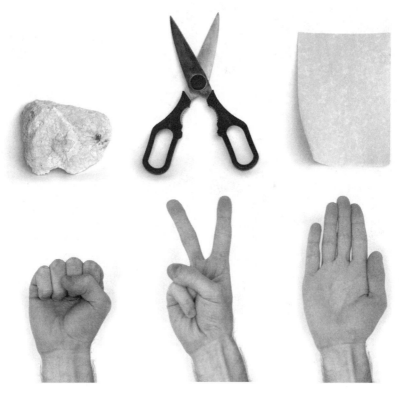

Figure 1.3. Depending on our definition of similarity, we may recognize hand postures used in the Rock, Paper, Scissors game either as the "objects" they mean to imitate (top) or as postures of a hand (bottom).

literally available in our hands: the human hand. Most of the readers are probably familiar with *Rock, Paper, Scissors*, a hand game played by children in many countries (Figure 1.3). In this game, at count, the players simultaneously change their hands into any of three "objects": rock (represented by a closed fist), scissors (two extended fingers) or paper (open palm). The definition of shape similarity in this example is ambiguous, as we find disagreement about how to consider our objects. As children playing the Rock, Paper, Scissors game, we recognize the hand shapes as the objects they intend to mimic. As adults, we say that all these "objects" are nothing but deformations of the same human hand. The reason for this difference is due to the fact that in the first case, we consider the postures of the hands as stand-alone rigid shapes, whereas in the second case, we consider them as deformations of a non-rigid shape.

Defining similarity of non-rigid shapes, we are looking for properties that distinguish between what really characterizes the object and what can

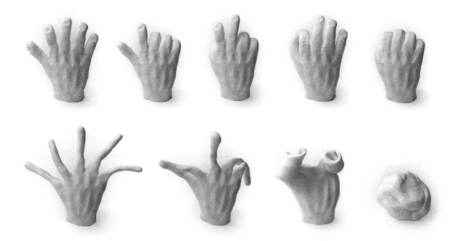

Figure 1.4. Clay figures deformed nearly isometrically (top) and non-isometrically (bottom).

be attributed to its deformations. Such properties are called *deformation-invariant*, and a similarity criterion based on these properties is called *deformation-invariant similarity*. In the human hand example, the length of the fingers, which always remains the same no matter how we articulate them, is one of the deformation-invariant characteristics.

Saying that a property is deformation-invariant, we need to specify what type of deformation is considered. A human hand sculpted out of clay will deform differently than will a real hand of flesh and bones: for example, the length and the width of the fingers of a clay hand may change almost without any restriction. Moreover, because a piece of clay can be torn apart, pierced, and bent, we can create almost any shape out of it (Figure 1.4). In this case, it will be hard to find any properties invariant under such a wide class of deformations.

If, however, we restrict ourselves to deformations similar to those of a human hand, we are in a much better situation. In geometric jargon, we call such deformations *articulations* or more generally *isometries*, and say that they preserve the *intrinsic geometry* of the shape. It appears that deformations of many natural objects can be modeled as isometries. Human and animal bodies deform approximately isometrically: we can flex our hands and legs to some extent but cannot stretch or shrink them. Of course, as in any model, there is a certain degree of inaccuracy in this assumption. At the same time, the benefit of limiting our discussion to the class of isometric deformations is that it leads to a well-defined geometric criterion of similarity based on the comparison of intrinsic geometry, which we refer to as the *intrinsic similarity*.

Such a similarity is invariant to isometries and therefore allows us to compare shapes no matter how they are bent.

1.2 Correspondence problems

Another important class of problems in shape analysis is known under the generic name of *correspondence problems*. When we put on a glove, we unconsciously solve a problem of great complexity: how to wear the glove such that it best fits the hand. For this purpose, we obviously need to insert each finger into its corresponding location in the glove. By saying "corresponding location," we implicitly assume that there is a natural correspondence between the hand and the glove – the thumb goes into the thumb of the glove, the index finger into the index finger, and so on. However, an automatic computation of such a "natural" correspondence is by no means trivial. Because both the hand and the glove are non-rigid objects, the correspondence must be independent of the deformations, that is, deformation-invariant.

The fact that we encounter deformation invariance again in this context suggests that correspondence and similarity problems are intimately related. In the glove fitting example, we can regard the "easiness" of putting on the glove as a criterion of similarity: if in order to fit the glove we make an effort to stretch it significantly (thus changing the intrinsic geometry), this means that the two objects are dissimilar (Figure 1.5). Had the jury in the United States court spoken in our terms, it would formulate the reason to acquit O. J. Simpson in his controversial murder trial as "intrinsic dissimilarity.[2]" The same criterion can be used for the definition of correspondence: we would like the fitting to be performed the easiest way. Trying to fit the thumb into the index finger of the glove is not an easy job. Therefore, like in similarity problems, we can speak about *intrinsic correspondence* between two objects and apply similar numerical tools for its computation.

Similarity and correspondence are two archetype problems used for shape *analysis* and *synthesis* applications. Many applications in the non-rigid world can be seen through the glasses of these two problems. As an example, we consider the problem of face recognition, dealing with the question of how to distinguish between the faces of two different people. Because of the flexibility of facial tissues and our ability to express a wide range of emotions, the face is a non-rigid object. Therefore, face recognition falls into the category of non-rigid shape similarity problems. Modeling facial expressions as deformations of the facial surface and using intrinsic geometric similarity criteria, we can distinguish between features resulting from expressions and those characterizing the person's identity, or in other words, make our face recognition *expression-invariant*.

Looking at the same application from a different perspective, we can ask the question how to find the same facial features in two different expressions

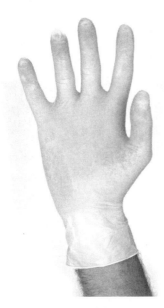 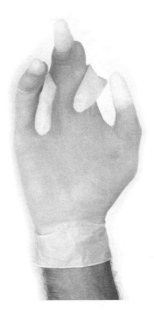

Figure 1.5. Glove fitting is an example of non-rigid shape similarity and correspondence problems. Correct correspondence (left) does not require significant stretching of the glove. Incorrect correspondence (right) results in a significant deformation.

of a face. This is exactly the problem of finding a deformation-invariant correspondence between two non-rigid shapes. The knowledge of the correspondence between faces allows us to perform different manipulations of them, including exaggeration of expressions, expression-invariant texture mapping, and *morphing* between faces.

In general, looking at the non-rigid world from the perspective of correspondence problems, we find applications more related to computer graphics and geometry processing (dealing mostly with synthesis), rather than to computer vision and pattern recognition (traditionally dealing with analysis problems).

1.3 A landscape of problems

To conclude our brief introduction to the forthcoming journey to the non-rigid world, let us overview the landscape of related fields and try to position the problems we will encounter in this book. In a broad sense, our problems belong to the realm of computer vision. As we mentioned, computer vision deals with extracting information about objects surrounding us from their visual representation. Traditionally, the most commonly used representation

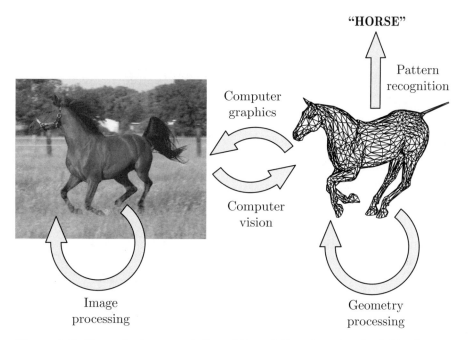

Figure 1.6. Conceptual representation of the relations between the fields of computer vision, computer graphics, image processing, pattern recognition, and geometry processing.

of the world is a two-dimensional image and the objects are geometric models, therefore, computer vision can be thought of as a "geometry from image" problem. A classic example is reconstructing the geometry of a scene from images taken by multiple cameras. The converse problem is addressed in the field of *computer graphics*: how to realistically and aesthetically render the description of the world, or how to produce an image from a geometric model. *Geometry processing* works with geometric models, trying to improve their quality or make their handling easier. *Image processing*, at the other end, operates on the images themselves, getting an image as the input and producing its "better" version as the output (Figure 1.6).

This division, once so clear to a specialist in each of the above fields, is becoming less obvious and less relevant in our days. For example, research results from the past decade have revealed a profound relation between geometry and image processing.[3] Considering images as geometric objects and operating on them using geometric tools created a revolution in image processing. The opposite process is taking place in the geometry processing community: it appears that by representing geometric objects as images, many efficient and powerful methods can be borrowed from image processing and adapted for geometry processing.

A similar situation seems to be happening in the analysis of non-rigid shapes: as we heavily rely on the model of the non-rigid world, our problems and methods reach far beyond "classic" computer vision. The tools we will use in this book range from theoretical geometry, used to model non-rigid objects, to methods used in manifold learning and artificial intelligence, which we employ for invariant representation of intrinsic geometry. In the next chapters, we will meet many disciplines and approaches brought together by problems of non-rigid shapes analysis. Such a diversity, in our opinion, makes research in this fascinating new field interesting and attractive.

Notes

[1] This citation from the Bard also appears as an epigraph in the Ph.D. thesis of Dragomir Anguelov [8].

[2] The moment at which the prosecution asked O. J. Simpson to put on a glove that allegedly had been used at the crime scene was the turning point of the trial, as the glove appeared too tight for Simpson to put on. The defense reflected this fact in the phrase "If it doesn't fit, you must acquit." As a result, the jury acquitted Simpson from murder charges.

[3] This relation is addressed in the book *Numerical Geometry of Images* [225].

2

A Taste of Geometry

> Μεδεις αγεωμέτρητος εισιτω μου τήν στήγων.
> Let none ignorant of geometry enter my door.
>
> Legendary inscription over
> the door of Plato's Academy.[1]

Geometry is probably one of the oldest fields of mathematics, going back to as early as 3000 BC. The ancient Greek venerated geometry and considered it the crown jewel of science up to the level that, according to a legend, the entrance to the school of Plato bore the inscription "let none ignorant of geometry enter my door" [332]. At the same time, geometry was not only a theoretical science and the domain of philosophers; the very fact that the Greek word γεωμετρια literally means "measurement of the Earth" implies that ancient engineers were skilled in geometry and employed it for practical purposes, mainly in construction and urban planning.

Though modern geometry has significantly evolved since Euclid and Archimedes (Figure 2.1), many of the basic terms were well-known, maybe in different formulation, to the ancient mathematicians. Today, engineers still use geometry for modeling of physical objects. For us, geometric tools are necessary for the description of the non-rigid world. In this chapter, we introduce a geometric vocabulary that will allow us to formulate properties of non-rigid objects.

2.1 Basic terms in metric geometry and topology

One of the most fundamental concepts in geometry is the notion of *distance*. From our experience of the three-dimensional Euclidean world in which we live, we are quite familiar with the intuitive way of measuring the distance between two points as the length of the straight line connecting them. However, the three-dimensional space with the Euclidean distance is simply a particular instance of a more general notion of a *metric space*. Formally, a set X equipped with a function $d: X \times X \to \mathbb{R}$ is said to be a metric space if the following axioms hold for all $x_1, x_2, x_3 \in X$:

(M1) *Non-negativity:* $d(x_1, x_2) \geq 0$.
(M2) *Indistinguishability:* $d(x_1, x_1) = 0$ if and only if $x_1 = x_2$.

Figure 2.1. A portion of Raffaello Santi's fresco *The School of Athens* (1509), supposedly depicting Euclid of Alexandria, one of the founding fathers of geometry.

(M3) *Symmetry:* $d(x_1, x_2) = d(x_2, x_1)$.
(M4) *Triangle inequality:* $d(x_1, x_3) \leq d(x_1, x_2) + d(x_2, x_3)$.

Elements of X are called *points* of the metric space, and the *metric* $d(x_1, x_2)$ is attributed the sense of distance between a pair of points. For brevity, we will often omit an explicit reference to the metric and write "metric space X," instead of the more rigorous form (X, d).

Example 2.1 (Euclidean space). The Euclidean space $\mathbb{R}^m$, with the metric defined using the Euclidean norm $d(x_1, x_2) = \|x_1 - x_2\|_2$ is a metric space. In general, any normed vector space can be thought of as a metric space in which the metric is induced by the norm.

The metric quantifies the concepts of "near" and "far." For example, in the Euclidean space we can pick a point and say that all points located less than one meter from it are "near" and the rest are "far." Obviously, the set of all near points falls into a *ball* (the interior of a sphere) of radius one meter. This idea can be generalized for arbitrary metric spaces: let $x \in X$ and $r > 0$; the set $B_r(x) = \{x' \in X : d(x, x') < r\}$ of points at a distance less than r from x is called an *open metric ball* (or simply an *open ball*) of radius r centered at x. Similarly, the set $\overline{B}_r(x) = \{x' \in X : d(x, x') \leq r\}$ is called a *closed ball*. The notion of a ball allows defining *open sets*: a set $A \subset X$ is said to be open if for every point $x \in A$ there exists $\epsilon > 0$ such that $B_\epsilon(x) \subset A$. Informally, this means that A has no "boundaries" and every point in it has at least a small neighborhood entirely contained in A. Closed sets are defined as complements of open sets. Through the definition of open and closed sets, the metric induces a *topology* on X, which is required to give a rigorous formulation of important properties such as convergence, connectedness, compactness, and continuity. Let us briefly overview each of them.

A space is said to be *disconnected* if it is the union of two disjoint, nonempty, open sets; otherwise, it is said to be *connected*. Our everyday lives give plenty of intuitive examples of connected and disconnected objects: a piece of paper is connected, because we can draw a line between every pair of points on it without raising the pen.[2] On the other hand, a plate and a fork placed apart are disconnected because they have no common points of intersection (hence are disjoint). A metric space X is said to be *compact* if any collection of open sets $A_i \subset X$ that covers X (i.e., $\bigcup A_i = X$) has a finite sub-collection that still covers X. This formal definition generalizes the simple Euclidean intuition that attributes compactness to every subspace of $\mathbb{R}^n$ that is *closed* (has a boundary) and *bounded* (is contained in a ball of a finite radius).

A sequence $\{x_n\}_{n=1}^\infty$ of points in X is said to *converge* to a point $x \in X$ if for every $\epsilon > 0$, there exists some N such that $x_n \in B_\epsilon(x)$ for all $n \geq N$. In other words, every neighborhood of x contains all but a finite number of points of the sequence. The point x is called a *limit* of the sequence and is denoted as $x_n \to x$ (as $n \to \infty$). Note that convergence is a topological definition rather than a metric one, meaning that we do not need the metric in order to define a convergent series. In a metric space, $x_n \to x$ if and only if $d(x, x_n) \to 0$, which is sometimes referred to as *convergence in the metric*.

2.2 Isometries

Given two metric spaces (X, d_X) and (Y, d_Y), we often look for some correspondence between their points, which can be expressed by a function $f : X \to Y$. Clearly, the distance $d_X(x_1, x_2)$ between a pair of points in

X is mapped to the distance $d_Y(f(x_1), f(x_2))$ between the images of x_1 and x_2 under f. Let us explore how d_X changes under f. The most familiar way to characterize $d_Y(f(x_1), f(x_2))$ in terms of $d_X(x_1, x_2)$ is the notion of *continuity*: f is said to be continuous at a point x if for every sequence $\{x_n\}$ in X with the limit $x_n \to x$, the sequence $\{f(x_n)\}$ converges to $f(x)$. Informally, a continuous function transforms limits to limits, or maps nearby points to nearby points. Continuous functions are known to preserve other topological properties as well. In particular, for every open set $A \subset Y$, its preimage[3] $f^{-1}(A)$ in X is open.[4] Compactness and connectedness are also preserved under continuous maps. A bijective continuous map f with a continuous inverse f^{-1} is called a *homeomorphism* and two spaces related by such a map are said to be homeomorphic. Homeomorphic spaces are indistinguishable in terms of topology.

A property stronger than continuity is *Lipschitz continuity*: f is said to be Lipschitz if there exists $C \geq 0$ such that $d_Y(f(x_1), f(x_2)) \leq C \cdot d_X(x_1, x_2)$ for all $x_1, x_2 \in X$. Being Lipschitz adds uniformity to the basic notion of continuity in the sense that a small change in x results in a small change in $f(x)$, whose size depends only on the size of change in x but not on x itself. For this reason, Lipschitz continuity is a global property in contrast with continuity. The smallest admissible C is called the *Lipschitz constant* or the *dilation* of f and is denoted by

$$\operatorname{dil} f = \sup_{x_1 \neq x_2 \in X} \frac{d_Y(f(x_1), f(x_2))}{d_X(x_1, x_2)}.$$

Dilation is a quantitative measure of the maximum relative change of distances in X under f. A map with $C \leq 1$ is called *nonexpanding* and a map with $C < 1$ is called a *contraction*.

Example 2.2 (continuity). A function $f : (0, 1) \to (1, \infty)$ defined by $f(x) = 1/x$ is continuous at every point. To prove it, observe that every open interval (a, b) is mapped to an open interval $(1/b, 1/a)$. On the other hand, f is not Lipschitz continuous. To verify this, consider two points x and x' in $(0, 1)$. Obviously,

$$C \geq \frac{|1/x - 1/x'|}{|x - x'|} = \frac{1}{x|x - x'|} \geq \frac{1}{x}.$$

For $x \to 0$, the Lipschitz constant C is unbounded.

Example 2.3 (Lipschitz constant of smooth functions). For real valued differentiable functions, a function $f : [0, 1] \to \mathbb{R}$ is Lipschitz if its derivative $|f'(x)|$ on $[0, 1]$ is bounded. Moreover, $\operatorname{dil} f \leq \sup_{x \in [0,1]} |f'(x)|$. To show this property, let $|f'(x)|$ be bounded by C and assume by contradiction that for some $x_1, x_2 \in [0, 1]$,

$$\frac{|f(x_2) - f(x_1)|}{|x_2 - x_1|} > C.$$

Then, by the mean value theorem, it follows that there exists $x_3 \in [x_1, x_2]$ such that $|f'(x_3)| > C$, which contradicts the bound. Hence, f is Lipschitz continuous and its dilation is bounded by C.

Going one step further, we can bound the minimum relative change of distances that f introduces, defining a yet stronger property of *bi-Lipschitz continuity*: f is said to be bi-Lipschitz if there exists a constant $C \geq 1$ such that $C^{-1} \cdot d_X(x_1, x_2) \leq d_Y(f(x_1), \text{ and } f(x_2)) \leq C \cdot d_X(x_1, x_2)$. Clearly, every bi-Lipschitz function is injective, i.e., there exists an inverse map $f^{-1} : f(X) \to X$, which is also Lipschitz continuous (the proof is left to the reader as Problem 2.1).

A bi-Lipschitz function f with $C = 1$ is called *distance preserving*, because $d_Y(f(x_1), f(x_2)) = d_X(x_1, x_2)$ for all $x_1, x_2 \in X$. A bijective distance preserving map is called an *isometry*, and two metric spaces related by such a map are referred to as *isometric*. Being isometric is an equivalence relation: isometric spaces share all properties that can be expressed in terms of distances and, consequently, are indistinguishable from the point of view of metric geometry.

Example 2.4 (Lipschitz, bi-Lipschitz, and isometric functions).

1. The function $f(x) = x^2$ on $(0, 1)$ is Lipschitz with the constant $C = 2$, but not bi-Lipschitz, because its inverse $f^{-1}(y) = \sqrt{y}$ has an unbounded derivative at $y = 0$.
2. The function $f(x) = 2x$ on $(0, 1)$ is bi-Lipschitz with the constant $C = 2$, but clearly not distance preserving.
3. The function $f(x) = 1 - x$ on $(0, 1)$ is an isometry.

A particular case of isometries are maps of X onto itself called *self-isometries*.[5] Self-isometries form a group with the function composition operator, as a composition of two self-isometries is also a self-isometry. This group is usually referred to as the *isometry group* of the space X and denoted by Iso(X). Iso(X) is trivial for a "generic" space, containing only the identity function. A non-trivial isometry group is often expressed by saying that the space has (intrinsic) *symmetries*. There is a remarkable class of spaces with a high degree of symmetry called *homogenous*. A metric space is said to be homogenous[6] if for every $x_1, x_2 \in X$ there exists a self-isometry $f : X \to X$ such that $f(x_1) = x_2$. Informally, this means that no place in the space is "privileged" and it is impossible to distinguish a particular element of X from any other element.

Example 2.5 (isometry groups).

1. In a planar triangle with unequal sides equipped with the Euclidean metric, the isometry group is trivial.
2. In a planar triangle with two equal sides unequal to the third equipped with the Euclidean metric, the isometry group is the cyclic group $Z/2Z$. This group contains only two elements: the identity transformation and

the reflection transformation, which flips the triangle about its symmetry axis. The space is not homogenous, because there exists no isometry that maps the center of one of the equal sides to the center of the unequal one.
3. In the Euclidean plane, the isometry group is the Euclidean group $E(3)$ (the group of rotations, translations, and reflections). The space is homogenous, because for every $x_1, x_2 \in \mathbb{R}^2$ there exists a planar translation by $v = x_2 - x_1$ such that $x_1 + v = x_2$.

In the real world, where many measurement devices have a finite precision, the notion of isometry might appear too restrictive. Consider for example an instrument capable of measuring distances with a relative error smaller than, say, 1%. Equipped with such a tool, we are unable to discern between an object and its scaled replica whenever the scaling factor falls between 0.99 and 1.01. We may say that two metric spaces related by a bi-Lipschitz function with dil $f \leq 1.01$ are indistinguishable due to the limited accuracy of our measurements, and, hence, such bi-Lipschitz maps can be considered isometries for every practical purpose.

Relative errors are what we usually care about concerning the metric of physical objects. For example, finding the radius of the Earth with an error of one centimeter is a formidable achievement, but measuring the width of conductors on a silicon chip with such a precision is of little use. However, some types of measurement devices can guarantee a certain precision no matter how large the measured distance is; in this case, it is usually said that the tool has an *absolute* error rather than a *relative* one. A natural relaxation of isometry in this case is the notion of *almost isometry*: a function $f : X \to Y$ is said to be an almost isometry (or ϵ-isometry) if there exists some $\epsilon \geq 0$ such that $|d_Y(f(x_1), f(x_2)) - d_X(x_1, x_2)| \leq \epsilon$ for all $x_1, x_2 \in X$, and for every $y \in Y$ there exists $x \in X$ such that $d_Y(y, f(x)) \leq \epsilon$ (the latter property implies that f is almost surjective in Y). Clearly, a 0-isometry is a true isometry. It is also convenient to define the *distortion* of a map

$$\text{dis } f = \sup_{x_1, x_2 \in X} |d_Y(f(x_1), f(x_2)) - d_X(x_1, x_2)|.$$

This definition resembles that of the dilation of Lipschitz maps. The only difference is that dilation measures the relative change of distances under the map f, whereas distortion measures absolute ones. In terms of distortion, an ϵ-isometry is an ϵ-distance-preserving (dis $f \leq \epsilon$) ϵ-surjective map. Returning to our example, a device measuring distances with an absolute precision of 1 mm cannot distinguish between two 1 mm-isometric objects and, hence, 1 mm-isometries are as good as isometries for any practical purpose.

So far, we have seen two natural ways of relaxing the notion of isometry. The first approach consisted of replacing distance preservation by the less restrictive bi-Lipschitz continuity with dil $f > 1$. In the second, we replaced distance preservation by ϵ-distance preservation and surjectivity by ϵ-surjectivity. Intuitively, one expects that the latter relaxation would have

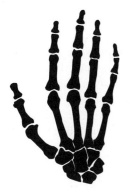  

Figure 2.2. Visualization of the difference between bi-Lipschitz maps and almost-isometries. The hand in the middle can be mapped to the smaller hand (right) using a bi-Lipschitz map, yet the absolute distance distortion of such a function may be very large. The hand can also be copied to the skeleton (left), this time with small absolute distortions, yet with no good local properties. Note that the topology of the skeleton and the hand is extremely different.

little effect on the good properties of a true isometry. In reality, it appears that passing from zero distortion to "almost zero" distortion leads to dramatic changes. In fact, almost isometries are not even necessarily continuous and may have very bizarre local behavior (Figure 2.2). Yet, if we were given the possibility of observing their behavior from a distance, we would see that on scales of distances significantly larger than ϵ, ϵ-isometries behave similarly to true isometries. Such a behavior may appear natural in our physical universe. In fact, at microscopic levels, the fabric of matter may be very non-smooth with singularities and bizarre foam-like fluctuations [186]. However, observed at a sufficiently coarse resolution at which we live, these details are hidden from our eyes. We will use the notion of almost isometry throughout the book. For example, it provides the basis for the definition of the Gromov-Hausdorff distance.

2.3 Length spaces

Metric spaces allow the definition of an abstract distance function between points. However, in many cases the notion of distance is somehow ambiguous. To illustrate this, we borrow the following example given by Burago *et al.* [88]. With a certain degree of precision, the Earth can be considered a sphere immersed into the three-dimensional Euclidean space representing our universe. Using the Euclidean distance, one can claim that the distance between New York and Sydney is about eight thousand miles; however, this number is of little use to an aircraft pilot,[7] as it assumes that we travel along a straight tunnel dug through the Earth between the two cities. In a similar way, a distance measured between two mountains by an optical instrument as the "bird flies" (i.e., using the Euclidean distance) is useless to an alpinist. Confined to the ground, our journey may be much longer than that of a winged creature.

The conclusion is that in many cases, the notion of distance is born from the lengths of *paths*. We define a path as a continuous map $\Gamma : [a, b] \to X$ of an interval. It is natural for us to associate *length* with a path Γ, i.e., to define a functional $L(\Gamma)$ that assigns a nonnegative number to every path.[8] Using these two concepts, we can readily define a metric induced by the length as

$$d_L(x, y) = \inf_{\Gamma} \{L(\Gamma) \text{ s.t. } \Gamma : [a, b] \to X, \Gamma(a) = x_1, \Gamma(b) = x_2\}. \quad (2.1)$$

The distance between two points is thereby the infimum of lengths of (admissible) paths connecting between them. Such a metric is called a *length metric* and a metric space (X, d_L) is called a *length space* (the proof that d_L is a metric is left as Problem 2.2). Length metrics are not necessarily finite; in fact, two disjoint sets are not path connected, as there exists no continuous path joining a point from one set with a point from the other, and therefore the distance between two points belonging to different disconnected components is infinite.

Example 2.6 (undirected graph). The prototype example of a length space is, without any doubt, an *undirected graph* comprising a set V of points (vertices), a set E of unordered pairs of distinct vertices (edges), and a function $L : E \to [0, \infty)$ assigning length to each edge. An ordered set of edges $\Gamma = \{(v_1, v_2), (v_2, v_3), \ldots, (v_{n-1}, v_n)\}$ constitutes a path between the vertices v_1 and v_n, whose length is given by $L(\Gamma) = L((v_1, v_2)) + L((v_2, v_3)) + \ldots + L((v_{n-1}, v_n))$. The length metric $d_L(v_1, v_2)$ between two vertices v_1, v_2 in the graph is defined as the length of the shortest path between them. We will encounter undirected graphs in the next chapters.

Note that in the definition of d_L, infimum is used instead of minimum because the shortest path between two points may not exist. For example, consider the Euclidean plane, from which the open interval $(0, 1)$ on the horizontal axis has been removed. The points $(0, 0)$ and $(1, 0)$ can be connected by a path, yet the shortest path passes through the removed interval, which is not a part of the space. A length space is said to be *complete* if for every two points x_1, x_2 there exists a path Γ joining them such that $L(\Gamma) = d_L(x_1, x_2)$; that is, there exists a shortest path (not necessarily unique) between every two points. Because all tangible objects appearing in most real-life situations are complete and compact, we will usually assume the existence of shortest paths, which will avoid unnecessary complications.

Recalling our alpinist example, we may notice a strange fact: the Euclidean distance was useless for measuring distances in mountains, because there were no paths passing on the surface of the Earth realizing this distance. This, however, means that before defining the length metric associated with a length function, we already knew how to measure length – this knowledge was derived from the Euclidean distance! Indeed, it is intuitive to measure the length of a path in Euclidean space by approximating it by a sequence of line segments and define the path length as the limit of their lengths. For a differentiable path Γ, this gives rise to the known formula

$$L(\Gamma) = \int_a^b ||\dot{\Gamma}(t)||_2 dt, \qquad (2.2)$$

where $\dot{\Gamma}(t)$ is the derivative of $\Gamma(t)$ with respect to the parameter t. This observation touches upon the important distinction between the concepts of *intrinsic* and *extrinsic* geometry, which we will often encounter in our exploration of non-rigid surfaces. The meaning of these two terms will become clearer after the discussion of Riemannian geometry later in this chapter. Formally, we say that a metric (the Euclidean metric in our example) *induces* a length structure. The latter, in turn, gives rise to a length metric d_L. One may ask what happens if we continue this process, i.e., use the length metric to define a length structure to induce another length metric. It appears, however, that this new metric will always be identical to the length metric from which it was created.[9] It is worthwhile mentioning that the terms *intrinsic metric* and *induced metric* are often used as synonyms referring to a length metric. In this book, we prefer to use the first term.

Example 2.7 (restricted vs. intrinsic metric). Consider the Euclidean plane equipped with the standard Euclidean metric and a connected region $X \subset \mathbb{R}^2$ thereof. The most straightforward way to define a metric on X is simply to assign a pair of points a distance equal to the distance between the same pair of points in $\mathbb{R}^2$. Such a metric is called a *restricted metric* and is denoted by $d_{\mathbb{R}^2}|_X$. Restriction of a metric is natural from the point of view of topology, as a subspace with a restricted metric inherits the topology of the space from which it was created. Another way to define a metric on X is associated with length structures. Admissible paths are all piecewise smooth paths contained in X and their length is measured using integration. If the region X is convex, then all such paths are straight lines and the intrinsic metric d_X coincides with the usual (restricted) Euclidean distance. Non-convex regions give rise to different metrics (compare between them in Figure 2.3).

This example brings up the important notion of convexity. Given a subset X' in a complete length space (X, d_L), we may define a metric on X' as the *restriction* of the length metric d_L. Another way to measure distances in X' is by using the length structure L to induce an intrinsic metric on X'. A set for which the two metrics coincide is called *convex*. An analogous definition is that for each pair of points in X', the shortest path between them entirely belongs to X' (the reader is invited to prove the equivalence of these two definitions in Problem 2.6).

Analogously to isometries that preserve metric structures, we may define a class of maps preserving length structures. An injective map $f : X \to Y$ is

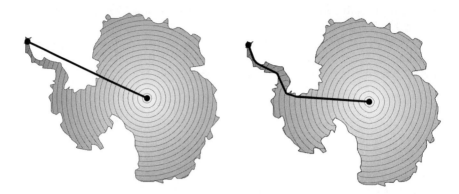

Figure 2.3. Birds flying over Antarctica travel along straight lines (left) and thus measure distances using the standard Euclidean metric restricted to the non-convex "island" in the Euclidean plane. People and quadrupeds, on the other hand, are confined to the glacial surface and have to travel distances measured by the intrinsic metric (right). Shades of gray visualize the metric balls produced by each metric.

called an *arcwise isometry* or an *isometric embedding* if $L(\Gamma) = L(f(\Gamma))$ for every path Γ. A bijective arcwise isometry is nothing but an isometry. However, unlike isometries, surjective arcwise isometries may have bad behavior. We leave the comparison between these different types of maps as an exercise in Problem 2.5.

2.4 Manifolds

So far, we have encountered objects like metric and length spaces, which allowed us to abstract the notions of distance and path length. Both concepts belong to the realm of metric geometry. Let us now take a step backward to introduce an important class of topological objects called *manifolds*. A space[10] is said to be an n-dimensional manifold if every point in it has a neighborhood homeomorphic to an open subset of $\mathbb{R}^n$. In other words, for every point x on the manifold, there exists a map $\alpha : U_\alpha \to \mathbb{R}^n$ from an open neighborhood U_α of x such that U_α is homeomorphic to $\alpha(U_\alpha)$. The map α is called a *chart* or a *system of local coordinates*. A collection of charts whose domains cover the entire manifold is called an *atlas*. These terms, borrowed from cartography, appear intuitive if we apply them to the most familiar example of a manifold: the Earth. Although we know that our planet is round, our everyday experience tells us that locally it is planar (that is why the erroneous belief that the Earth is flat took so much time to eradicate), meaning that a small region of the terrestrial surface can be charted or mapped on the plane. Similarly, in our everyday lives a world atlas is nothing but a book collecting several maps of different geographic zones that together cover the entire surface of our planet.

If we have two charts $\alpha : U_\alpha \to \mathbb{R}^n$ and $\beta : U_\beta \to \mathbb{R}^n$ with overlapping domains, a change of coordinates is possible by applying a *transition function* $\beta \circ \alpha^{-1} : \alpha(U_\alpha \cap U_\beta) \to \mathbb{R}^n$. Resorting again to our cartographic illustration, if a traveler has two different pages in a road atlas representing overlapping regions, he or she needs to know how to translate the coordinates of a point on one map into the coordinates on the other one. Thus, transition functions in some sense allow us to seamlessly "glue" together different charts in the atlas, such that we can plan our travel using local maps, passing from one map to another without too much hindrance. When all the transition functions are r-times continuously differentiable, the manifold is said to be $\mathcal{C}^r$. We will use a slightly relaxed terminology referring to a manifold as *smooth*, implying $\mathcal{C}^r$ with a sufficiently large r (it is common to assume $r = \infty$). We will often tacitly assume smoothness for our objects of interest.

There are a few important comments to be made at this point about manifolds. First, from the definition it is obvious that manifolds are topological objects and therefore are generally disconnected from the notion of metric and length. There may exist metric and length spaces that are not manifolds and manifolds without a metric. However, our discussion will focus on the intersection of these two worlds, i.e., manifolds that are also metric or length spaces. Second, we should distinguish between a manifold and a *manifold with boundary* (see Figure 2.4). Consider as an example a closed subset $A \subset \mathbb{R}^n$ of the Euclidean space. Points "inside" A have open neighborhoods homeomorphic to $\mathbb{R}^n$ and thus the interior of A is, indeed, a manifold. However, points belonging to the "edge" of A are clearly not homeomorphic to $\mathbb{R}^n$ but to the closed Euclidean half-space $[0, \infty) \times \mathbb{R}^{n-1}$. Being more precise, a manifold with boundary is a (non-empty) space X in which every point has a neighborhood that can be charted either in $\mathbb{R}^n$ or in $[0, \infty) \times \mathbb{R}^{n-1}$. The set of points in X that can be charted only in $[0, \infty) \times \mathbb{R}^{n-1}$ is called the *boundary* and denoted by ∂X, whereas their complement is referred to as the *interior* and is denoted by $\text{int}(X)$. The interior (always non-empty) is an n-dimensional manifold, and the boundary, when non-empty, is an $(n-1)$-dimensional manifold. With a slight abuse of notation, wherever no confusion may arise, we will abbreviate "manifold with boundary" to simply "manifold."

2.5 Embedded surfaces

Manifolds can be found everywhere: surfaces of all tangible objects can be represented as smooth two-dimensional manifolds residing in the ambient three-dimensional Euclidean space. Such manifolds are often referred to as *embedded surfaces*, as they constitute subspaces of a larger ambient space (in our case $\mathbb{R}^3$) and are therefore *embedded* into it.[11] Such surfaces can often be described by a smooth map $x : U \to \mathbb{R}^3$ from a

Figure 2.4. Example of a two-dimensional manifold without boundary (left); object that is not a manifold, as at one point it is not homeomorphic to $\mathbb{R}^2$ (center); and a manifold with boundary (right).

subset U of $\mathbb{R}^2$ to the three-dimensional Euclidean space. The set U is called a *parameterization domain* and the vector-valued function $x(u^1, u^2) = (x^1(u^1, u^2), x^2(u^1, u^2), x^3(u^1, u^2))$ a (global) *parameterization* or the *embedding* of the surface. The two-dimensional manifold formed by the image $X = x(U)$ of U in $\mathbb{R}^3$ is referred to as a *parametric surface*.

Example 2.8 (parameterization of the Earth). The Earth modeled as a two-dimensional sphere $\mathbb{S}^2$ with the radius $r \approx 3678\,km$ can be parameterized by $x : [-\frac{\pi}{2}, \frac{\pi}{2}] \times [0, 2\pi) \to \mathbb{R}^3$ in the following way,

$$x^1(u^1, u^2) = r\cos u^2 \cos u^1;$$
$$x^2(u^1, u^2) = r\sin u^2 \cos u^1;$$
$$x^3(u^1, u^2) = r\sin u^1,$$

where (u^1, u^2) are known as *latitude* and *longitude*, respectively.

The map x can be considered as a *global coordinate system*. A coordinate system assigns to each point on the surface an ordered pair of real numbers, whose meaning is usually associated with "location." Using the Earth as an illustration, location of geographical objects can be expressed using two global coordinates: latitude and longitude. Sometimes it may be impossible to provide one consistent smooth coordinate system for the entire surface. Surfaces for which a smooth $x : U \to \mathbb{R}^3$ does not exist are said not to admit a global parameterization. In such cases, local coordinate systems (smooth charts) are put together to form an atlas covering the entire surface. As a visualization, recall again the road atlas example: each page in it is equipped with a local Cartesian coordinate system, serving as a local parameterization of a portion of the surface.

When the derivatives $x_1 = \partial_{u^1} x$ and $x_2 = \partial_{u^2} x$ of x with respect to the coordinates are linearly independent for every $(u^1, u^2) \in U$, we say that x is *regular*. In such a case, the vectors x_1, x_2 span a two-dimensional space at $x = x(u^1, u^2)$, referred to as the *tangent plane* or the *tangent space* and denoted by $T_x X$. The name speaks for itself: the tangent space can be thought of as a local Euclidean approximation of the surface at x. A vector perpendicular to the tangent plane $T_x X$ is called the *normal* to the surface and is denoted by N.[12] If the normal direction depends smoothly on x, the surface is said to be *orientable*. Orientability means for example that if a traveler starts with a local right-handed system of coordinates, any round trip he makes on the surface would not change the orientation ("handedness") of his coordinates. We will assume our surfaces to be orientable to exclude pathologies like the Möbius stripe or the Klein bottle.

Let us select a point u in the parameterization domain and the corresponding point $x = x(u)$ on the surface. An infinitesimal displacement by $du = (du^1, du^2)$ around u will displace the point on the surface to

$$x(u+du) = x + x_1 du^1 + x_2 du^2 = x + Jdu. \quad (2.3)$$

Here, J is a 3×2 matrix having x_1 and x_2 as the columns; such a matrix is usually referred to as the *Jacobian* of the parameterization $x : U \to X$ at a point u (Figure 2.5, left). To quantify the length of the displacement $dx = Jdu$, we may write

$$d\ell^2 = \|dx\|^2 = \|Jdu\|^2 = du^T J^T J du = du^T G du, \quad (2.4)$$

where $G = J^T J$ is a symmetric 2×2 matrix, whose elements are the inner products $g_{ij} = \langle x_i, x_j \rangle$, depending on the choice of the local system of coordinates. Note that for a regular parameterization, $g_{11}g_{22} - g_{12}^2 > 0$, and hence G is positive definite. The quadratic form (2.4) is called the *first fundamental form* of the surface or the *Riemannian metric*.[13] The same definition extends to n-dimensional manifolds embedded into a finite-dimensional Euclidean space.

Example 2.9 (first fundamental form of a sphere). Let us consider the two-dimensional unit sphere with the parameterization $x^1 = \cos u^2 \cos u^1$, $x^2 = \sin u^2 \cos u^1$, $x^3 = \sin u^1$, as in the previous example. The basis vectors of the tangent plane of the sphere at the point (u^1, u^2) are given by

$$x_1 = \partial_{u^1} x = (-\cos u^2 \sin u^1, -\sin u^2 \sin u^1, \cos u^1)$$
$$x_2 = \partial_{u^2} x = (-\sin u^2 \cos u^1, \cos u^2 \cos u^1, 0).$$

The coefficients of the first fundamental form are given by the 2×2 matrix of inner products,

$$G = \begin{pmatrix} \langle x_1, x_1 \rangle & \langle x_1, x_2 \rangle \\ \langle x_1, x_2 \rangle & \langle x_2, x_2 \rangle \end{pmatrix} = \begin{pmatrix} 1 & 0 \\ 0 & \cos^2 u^1 \end{pmatrix}.$$

The form of G implies that a latitudinal displacement by du^2 translates to a displacement by $\cos u^1 \, du^2$ on the sphere. For this reason, a ship sailing between two islands 1° latitude apart in the equatorial zone needs to travel a larger distance than one navigating in the polar area.

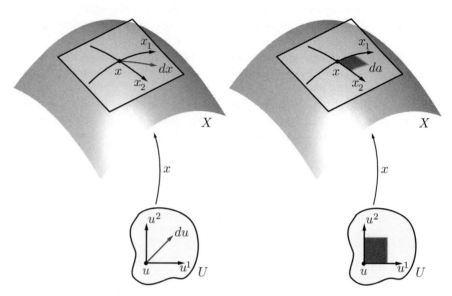

Figure 2.5. Left: an infinitesimal displacement by du in the parameterization domain is translated to a displacement by dx in the tangent space, whose length can be expressed using the first fundamental form. Right: an infinitesimal area element in the parameterization domain is translated to the area element in the tangent space, expressed in terms of $\det(G)$.

This example teaches us that the first fundamental form can be used to translate the measurement of length on the surface into measurement of length on its planar chart. To illustrate this better, imagine that a smooth curve $\gamma : [a, b] \to U$ describes the route of a traveler plotted on a map, such that $\Gamma = x \circ \gamma$ is the actual itinerary traveled on the surface. Equipped with the first fundamental form, we may say that the distance the traveler traverses by moving from t to $t + dt$ along the curve on the map is

$$d\ell = \sqrt{\dot\gamma(t)^\mathrm{T} G(\gamma(t)) \dot\gamma(t)}\, dt, \tag{2.5}$$

where

$$\dot\gamma = \left(\frac{d\gamma^1}{dt}, \frac{d\gamma^2}{dt}\right)^\mathrm{T}$$

denotes the derivative of γ with respect to the parameter t. Clearly, choosing a different chart will change the curve γ, and G will contain different coefficients to account for this change.

The quantity $d\ell$ is called the *differential arc length* of the curve Γ at the point t. The length of the entire curve can be obtained by simply integrating

$d\ell$ along Γ,

$$L(\Gamma) = \int_\Gamma d\ell = \int_a^b \sqrt{\dot\gamma(t)^{\mathrm{T}} G(\gamma(t))\dot\gamma(t)}\, dt. \qquad (2.6)$$

As we have already seen, $L(\Gamma)$ induces a length metric, which for a two-dimensional surface embedded in $\mathbb{R}^3$ is nothing but the intrinsic metric induced from the ambient Euclidean space. In the following, thinking of an embedded surface X as of a metric space, we will assume it to be equipped with the intrinsic metric, which will be denoted by d_X. We will refer to the geometry described by the intrinsic metric d_X as to the *intrinsic* geometry of the surface. The intrinsic geometry is completely described by the first fundamental form and is invariant to isometries.

Apart from length and distance, another geometric quantity that can be computed in the parameterization domain with the help of the first fundamental form is *area*. In order to understand how to measure area on the surface, consider an infinitesimal rectangle formed at a point u in the parameterization domain by du^1 and du^2. This rectangle is copied to the parallelogram formed at x by $x_1 du^1$ and $x_2 du^2$ in the tangent space of the surface, whose area is given by the length of the outer product of its sides,

$$da = \|x_1 du^1 \wedge x_2 du^2\| = \|x_1 \wedge x_2\| du^1 du^2 \qquad (2.7)$$

(Figure 2.5, right). Using simple identities, we can write

$$da = \sqrt{\|x_1\|^2 \|x_2\|^2 - \langle x_1, x_2\rangle^2}\, du^1 du^2$$
$$= \sqrt{g_{11}g_{22} - g_{12}^2}\, du^1 du^2 = \sqrt{\det(G)}\, du^1 du^2. \qquad (2.8)$$

The element da is called the *differential area element* of the surface at the point u. In order to compute the area of a subset $\Omega \subset X$ charted as $\omega = x^{-1}(\Omega)$, we integrate da, obtaining

$$\mu(\Omega) = \int_\Omega da = \int_\omega \sqrt{\det(G)}\, du^1 du^2. \qquad (2.9)$$

Similarly to length, area is also an intrinsic quantity, completely described by the determinant of G and invariant to isometries.

Although the function μ in (2.9) is probably the most natural way of assigning the notion of "size" to a subset of the surface, it is clearly not the only way to do so. For example, we can normalize μ by the area of the entire surface, obtaining the *relative area*

$$\nu(\Omega) = \frac{\mu(\Omega)}{\mu(X)} \qquad (2.10)$$

of the subset. Because the relative area of X is exactly one, we can associate $\nu(\Omega)$ with the probability of a point chosen at random from a uniform distribution on the surface to fall into Ω. It is worthwhile noting that area and relative area are particular cases of a more general concept of *measure*, which assigns non-negative values to subsets, quantifying their size.

2.6 Curvature and the second fundamental form

When analyzing embedded surfaces, it is often important to tell how different they are from a Euclidean space. The "non-Euclideness" can be expressed in terms of how fast the normal vector rotates as we move on a surface (or, informally, how "curved" is the surface). To quantify the rate of change of the normal, we may use the notion of *directional derivative* that should be familiar from multivariate calculus. Given a point $x \in T_x X$ and a direction $v \in T_x X$, the directional derivative of N is defined as

$$D_v N = \lim_{t \to 0} \frac{1}{t}(N(\Gamma(t)) - N(x)) = \frac{d}{dt} N(\Gamma(t)) \Big|_{t=0}, \quad (2.11)$$

where $\Gamma : (-\epsilon, +\epsilon) \to X$ is an arbitrary smooth curve with $\Gamma(0) = x$ and $\dot{\Gamma}(0) = v$. $D_v N$ is a vector in $\mathbb{R}^3$ measuring the change in N as we make a differential step in the direction v.

It is important to observe that $D_v N \in T_x X$. Indeed, differentiating the equation $\langle N, N \rangle = 1$ with respect to the direction v, we obtain $D_v \langle N, N \rangle = 2 \langle D_v N, N \rangle = 0$. Thus, $D_v N$ is perpendicular to the normal and lies in $T_x X$. The negative[14] directional derivative of the normal, $-D_v N$, is called the *shape operator* or the *Weingarten map* of the surface and is denoted by $S(v)$. The shape operator defines a linear map $S : T_x X \to T_x X$. For a surface admitting a parameterization $x : U \to \mathbb{R}^3$, this linear map can be expressed in the basis spanned by x_1, x_2, as a 2×2 matrix S satisfying $S(x_i) = Sx_i$, or

$$(S(x_1), S(x_2))^\mathrm{T} = S(x_1, x_2)^\mathrm{T}.$$

Multiplying the former equation by (x_1, x_2) from the right, we obtain

$$(S(x_1), S(x_2))^\mathrm{T} (x_1, x_2) = S(x_1, x_2)^\mathrm{T} (x_1, x_2) = SG,$$

where G is the first fundamental form matrix. The left-hand side is a 2×2 matrix $B = (S(x_1), S(x_2))^\mathrm{T} (x_1, x_2)$, whose elements are given by $b_{ij} = \langle S(x_i), x_j \rangle$ (using the simple fact that $S(x_i) = -\partial_{u^i} N$, we may also write $b_{ij} = -\langle \partial_{u^i} N, x_j \rangle$). The matrix B is called the *second fundamental form* of the surface and is expressed in the coordinates of the parameterization as the quadratic form

$$B(v, w) = \langle S(v), w \rangle. \quad (2.12)$$

The identity $S = BG^{-1}$ connects the shape operator with the first and the second fundamental forms.

Example 2.10 (second fundamental form of a sphere). We consider again the two-dimensional unit sphere with the parameterization $x^1 = \cos u^2 \cos u^1$, $x^2 = \sin u^2 \cos u^1$, $x^3 = \sin u^1$. The normal to the surface at a point (u^1, u^2) is given by

2.6 Curvature and the second fundamental form

$$N = x = (\cos u^2 \cos u^1, \sin u^2 \cos u^1, \sin u^1),$$

and its derivatives with respect to u are

$$\partial_{u^1} N = (-\cos u^2 \sin u^1, -\sin u^2 \sin u^1, \cos u^1)$$
$$\partial_{u^2} N = (-\sin u^2 \cos u^1, \cos u^2 \cos u^1, 0);$$

in this specific example, $\partial_{u^i} N$ coincide with x_i. The second fundamental form is given by the matrix of the inner products

$$B = -\begin{pmatrix} \langle \partial_{u^1} N, x_1 \rangle & \langle \partial_{u^1} N, x_2 \rangle \\ \langle \partial_{u^2} N, x_1 \rangle & \langle \partial_{u^2} N, x_2 \rangle \end{pmatrix} = -G = \begin{pmatrix} -1 & 0 \\ 0 & -\cos u^1 \end{pmatrix}.$$

The shape operator matrix is therefore $S = BG^{-1} = -I$.

It appears that the spectrum of the shape operator contains all information about the *curvature* of the surface. Indeed, observe that for a direction v, the second fundamental form $B(v,v) = \langle -D_v N, v \rangle$ is the projection of $-D_v N$ on v, which describes the change of the normal in the direction v caused by an infinitesimal step in that direction. In the coordinates of the parameterization, $B(v,v) = v^T S v$, from where it follows that the smallest and the largest eigenvalues of S express the smallest and the largest change of the normal, respectively (Figure 2.6). The eigenvalues of S are denoted by $\kappa_1 \leq \kappa_2$ and are referred to as *principal curvatures* of the surface at the point x. The corresponding eigenvectors are called the *principal directions*. An interesting fact is that κ_1 and κ_2 are the same regardless of the parameterization, meaning that the spectrum of the shape operator is invariant to the choice of coordinates.

Principal curvatures can be also interpreted in terms of curves passing on the surface through the point x. Let us develop this intuition by considering a smooth curve $\Gamma : [a,b] \to X$, parameterized by arc length. We can think of Γ as of a trajectory of a race car driving on the surface with a constant velocity $1\, m/sec$. From a physical point of view, the first derivative $\dot{\Gamma}(t)$ is the velocity vector, measuring the rate of change of the car position, and its direction is tangential to the curve. The second derivative $\ddot{\Gamma}(t)$ is the acceleration vector, measuring the rate of the change of the unit velocity vector direction. The direction of $\ddot{\Gamma}(t)$ is called the *principal normal* to the path, and as follows from the name, it is perpendicular to $\dot{\Gamma}(t)$. At tight turns, the driver feels that the direction of the velocity vector changes quickly, whereas when the car goes nearly straight, the change is close to zero. The speed of the velocity vector rotation at every point is called the *curvature* of the path and is denoted by $\kappa = \|\ddot{\Gamma}(t)\|_2$. Often, $\ddot{\Gamma}(t)$ is called the *curvature vector*. At a point x on the path, we can decompose the curvature vector into two components. The first component is the projection of the curvature vector onto the tangent plane $T_x X$, called the *geodesic curvature* and denoted by κ_g. The second is the projection of the curvature vector onto the surface normal, called the *normal curvature* and denoted by κ_n. Because the curve Γ lies on the surface,

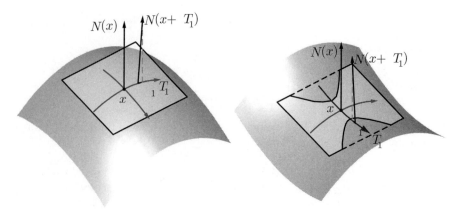

Figure 2.6. On an elliptic surface (left), both principal curvatures are positive, meaning that a small step in the principal direction rotates the normal to the surface toward the same direction. On the other hand, on a hyperbolic surface (right), the principal curvatures have different signs. This means that a step in the principal direction corresponding with κ_1 rotates the normal toward the opposite direction.

$\dot{\Gamma} \in T_x X$, and, consequently, $\langle \dot{\Gamma}, N \rangle = 0$. Differentiating this equation with respect to the parameter t yields

$$0 = \frac{d}{dt}\langle \dot{\Gamma}, N \rangle = \langle \ddot{\Gamma}, N \rangle + \langle \dot{\Gamma}, \frac{d}{dt} N \rangle,$$

from where

$$\kappa_n = \langle \ddot{\Gamma}, N \rangle = \langle \dot{\Gamma}, -D_{\dot{\Gamma}} N \rangle = B(\dot{\Gamma}, \dot{\Gamma}).$$

Because $\dot{\Gamma}$ is a unit vector, the normal curvature of any curve passing through x lies in the interval $\kappa_1 \le \kappa_n \le \kappa_2$. Curves following the principal directions realize the smallest and the largest κ_n. The mean value $H = \frac{1}{2}(\kappa_1 + \kappa_2) = \frac{1}{2}\text{trace}(S)$ is called the *mean curvature* and the product $K = \kappa_1 \kappa_2 = \det(S)$ is called the *Gaussian curvature* of the surface. The values of K and H define the local behavior of the surface, that is, how the surface is curved at the point x. For example, a plane has $K = H = 0$ at every point. A sphere has $K > 0$. A hyperbolic surface has $K < 0$ because one of the principal curvatures is positive and one is negative.

In the same way the first fundamental form of a surface completely describes its intrinsic geometry, the second fundamental form is responsible for the *extrinsic* one, that is, the way the surface resides in the ambient Euclidean space. Note that the second fundamental form is invariant to Euclidean isometries (i.e., a rigid motion of the surface in $\mathbb{R}^3$ does not affect its second fundamental form). Formally, a Euclidean isometry is often referred to as *congruence*, and two surfaces differing by a Euclidean isometry are said to be *congruent*. Congruence preserves both the intrinsic and the extrinsic geometries and thus the first and the second fundamental forms. The converse also appears to be true: any map preserving the first and the second fundamental forms is necessarily a congruence (or, said differently, an isometry[15] preserving the second fundamental form is a restriction of a Euclidean isometry). In simple words, this means that the first and the second fundamental forms completely describe the geometry of the surface, which is sometimes referred to as the fundamental theorem of the theory of surfaces.

2.7 Intrinsic view on geometry of surfaces

Embedded surfaces are important examples of manifolds, which have been the motivation that led to the development of non-Euclidean geometries during the nineteenth century. The German mathematician Berhnard Riemann (1826–1866) was among the first to notice that some of the properties we have defined so far can be described without the use of parameterization, but rather as properties intrinsic to the surface [324]. Such a viewpoint requires a certain degree of imagination, as it assumes that the surface has no ambient space, contrary to our everyday experience. Those familiar with general relativity know that one of the hardest things to comprehend is the idea of a four-dimensional space-time Riemannian[16] manifold as a model of our universe. Simply put, this manifold is the whole universe and therefore, there is no ambient space beyond it. In order to clarify the difference between the two viewpoints, recall again the Earth example. Astronauts on a space shuttle looking at the Earth and the space around it have an *extrinsic* point of view. However, an insect living on the surface of the Earth will not agree with this perception, as it has no idea about the three-dimensional space around the Earth. It is constrained to the two-dimensional Earth surface, and therefore has an *intrinsic* point of view.

From the intrinsic viewpoint, the tangent space T_xX is a Euclidean space associated with each point x of X. A Riemannian metric can be thought of as an abstract inner product $\langle \cdot, \cdot \rangle_x$ on the tangent space. Formally, a Riemannian metric is a bilinear symmetric positive-definite map $g_x : T_xX \times T_xX \to \mathbb{R}$, depending smoothly on x. The advantage of such a definition is that it is coordinate-free. If we select a basis for the tangent space T_xX, the Riemannian metric g_x can be expressed as a 2×2 matrix G and identified with the first fundamental form.

Using the intrinsic definition of the Riemannian metric, we can give a slightly different flavor to equation (2.6). Assuming $\Gamma : [a,b] \to X$ to be a smooth path on the surface X, the length of Γ is given by

$$L(\Gamma) = \int_a^b \sqrt{\langle \dot{\Gamma}(t), \dot{\Gamma}(t) \rangle_{\Gamma(t)}} \, dt = \int_a^b \sqrt{g_{\Gamma(t)}(\dot{\Gamma}(t), \dot{\Gamma}(t))} \, dt. \qquad (2.13)$$

This formula has the following physical meaning: the length of a path can be thought of as the time it takes to travel with a certain velocity from one point to another, and the distance we effectively pass depends on where and in which direction we travel. The vector $\dot{\Gamma}(t)$ can be interpreted as the velocity of movement along the path. The effective distance we travel at point $\Gamma(t)$ with the velocity vector $\dot{\Gamma}(t)$ is given by the Riemannian metric

$$\sqrt{g_{\Gamma(t)}(\dot{\Gamma}(t), \dot{\Gamma}(t))},$$

which is the length of the vector $\dot{\Gamma}(t)$ measured locally.

In Riemannian geometry, extrema (i.e., minimizers or maximizers) of the functional $L(\Gamma)$ are called *geodesics*. A path that minimizes $L(\Gamma)$ is called a *minimal geodesic*. An important question is whether for every pair of points, there exists a minimal geodesic that connects them, or in other words, whether the space is complete? As we have already seen, minimal geodesics do not necessarily exist in general length spaces. However, it appears that connected and compact Riemannian manifolds are complete, a fact known as the *Hopf-Rinow theorem*. We can therefore define the length metric, called the *geodesic metric*, as

$$d_X(x_1, x_2) = \min_{\Gamma} \left\{ L(\Gamma) \text{ s.t. } \Gamma : [a,b] \to X, \Gamma(a) = x_1, \Gamma(b) = x_2 \right\}.$$

Note that we arrived at the same expression for d_X using two different definitions. In the first one, the metric was induced by Euclidean geometry of the ambient space, whereas in the second one, it was induced by a local Riemannian structure. It may seem that the intrinsic definition is more general, as it is not restricted by a specific embedding: if in the extrinsic case the Riemannian metric is a by-product of the parameterization, in the intrinsic case, the Riemannian metric is an abstract positive-definite bilinear form. However, it appears that the intrinsic and the extrinsic views are equivalent. According to the *Nash embedding theorem*, any smooth Riemannian manifold can be realized as an embedded surface in a Euclidean space of sufficiently high, but finite dimension [290].[17] As a particular case, $\mathbb{R}^{17}$ is sufficient to realize any smooth two-dimensional manifold. It means that if we define an arbitrary Riemannian metric on our surface, we can always find an embedded surface in $\mathbb{R}^{17}$, such that the Riemannian metric induced on it by the Euclidean structure of the ambient space is equal to the Riemannian metric we have defined.

As for notation, we will henceforth denote by $(\mathcal{S}, d_{\mathcal{S}})$ an abstract metric space created by a Riemannian metric, and by $(X = x(\mathcal{S}), d_X)$ its realization as an embedded surface under the embedding $x : \mathcal{S} \to \mathbb{R}^3$. We will say that the intrinsic metric d_X induced by the length structure of $\mathbb{R}^3$ realizes the metric $d_{\mathcal{S}}$. When our interest will be the intrinsic geometry only, we will prefer the notation $(\mathcal{S}, d_{\mathcal{S}})$.

2.8 Bending and rigidity

Though the Nash embedding theorem guarantees that any Riemannian metric can be realized as an embedded surface, it says nothing about the uniqueness of such a realization. As a particular case, let us consider a Riemannian manifold $\mathcal{S}$, whose metric is realized in $\mathbb{R}^3$ by the embedding $x : \mathcal{S} \to \mathbb{R}^3$, that is, the intrinsic metric induced on the embedded surface $X = x(\mathcal{S})$ coincides with $d_\mathcal{S}$. Clearly, the embedding $x : \mathcal{S} \to \mathbb{R}^3$ is not unique, as for any congruence $h \in \text{Iso}(\mathbb{R}^3)$, $h \circ x(\mathcal{S})$ also realizes $d_\mathcal{S}$. This type of non-uniqueness is rather trivial, and we are interested in richer non-uniqueness going beyond Euclidean isometries. The question of existence of a non-unique embedding can be therefore posed as whether $d_\mathcal{S}$ can be realized by another embedding $y : \mathcal{S} \to \mathbb{R}^3$, such that $Y = y(\mathcal{S})$ is incongruent with $X = x(\mathcal{S})$. Said differently, we are looking for X and Y realizing the same intrinsic geometry while differing in their extrinsic geometries.

It appears that some surfaces have non-unique embeddings in $\mathbb{R}^3$. The simplest example is the plane, which can be bent and folded in many ways. A more sophisticated example is shown in Figure 2.7. Let us assume that $\mathcal{S}$ admits two embeddings x and y. The map $f : \mathbb{R}^3 \to \mathbb{R}^3$ defined by $f = y \circ x^{-1}$ describes the extrinsic deformation that we need to apply to X in order to obtain Y. Such a deformation is called *bending*, and a surface having a non-unique embedding is called *bendable*. Clearly, because $f \circ x(\mathcal{S}) = y(\mathcal{S})$ and X and Y are isometric, f must be distance preserving, that is,

$$d_X(x_1, x_2) = d_Y(f(x_1), f(x_2))$$

for every $x_1, x_2 \in X$. Here, d_X and d_Y denote the intrinsic metrics on the embedded surfaces X and Y, respectively, induced by the Euclidean metric in $\mathbb{R}^3$.

Considering the example in Figure 2.7, it is easy to transform one wine bottle into another by sawing off the bottle neck and welding it back "upside down." Now imagine that instead of glass the bottle is made of flexible, yet inelastic material. Trying to push the neck inside the bottle, we will soon realize that it is an impossible task, although the two versions of the bottle are isometric. That is, there is no way to turn the bottle neck upside down without distorting it – the only way to do so necessarily involves a cut. This experiment brings us to the notion of continuous bending. Two isometric surfaces X and Y are called *applicable* or *continuously bendable* if there exists a family of bendings $\{f_\lambda\}_{\lambda=0}^1$ continuous with respect to λ, such that $f_0(X) = X$ and $f_1(X) = Y$. Physically, applicability means that given a surface X realized as a thin shell of inelastic material, it can be pressed without tearing into a mold having the form of Y. Being continuously bendable is a stronger property than being bendable; in fact, the wine bottle example shows that incongruent embeddings of the same surface are not necessarily applicable to each other. A particularly interesting class of continuously bendable surfaces

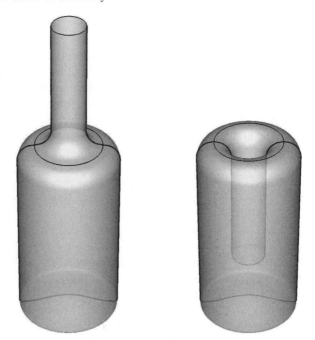

Figure 2.7. An example of non-rigid objects. If we cut the neck of the wine bottle on the left along the indicated circle and weld it back upside down, we will obtain the object on the right. The new bottle is isometric to the original one, yet the two objects are incongruent. Note that while the bottle is bendable, it is not continuously bendable.

are *flat* surfaces, which are applicable to a subset of the plane.[18] Such surfaces can be flattened onto a plane without distortion and thus can be constructed by cutting, folding, and bending a sheet of paper (Figure 2.8). Flat surfaces are important in manufacturing, especially in shipbuilding, where different parts of a ship are created from sheet steel.

Along with bendable and continuously bendable surfaces, many other surfaces admit a unique embedding into $\mathbb{R}^3$ (of course, up to a Euclidean isometry). Such surfaces, whose extrinsic geometry is completely determined by the intrinsic one, are called *non-bendable* or *rigid*.

Example 2.11 (rigidity of planar shapes). Let X and Y be two subsets of the plane with the intrinsic metrics d_X and d_Y, respectively, induced by the Euclidean metric. Because both X and Y are restricted to the plane, their second fundamental forms are identically zero. As a consequence, if X and Y are isometric, they are necessarily congruent. This result implies that planar shapes are rigid, as their geometry is completely determined by the first fundamental form. This fact has an interesting consequence. If $\mathcal{S}$ is a flat surface, one of its embeddings is a planar shape $X = x(\mathcal{S}) \subset \mathbb{R}^2$. This means that the

Figure 2.8. Left: origami, the art of paper folding, consists of creating intricate curved shapes from a flat piece of paper (image courtesy Alex Hubris). Right: approximation of the Stanford Bunny as developable patches glued together (reproduced from [354] with permission).

isometry group of $\mathcal{S}$ is isomorphic to the isometry group of X, comprising all distance preserving mappings $h : X \to X$. However, we have seen that X is rigid, which implies that every such h must be a restriction of a Euclidean isometry to X. Hence, any $g \in \text{Iso}(\mathcal{S})$ can be represented as the composition $g = (x|_X)^{-1} \circ h \circ x$, where h is an Euclidean isometry obeying $h(X) = X$. Intuitively, this means that studying the isometry (intrinsic symmetry) group of $\mathcal{S}$ can be replaced by applying the surface to the plane followed by studying the extrinsic symmetries of the obtained planar shape.

The question of rigidity interested many mathematicians for centuries, who focused mostly on the class of *polyhedral* surfaces. Let us briefly review the history of some dramatic developments in the field. As a starting point we should probably consider 1766, when Euler proposed his renowned rigidity conjecture, stating that all closed polyhedra embedded in $\mathbb{R}^3$ are rigid. In 1813, Cauchy (then only 24 years old) proved that *convex* polyhedra are, indeed, rigid [94].[19] In 1974, Gluck showed that almost all triangulated simply connected closed surfaces are rigid, remarking that Euler was right "statistically" [174]. Informally, Gluck's result implies that picking a closed polyhedron "at random," the probability that it will bend is zero. In 1977, Euler's conjecture was finally disproved by Connelly [114], who found a simple bendable closed polyhedron (Figure 2.9, right), sometimes referred to as the *Connelly sphere* [113, 115]. Unlike polyhedra, much less is known about the rigidity of smooth surfaces. One of the main results was proved in 1927 by Cohn-Vossen

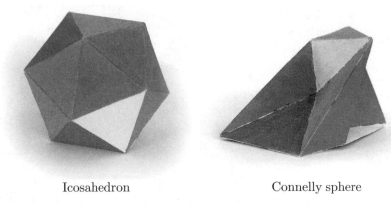

Icosahedron Connelly sphere

Figure 2.9. A paper realization of a convex polyhedron with twenty faces (left), and a variant of the Connelly sphere (right). Unlike the icosahedron, the Connelly sphere is non-rigid and can be bent along some of its edges.

[108], who showed that all surfaces with strictly positive Gaussian curvature are rigid. Cohn-Vossen's rigidity theorem can be thought of as a continuous analog to Cauchy's theorem for convex polyhedra.

These results may give the impression that most surfaces around us are rigid, which is probably true if rigidity is considered in a strict sense. The situation changes dramatically if small distortions are allowed. Although, to the best of our knowledge, the question of whether or not two rigid surfaces can be bent by an ϵ-isometric bending has not been addressed in the literature, practice shows that a great variety of surfaces appearing rigid in the strict sense, can be bent *almost isometrically*, or at least, can be realized with low distortion by many incongruent surfaces in $\mathbb{R}^3$. In a sense, a significant part of this book is motivated by this astonishing fact.

2.9 Intrinsic invariants

So far, we have seen that the Gaussian curvature of a surface can be defined as a product of the two principal curvatures or the determinant of the shape operator. Apart from these two definitions, there exists yet another one. Imagine an insect tied with a thread of length r to some point x on the surface. The insect makes a round trip around x while the thread is completely stretched on the surface, and we measure the perimeter $P(r)$ of the closed curve it describes. For a flat surface, we would obviously get $P(r) = 2\pi r$. When we repeat the experiment on a curved surface, $P(r)$ will differ from $2\pi r$, and measuring this discrepancy, we hope to realize how our surface is different from a plane. It appears that for a sufficiently small r, we will measure

$$P(r) = 2\pi r - \frac{\pi}{3} K r^3 + \mathcal{O}(r^4),$$

where $\mathcal{O}(r^4)$ is a fourth-order term. Thus, a Riemannian metric is locally Euclidean up to a third-order error, and the third-order distortion is controlled by K. This imaginary experiment leads to the following alternative definition of the Gaussian curvature,

$$K = \lim_{r \to 0} (2\pi r - P(r)) \frac{3}{\pi r^3},$$

where the term $2\pi r - P(r)$ measures the "defect" of the metric ball perimeter compared with its Euclidean counterpart.

Note that the way we define the Gaussian curvature does not rely on the ambient spaces in which the surface resides. The Gaussian curvature appears to be an intrinsic quantity, that is, a two-dimensional creature living on the surface is capable of measuring it. This striking result was probably accidentally discovered by Gauss, who was so astonished that he labeled it as *theorema egregium* (Latin for "Remarkable Theorem") [169]. Gauss observed that the Gaussian curvature can be expressed solely in terms of the first fundamental form coefficients and is therefore an intrinsic property.[20] This implies that if we bend the surface in such a way that the first fundamental form is preserved, *theorema egregium* guarantees that the Gaussian curvature at every point is also preserved. Put differently, the Gaussian curvature is invariant under isometries.

In our everyday lives, we encounter the *theorema egregium* in a pizzeria. An Italian pizza is topped with tomato sauce, mozzarella cheese, and olive oil, which can easily fall off and spot our clothes. Most people bend the slice across the radius, which creates non-zero principal curvature along the fold. Because the slice is a patch of the plane and has zero Gaussian curvature, the other principal curvature must remain zero in order to preserve the Gaussian curvature of the bent surface – otherwise stretching will be inevitably introduced (we of course assume that the bend is an isometry, which makes the slice a flat surface). This maintains rigidity in the direction perpendicular to the fold, such that the topping does not fall off.

Gauss' Remarkable Theorem gives us a local invariant, which could be potentially employed as a recipe for comparison of non-rigid surfaces insensitively to isometric deformations: Given two surfaces, compare their Gaussian curvature at every point and conclude whether they are similar or not (Figure 2.10). Unfortunately, in practice Gaussian curvature is a second-order differential quantity and, thus, is sensitive to noise. Moreover, even if we were able to compute it perfectly, there still remains the problem of identifying the corresponding points on two different surfaces. In the following chapters, we provide an answer to this question.

A possibility to overcome the correspondence problem is a trivial consequence from the *theorema egregium*. If we integrate the Gaussian curvature on the whole surface, such an integral will be invariant to isometries. Surprisingly, the invariance is much stronger. It appears that deformations that preserve a topological property of the surfaces known as the *Euler charac-*

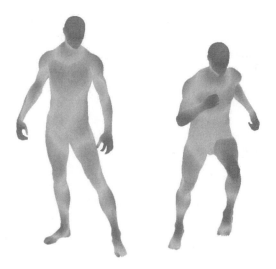

Figure 2.10. Gaussian curvature computed on two approximately isometric surfaces. *Theorema egregium* guarantees the invariance of Gaussian curvature under isometries.

teristic [153] also preserve the integral of the Gaussian curvature. This result is known as the *Gauss-Bonnet theorem* [43], which formally says that for a compact orientable Riemannian surface X, the following relation holds[21]:

$$\int_X K da = \chi_X,$$

where da is an area element and χ_X is the *Euler characteristic*. We will not extend our discussion on the meaning of χ_X. For polyhedra, it is defined as $\chi_X = N_F - N_E + N_V$, where N_F is the number of faces, N_E is the number of edges, and N_V is the number of vertices in the polyhedron. For smooth surfaces, in a sense, the Gauss-Bonnet theorem can be thought of as a definition of χ. It is worthwhile mentioning that the Euler characteristic is related to another topological invariant called *genus*. Genus represents the largest number of cuts along nonintersecting closed simple curves that leave the manifold connected, and can be intuitively interpreted as the number of "handles" or "holes" a surface has. Genus and the Euler characteristic are related by the formula $\chi_X = 2 - 2\operatorname{genus}(X)$.

The Gauss-Bonnet theorem allows us to compute a global invariant that is robust but at the same time very crude – surfaces as different as those depicted in Figure 2.11 may have the same Euler characteristic. This is not a very useful result for our applications, as for example, all surfaces homeomorphic to a sphere have $\chi = 2$ and are thus indistinguishable. In general, global topological characteristics usually do not give a sufficient level of "resolution"

2.9 Intrinsic invariants

Figure 2.11. Though the depicted objects look different, they are all homeomorphic to a sphere and have the same Euler characteristic $\chi = 2$.

necessary to distinguish between non-rigid surfaces. This will motivate us to look for better methods for comparing surfaces in the next chapters.

Suggested reading

An excellent introduction to metric geometry is given in the textbook *Course on Metric Geometry* by Burago *et al.* [88]. A discussion of isometries can be found in Chapter 11 in *Geometry: a Metric Approach with Models* by Millman and Parker[275]. Good guides to differential geometry and analytic machinery are the books *Differential Geometry of Curves and Surfaces* by do Carmo[139], *Elementary Differential Geometry* by O'Neil [300], and *Curves and Surfaces* by Montiel and Ros [281]. A great overview of Riemannian geometry is Berger's *Panoramic View of Riemannian Geometry* [26], which touches virtually every topic in modern differential and Riemannian geometry. Particularly, the reader may find interesting the two proofs of *theorema egregium* given by Berger.

Problems

2.1. Show that a bi-Lipschitz function is injective (i.e., there exists an inverse from its image) and its inverse is Lipschitz continuous.

2.2. Describe the isometric group for the following metric spaces:

1. Equilateral triangle in the plane with the standard Euclidean metric.
2. Perfect n-gon in the plane with the standard Euclidean metric.

3. A unit two-dimensional sphere $\mathbb{S}^2$ with the intrinsic metric induced from the ambient Euclidean space $\mathbb{R}^3$.

2.3. Show that
$$d_L(x,y) = \inf_{\gamma} \left\{ L(\gamma) \text{ s.t. } \gamma : [a,b] \to X, \gamma(a) = x_1, \gamma(b) = x_2 \right\}.$$
stemming from a length structure is a metric.

2.4.✓ Show that the length structure induced by the intrinsic metric d_L induces the same intrinsic metric d_L.

2.5. Articulate the difference between isometries and arcwise isometries (maps preserving length structures) by showing examples of arcwise isometries, which are not isometries.

2.6. Show that in a set A of a metric space (X, d_X), $d_X|_A$ coincides with the intrinsic metric on A if and only if the shortest path between every pair of points in A lies completely inside A.

2.7. Prove that the quadratic form $d\ell^2 = g_{11}(du^1)^2 + 2g_{12}du^1 du^2 + g_{22}(du^2)^2$ is positive definite if and only if the parameterization $x : U \to \mathbb{R}^3$ is regular.

2.8. Prove that $S(x_i) = -\partial_{u^i} N$.

2.9. Prove that the second fundamental form is symmetric, i.e., $B(v,w) = B(w,v)$.

2.10. Prove that the elements of B are also given by $b_{ij} = \langle N, \frac{\partial^2 x}{\partial u^i \partial u^j} \rangle$.

2.11. Prove that the spectrum of the shape operator is invariant to the choice of coordinates.

2.12.✓ Express the first and the second fundamental forms of the surface given as the graph of a function, $(x, y, z(x, y))$. Give an expression for the shape operator and the mean and Gaussian curvatures.

2.13. Prove that a curve with everywhere vanishing geodesic curvature κ_g is a geodesic.

2.14. Show surfaces with identical second fundamental forms, yet different first fundamental forms.

2.15. Show that the Gaussian curvature is an intrinsic quantity by expressing the determinant of the shape operator in terms of the first fundamental form and its derivatives.

2.16. Prove equivalence of the intrinsic and the extrinsic definitions of the Gaussian curvature.

2.17. The Gaussian curvature has an alternative intrinsic definition: consider a geodesic triangle ABC on the surface, i.e., a triangle composed of the intersection of three geodesics. The angles in the triangles are measured between the vectors in the tangent space at the three vertices. Define the *defect* of the triangle as $\delta(ABC) = \alpha + \beta + \gamma - \pi$, i.e., how the sum of angles in the triangle is different from the 180° obtained on the plane. The limit

$$K = \lim_{ABC \searrow x} \frac{\delta(ABC)}{\text{Area}(ABC)}$$

defines the Gaussian curvature at the point x, where $ABC \searrow x$ means that the triangle converges to a point. Prove equivalence of the two intrinsic definitions.

Notes

[1] A version of this Greek phrase decorates the logo of the American Mathematical Society.

[2] The piece of paper enjoys an even stronger property of path-connectedness, which means that every two points on it can be joined by a path.

[3] The notation of preimage is sometimes misleading. The symbol $f^{-1}(A)$ does not assume that f is bijective, but rather denotes the set $\{x \in X : f(x) \in A\}$ of points in X that are mapped to A by f.

[4] This property is often used as an alternative, "purely topological," definition of continuity.

[5] We will slightly abuse the notation, referring to *self-isometries* simply as *isometries*. Sometimes we will use the term *isometry* to denote the image of such a transformation, i.e., $f(X)$. For example, a smiling face will be called an isometry of the same face with neutral expression.

[6] In the group theory jargon, we say that the concept of homogenous space is equivalent to the concept of transitive group action.

[7] Though, knowing the radius of the Earth, the pilot can compute the geodesic distance from the Euclidean one.

[8] A space equipped with the a length is usually referred to as a *length space*. For a formal definition, the reader is referred to [88].

[9] More formally, we can say that inducing a length metric is an idempotent operation.

[10] Being completely rigorous, a manifold has to be a Hausdorff space, a topological property implying that for every two distinct points, there exist two disjoint open sets containing one of the points (or, in a more humorous version, in a Hausdorff space, points can be "housed off" from one another by open sets).

[11] In some literature, the term *immersed* is used instead.

[12] Because N is a unit vector in $\mathbb{R}^3$, it can be represented as a point on the two-dimensional unit sphere $\mathbb{S}^2$. The map $N : X \to \mathbb{S}^2$ is called the *normal* or *Gaussian map*.

[13] Traditionally, in the literature, the term *metric* or *Riemannian metric* is often used along with the term *first fundamental form*. This creates confusion with the notion of abstract metrics that we defined in the beginning of this chapter, although

the two notions are intimately related. Readers familiar with tensor notation may observe that the quadratic form can also be written as $d\ell^2 = g_{ij}du^i du^j$. The tensor g represented as the 2×2 matrix G once the coordinate system is chosen is called the *Riemannian metric tensor*.

[14] Sometimes, the shape operator is defined as the negative directional derivative.

[15] To be precise, an arc-wise isometry is sufficient.

[16] Being completely rigorous, general relativity models the space-time as a *pseudo-Riemannian manifold*.

[17] The Nash embedding theorem guarantees that an n-dimensional $\mathcal{C}^3$ Riemannian manifold is Riemann-isometric to an n-dimensional Riemannian surface embedded in $\mathbb{R}^m$ for $m \geq n^2 + 5n + 3$. Two Riemannian manifolds (X, g) and (Y, h) are Riemann-isometric if there exists a diffeomorphism $f : (X, g) \to (Y, h)$ with the associated differential $df_x : T_x X \to T_{f(x)} Y$, such that for all $x \in X$ and $v_1, v_2 \in T_x X$, $g_x(v_1, v_2) = h_{f(x)}(df_x(v_1), df_x(v_2))$. Therefore, Nash's theorem says that there exists a smooth injective map $f : (X, g) \mapsto \mathbb{R}^m$ such that for all $x \in X$ and $v_1, v_2 \in T_x X$, $g_x(v_1, v_2) = \langle df_x(v_1), df_x(v_2) \rangle_{\mathbb{R}^m}$.

[18] Flat surfaces are a particular case of a wider family of the so-called *developable* surfaces, which can be applied to a plane, a cylinder, or a cone.

[19] In fact, Cauchy's result is even stronger: it appears that a convex polyhedron is *completely rigid*, that is, cannot be subject to small perturbations without distortion of the intrinsic geometry.

[20] Gauss himself presented his theorem this way:

> Formula itaque art. praec. sponte perducit ad egregium theorema: Si superficies curva in quamcunque aliam superficiem explicatur, mensura curvaturae in singulis punctis invariata manet.

which can be translated as

> Thus, the formula of the previous article leads itself to a remarkable theorem: If a curved surface is developed onto any other surface, the measure of curvature at each point remains unchanged.

[21] The Gauss-Bonnet theorem can be extended to manifolds with boundary with the addition of a term: $\int_X K da + \int_{\partial X} \kappa_g d\ell = \chi_X$, where κ_g is the geodesic curvature, integrated along the boundary.

3
Discrete Geometry

> The world is continuous, but the mind is discrete.
>
> D. MUMFORD

Surfaces we have encountered so far had the property of varying continuously: for example, we could give the coordinates of a point on the surface in the three-dimensional Euclidean space for every pair of real-valued coordinates in the parameterization domain. In other words, our objects belonged to the *continuous* world. All physical objects that surround us are continuous (at least, up to a very fine resolution level where quantum phenomena break this nice picture).

Unfortunately, digital computers can only work with *discrete* data, i.e., data that can assume only distinct, separated values. In order to perform any computation on a surface, we have first to approximate it by some discrete representation. In this chapter, we focus on building discrete approximations to surfaces. In mathematics, properties of discrete geometric objects are studied by *combinatorial topology* and *geometry*. The branch of computer science that studies representations, data structures, and algorithms for problems stated in terms of geometrical objects is called *discrete* or *numerical geometry*. The two main problems we will encounter in this chapter are how to discretize surfaces and how to approximate geometric quantities using discrete representations.

3.1 Point clouds and sampling

The most basic problem in discrete surface representation is *sampling*. When we say that a surface X is sampled, we imply a finite discrete set of points $X' = \{x_1, \ldots, x_N\} \subseteq X$, called a *point cloud* (Figure 3.1, top left). Obviously, there are infinitely many ways to produce a point cloud out of X, and the natural question is how to decide whether one sampling is better than another. Intuitively, we wish the sampling to be as dense as possible, in order to better represent the underlying surface. On the other hand, we need to keep in mind that the discrete representation is used by computer algorithms, and every additional point increases storage and computational complexity costs.

42   3 Discrete Geometry

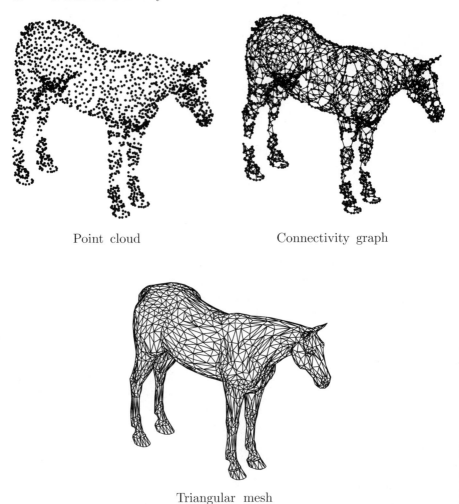

Point cloud   Connectivity graph

Triangular mesh

Figure 3.1. Different representations of a surface that we will encounter in this chapter: a point cloud (top left), a local connectivity graph (top right), and a triangular mesh (bottom).

As an illustration to the situation where trade-offs of this kind are encountered in real life, let us recall Example 2.7 from Chapter 2. Imagine that a scientific expedition is sent to Antarctica. The researchers establish a number of stations, spread all over the continent, which collect ice samples and study the glacial surface in a certain region around the station. On one hand, the stations have to be placed sufficiently densely to provide a good coverage of the studied area. On the other hand, installation of each station costs millions of dollars, therefore, the scientists are interested in having their network as sparse as possible.

In the above example, when we say that the stations are "spread all over the continent," we imply that such a sampling leaves no large uncovered regions or "holes." In order to give a quantitative interpretation to this phrase, we will say that a subset $X' \subseteq X$ is an *r-covering* of (or an *r-net* in) X if

$$\bigcup_{x' \in X'} B_r(x') = X,$$

(here B_r denotes a metric ball of radius r). In other words, an r-covering does not leave uncovered regions of diameter larger than r in the sense of the metric d_X. An alternative to express this is by saying that $d_X(x, X') \leq r$ for all $x \in X$, where $d_X(x, X') = \inf_{x' \in X'} d_X(x, x')$ is an abbreviated notation for the *point-to-set distance*.

Now, let use address the opposite problem: are all the points in the sampling necessary? An r-covering is guaranteed to cover the whole surface, however, it is not necessarily the best way to do it, as an r-covering needs not be a discrete set. Hence, we need some measure of how well the samples are separated. We say that X' is r'-*separated* if $d_X(x, x') \geq r'$ for all $x, x' \in X'$. If the surface X is compact, a set that is r'-separated for some $r' > 0$ is always finite.

The combination of these two criteria defines a good sampling: the whole surface is covered on one hand, and the sampling is sparse enough on the other. In our terminology, the expedition wants to sample the Antarctic area (modeled as a closed[1] metric space (X, d_X), see Figure 3.2), such that the sampling is an r-covering and an r-separated set at the same time. Depending on the sampling radius r, denser or sparser point clouds can be produced. At this point, we leave open the question how to select r in such a way that the surface is well-sampled and will return to it at the end of the chapter.

3.2 Farthest point sampling

Suppose that we start with placing the first station at a point x_1. The second station should be placed as far as possible from the first one, in order to provide the best coverage:

$$x_2 = \operatorname*{argmax}_{x \in X} d_X(x, x_1).$$

Obviously, the points x_1, x_2 are a $d_X(x_1, x_2)$-separated set and an r-covering of X with $r \leq d_X(x_1, x_2)$. The third station will be placed at the maximum distance from x_1 and x_2, i.e.,

---
input : metric space (X, d_X), some initial point $x_1 \in X$ and the desired sampling radius r_0.
output : a sampling $X' = \{x_1, \ldots, x_N\}$.
initialization: $X' = \{x_1\}$, $d(x) = d_X(x, x_1)$.

1 **while** $r > r_0$ **do**
2   Find the farthest point from X', $x' = \mathrm{argmax}_{x \in X}\, d(x)$.
3   Update the set of selected samples: $X' \longleftarrow X' \cup \{x'\}$.
4   Update the distance function $d(x)$,

$$d(x) \longleftarrow \min_{x \in X}\{d(x), d_X(x, x')\};$$

$$r \longleftarrow \max_{x \in X} d(x).$$

5 **end**

Algorithm 3.1. Farthest point sampling algorithm.

$$x_3 = \underset{x \in X}{\mathrm{argmax}}\, d_X(x, \{x_1, x_2\}),$$

and so on. After repeating the procedure $N-1$ times, we will end up with N samples $X' = \{x_1, \ldots, x_N\}$, which by construction constitute an r-covering and an r-separated set in X with

$$r = \max_{i=1,\ldots,N} \min_{k=1,\ldots,N} d_X(x_i, x_k).$$

We leave the formal proof as an exercise (Problem 3.3). The described strategy can be summarized in Algorithm 3.1.

Algorithm 3.1 is known as *farthest point sampling* (FPS) [179, 151] and is a way to obtain the "good" sampling we have defined. The FPS method is generic, as we have only assumed X to be a closed metric space, without adding any other restrictions to it. We have also tacitly assumed that the metric d_X is available, whereas in practice it must be approximated numerically. We defer the discussion on metric approximation to the next chapter.

In many practical applications, farthest point sampling is employed to *subsample* a discrete surface with a given dense but not necessarily uniform sampling. Given a point cloud X' containing M points, we would like to produce a subsampling $X'' \subset X'$ with $N < M$ points. For this purpose, Algorithm 3.1 is applied to the discrete metric space $(X', d_X|_{X'})$ with the stopping condition (Step 1) replaced by $|X''| < M$. Implemented straightforwardly, the computational complexity of the farthest point sampling in this case is $\mathcal{O}(NM)$, as every iteration of Algorithm 3.1 requires $\mathcal{O}(M)$ operations. Using efficient data structures, this complexity can be reduced to $\mathcal{O}(N \log M)$ [151].

Note that in the farthest point sampling strategy we are not allowed to move the samples already placed and at each iteration can only add one point in the best way. Algorithms with such a behavior, unable to undo what was

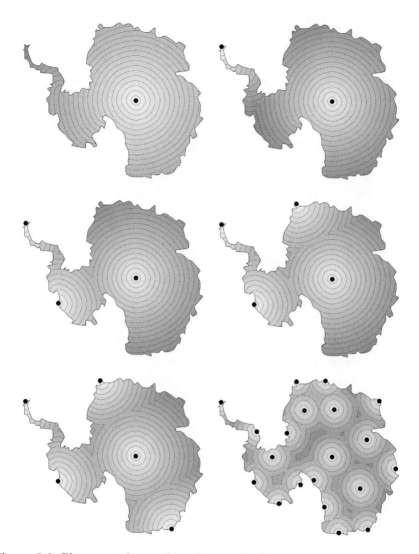

Figure 3.2. Placement of research stations on the Antarctic continent according to the farthest point strategy with the intrinsic metric d_X (progress of the algorithm is shown left to right, top to bottom after 1, 2, 3, 4, 5, and 20 iterations). Shades of gray and contours visualize the distance function $d(x)$ from the samples after each iteration.

done on previous iterations, are called *greedy*. As a consequence, FPS does not necessarily produce the optimal sampling, in the sense that there may be another r-covering and r-separated set with the same radius containing less points.

3.3 Voronoi tessellation

The sub-optimality of FPS leads us to the conclusion that we need a finer criterion for the sampling quality than just saying that X' is an r-separated r-covering. A somewhat better point of view is to think of sampling as *representation*, consisting of replacing continuous regions on the surface by single representative points. We are now going to embark on a discussion that will lead us to a quantitative definition of the representation error, and will allow us to construct a sampling strategy minimizing it.

Taking x_i from the sampling X', we may say that x_i represents all the points on X that are closer to it than to any other point in X'. Consequently, the sampling X' partitions the surface X into N regions,

$$V_i(X') = \{x \in X : d_X(x, x_i) < d_X(x, x_j), x_j \in X'\}. \tag{3.1}$$

The open set $V_i(X')$ is called the *Voronoi region* of x_i with respect to X', and the collection of all Voronoi regions is called the *Voronoi decomposition*[2] of the surface generated by X' [390]. For brevity, we will omit X' and write simply V_i wherever possible.

To avoid degenerate cases, we will henceforth assume that no three points in X' lie on the same geodesic, and no four points lie on the boundary of the same metric ball in X. In the Euclidean case, where d_X is the restriction of $d_{\mathbb{R}^2}$, the latter means that no three points in X' are *collinear* (i.e., belong to the same line), and no four points are *cocircular* (i.e., belong to the same circle). We refer to these two condition by saying that the points X' are in *general position*. General position can be given a probabilistic interpretation by saying that if we randomly distribute a set of points on X, the probability of them not being in a general position is zero.

Given a point x on X, we may construct a closed metric ball of some small radius ρ around it, and gradually increase ρ until the ball intersects X' for the first time. If x belongs to the Voronoi region V_i, the intersection will occur at one point, x_i (Figure 3.3, left). Another possibility is that x is *equidistant* from two points x_i and x_j, in which case the intersection will be at x_i and x_j (Figure 3.3, middle). In this case, we say that x belongs to the *Voronoi edge* $V_{ij} = \overline{V}_i \cap \partial \overline{V}_j$ separating the regions V_i and V_j ($\overline{V}_i$ denotes the *closure* of V_i, i.e., $\overline{V}_i = V_i \cup V_i$). Voronoi regions separated by a Voronoi edge are said to be *adjacent*. A third possibility is that x is equidistant from three points x_i, x_j, x_k (Figure 3.3, right). We will say that such x is the *Voronoi vertex* $V_{ijk} = \overline{V}_i \cap \overline{V}_j \cap \overline{V}_k$ adjacent to the three Voronoi regions V_i, V_j, and V_k.[3]

Shortly, a point on X can lie either in the interior of a Voronoi region, or on a Voronoi edge, or be a Voronoi vertex. This brings us to an alternative way of defining a Voronoi decomposition. Consider the open set of all points, which are closer to x_i than to x_j,

$$D_{ij} = \{x \in X : d_X(x, x_i) < d_X(x, x_j)\}.$$

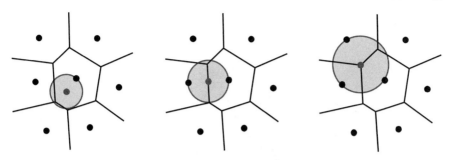

Figure 3.3. Voronoi diagram in the Euclidean plane. For each point $x \in X$, the following three cases can be distinguished: x belongs to the interior of a Voronoi region (left), x belongs to an edge shared by two Voronoi regions (center), and x is a vertex shared by three Voronoi regions (right).

Such a set is called the *domain of dominance* of x_i over x_j, and its boundary, $\partial D_{ij} = \{x \in X : d_X(x, x_i) = d_X(x, x_j)\}$, describes the locus of all points on the surface that are *equidistant* from x_i and x_j. In the case $X \subset \mathbb{R}^2$ and d is a Euclidean metric restricted to X, ∂D_{ij} is nothing but a straight line, bisecting the segment $x_i x_j$, and D_{ij} is an open half-plane. The interior of the Voronoi region of x_i can be described as the intersection of the domains of dominance of x_i over other points in X',

$$V_i = \bigcap_{i \neq j} D_{ij}.$$

In the Euclidean case, V_i is the intersection of $N - 1$ open half-planes containing x_i, which is a polygon without a boundary. Moreover, because the half-planes are convex and convexity is preserved under intersection, the resulting polygon is convex and thus homeomorphic to an open disk (we leave it to the reader to prove that a convex set is homeomorphic to a disk).

Observe that Voronoi regions are disjoint, yet their closures cover the entire surface. This means that if we cut X along Voronoi edges, it will fall apart into tiles, which are precisely the Voronoi regions. As we said before, in the Euclidean case, these tiles are convex polygons. A finite collection of disjoint open topological disks (usually referred to as *cells*), whose closures cover the entire surface, is called a *cell complex* or a *tessellation* of X (from *tessella* meaning "small tile" in Latin). A particular tessellation created by decomposition of X into Voronoi cells $\{V_i\}$ is called a *Voronoi tessellation* (Figure 3.4).

The situation changes significantly when considering a non-Euclidean metric. Figure 3.5 shows that on a general surface, it may happen that Voronoi regions are neither convex nor homeomorphic to a disk. Although Euclidean Voronoi decompositions are well-studied and understood, much less results are available for the general case. An important result is the paper by Leibon

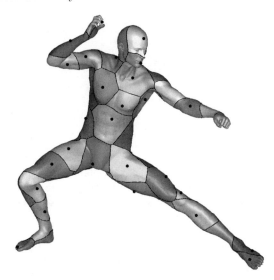

Figure 3.4. Voronoi decomposition of a surface with a non-Euclidean metric.

and Letscher [248], showing the existence of a Voronoi tessellation of a general Riemannian manifold sampled with "sufficient density." To give a quantitative definition of what is meant by "sufficient density," the authors resort to the notion of convexity radius. Recall that in Chapter 2, we defined a convex set in X as a set in which the minimal geodesic between each pair of points lies inside the set. The *convexity radius* of the surface X at a point x is the largest r for which the closed ball $\overline{B}_r(x)$ is convex in X. The convexity radius of the entire X is simply the infimum of the convexity radii over all points in X. Informally, we may say that on the scale below the convexity radius, the surface behaves very much like a Euclidean space – any interesting topological and geometric properties, like the one in Figure 3.5, appear on larger scales. Leibon and Letscher prove that if X' is an r-separated r-covering with r smaller than $\frac{1}{5}$ the convexity radius of X, then the Voronoi regions are topological disks and the Voronoi decomposition generated by X' is a valid tessellation of the surface [248, 301].

3.4 Centroidal Voronoi sampling and the Lloyd-Max algorithm

The notion of Voronoi tessellation allows us to express the sampling of a surface as a mapping $y^* : X \to X'$, copying the interior of each V_i to its *closest point* x_i in X' (the boundaries shared by more than one Voronoi cell can be copied to x_i belonging to any of the intersecting cells). Thinking of sampling in these terms, it is natural to quantify the error introduced by

3.4 Centroidal Voronoi sampling and the Lloyd-Max algorithm

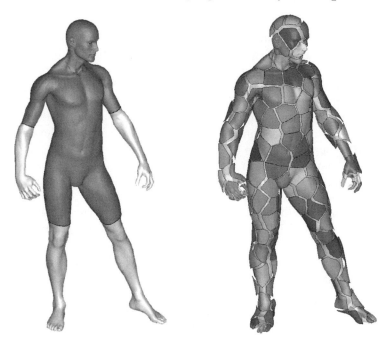

Figure 3.5. Left: Voronoi decomposition of a surface with insufficient sampling density. The shaded Voronoi region is not homeomorphic to a disk. Right: increasing the sampling density produces a valid tessellation. Observe that now the body is decomposed into topological disks.

replacing x with its representation $y^*(x)$. For that purpose, let us assume x is chosen at random with uniform distribution on X, where by *uniform distribution* we mean that the probability that x falls into a subset A on X is proportional to the area of A. Formally, this can be written as

$$\mathbb{P}(x \in A) = \frac{\mu(A)}{\mu(X)} = \frac{1}{\mu(X)} \int_A da,$$

where $\mu(X)$ is the area of X, da is the differential area element, and $\mathbb{P}$ stands for probability. The representation error associated with a sampling X' can be expressed as the *variance* of the random variable $d_X(x, y^*(x))$,

$$\varepsilon(X') = \operatorname{Var}(d_X(x, y^*(x)))$$
$$= \frac{1}{\mu(X)} \int_X d_X^2(x, y^*(x)) = \frac{1}{\mu(X)} \sum_{i=1}^N \int_{V_i(X')} d_X^2(x, x_i) da. \quad (3.2)$$

In the Euclidean case, the latter expression becomes

$$\varepsilon(X') = \frac{1}{\mu(X)} \sum_{i=1}^N \int_{V_i(X')} \|x - x_i\|_2^2 dx,$$

which can be interpreted as the *mean squared error* of the representation.

The representation error $\varepsilon(X')$ gives a quantitative measure of the sampling quality and leads to the natural question of finding the best sampling. Formally, we are looking for a set X' of N points on X, bringing the error $\varepsilon(X')$ to minimum among all sets of N points on X. A related problem is finding the smallest sampling X' having $\varepsilon(X')$ below some predefined value. It appears that a similar question arises in a variety of fields. For example, in image processing, a continuous domain of vectors encoding the intensities of different colors in an image often need to be represented by a finite set of symbols. Computation of such a representation is known as *vector quantization* [185, 172]. In machine learning, pattern recognition, and data mining, we encounter objects represented as vectors of features, and it is often required to aggregate them into groups sharing some common trait. The process of partitioning a space into such groups is termed *clustering* or *unsupervised learning* [141].

It can be shown (see Problem 3.5) that in order for a sampling X' to minimize $\varepsilon(X')$, each point x_i has to satisfy

$$x_i = \arg\min_{x \in V_i} \int_{V_i} d_X(x, x') da.$$

A point minimizing[4] the latter integral is called the *intrinsic centroid*[5] of V_i [140]. To understand the origin of this name, note that in the Euclidean case the condition becomes

$$x_i = \arg\min_{x \in V_i} \int_{V_i} \|x - x'\|_2 dx' = \frac{\int_{V_i} x\, dx}{\int_{V_i} dx},$$

which is simply the *centroid* (or *center of mass*) of the set V_i. A Voronoi tessellation generated by a set of points x_i, which are themselves the intrinsic centroids of the corresponding Voronoi cells V_i, is called a *centroidal Voronoi tessellation*. We refer to a sampling associated with a centroidal Voronoi tessellation as to a *centroidal Voronoi sampling*. Such a sampling is optimal in the sense of $\varepsilon(X')$ and is not unique.

A simple way to compute a centroidal Voronoi sampling starts by picking up an arbitrary sampling X' (produced, for example, using the farthest point sampling algorithm), and computing the associated Voronoi tessellation. Next, we compute the intrinsic centroids for each Voronoi cell and use them as a new sampling X'. Clearly, the Voronoi tessellation has changed and needs to be recomputed. Repeating this process several times gives a reasonable

3.4 Centroidal Voronoi sampling and the Lloyd-Max algorithm

approximation of a true centroidal Voronoi tessellation. The entire procedure can be summarized as shown in Algorithm 3.2.

---
input : metric space (X, d_X), initial set of points X'.
output : optimal set of points X', minimizing $\epsilon(X')$.
1 **repeat**
2   Construct the Voronoi tessellation associated with X'.
3   Compute the intrinsic centroids of V_i and set X' to be these points.
4 **until** X' *stops changing significantly*

Algorithm 3.2. Lloyd-Max algorithm.

This procedure is known as *Lloyd-Max algorithm*[6] in signal and image processing [255, 263, 256] and *k-means* in statistics [260]. We are not going to explore the full details of this algorithm and refer the reader to [312] for additional information. We only mention that Lloyd-Max algorithm is a very simple *alternating minimization* procedure, which attempts to produce a sequence of samplings that decrease the value of $\varepsilon(X')$ [107]. We will encounter more sophisticated numerical recipes for minimization of functions in Chapter 5.

Compared with the greedy farthest point sampling, which never changes the locations of the points previously added to the sampling and optimizes only the location of the next point, centroidal Voronoi sampling allows us to change the locations of all the points. As a consequence, the produced sampling is more uniform and may require less samples than one produced by FPS. In fact, as the number of samples grows asymptotically, all Voronoi cells converge to a hexagonal shape, which is known to produce the densest possible tessellation ("honeycomb" tiling, called this way because it is often encountered in Nature, see Figure 3.6). This result was known for a while in the Euclidean case and has been proved only recently for two-dimensional Riemannian manifolds [191].

Figure 3.6. Nature surprises us by producing a variety of living examples of Voronoi tessellations, including the spot-shaped coloring of an African giraffe (left), the pattern on the shell of a *testudo hermanni* turtle (middle), and the hexagonal wax honeycomb cells built by honey bees (right).

3.5 Connectivity

The sampling techniques we have been discussing so far approximate a continuous surface by a point cloud. Point clouds are the most primitive representations of surfaces at the level of set theory and, in fact, capture only the extrinsic geometry. In many cases, such a representation is too crude as it does not convey any information about the relations between different points. In our Antarctica exploration example, the only information a point cloud X' gives us is the coordinates of the stations. Yet, such an information may be insufficient. Suppose that an accident happens at station x_i, and the researchers request help from nearby stations. For that purpose, they need to know which stations are located near x_i, in order to get the aid as promptly as possible.

More formally, we say that we want to know the *neighborhood* of x_i in X', which is denoted by $\mathcal{N}(x_i)$. In our example, if the accident is so serious and the aid is required so urgently that only people from stations in a radius of two miles could arrive in time, we will define $\mathcal{N}(x_i)$ as the ball of radius $r = 2$ around x_i in X',

$$\mathcal{N}(x_i) = \{x_j \in X' : d_X|_{X'}(x_i, x_j) \leq r\}.$$

Alternatively, we can think of a situation where in case of a disaster, aid from K *nearest neighbors* is required, and define the neighborhood $\mathcal{N}(x_i)$ as the set of K points closest to x_i.

It is worthwhile noting that though in our example the neighborhood $\mathcal{N}(x_i)$ is related to the metric d_X, in general it can be defined independently. The notion of neighborhood is *topological* rather than geometric and can therefore exist in spaces even not equipped with a metric. However, because we are dealing with surfaces that are Riemannian manifolds and thus equipped with a metric structure, the most natural way to define the concept of neighborhood is in the way presented above.

We say that two points are *adjacent* or *directly connected* if they belong to the same neighborhood (it is common to exclude the point itself, i.e., x_i is not adjacent to itself). A natural requirement is that the adjacency relation is symmetric, i.e., if x_i is adjacent to x_j, then x_j is adjacent to x_i. The connectivity structure can be represented as an undirected graph with vertices representing the samples and edges telling us which samples are adjacent (see an example in Figure 3.1, top right). Because connectivity is a purely topological notion, we actually do not need the coordinates of the vertices x_i to define this graph. Therefore, the *vertices* of the graph are indices $I = \{1, \ldots, N\}$ representing the corresponding samples and the *edges* are pairs of indices $E = \{(i,j) \in I \times I : x_j \in \mathcal{N}(x_i)\}$. Numerically, the connectivity graph can be represented as an $N \times N$ matrix with elements

$$e_{ij} = \begin{cases} 1 & x_i \text{ and } x_j \text{ are connected;} \\ 0 & \text{otherwise,} \end{cases}$$

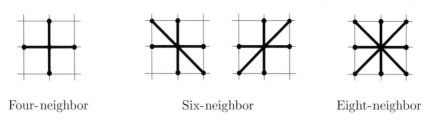

Four-neighbor   Six-neighbor   Eight-neighbor

Figure 3.7. Connectivity patterns on a Cartesian grid with four (left), six (center), and eight (right) neighbors.

E is usually referred to as *adjacency* or *connectivity matrix* and is typically sparse.

Given the connectivity E, we can assign a length function $L(x_i, x_j) = d_X(x_i, x_j)$ between all adjacent points $(i,j) \in E$. Edge lengths can be approximated as the Euclidean distances $L(x_i, x_j) \approx \|x_i - x_j\|_2$. If the sampling is sufficiently dense, L is a good approximation to the local distances measured on the surface using d_X, due to the fact that a Riemannian surface can be locally approximated as a Euclidean space. In order to measure distance between non-adjacent points in X, we may use the length metric d_L (often called the *graph distance*) induced by the length structure L of the graph. A legitimate question is whether d_L is a good approximation to d_X. We defer the answer to the next chapter.

Example 3.1 (four-, six-, and eight-neighbor connectivity). In numerical analysis application, we often encounter two important connectivity patterns of a regular planar Cartesian grid $\mathbb{Z} \times \mathbb{Z}$. When each point of the grid is directly connected to its top, bottom, left, and right neighbors, the connectivity is referred to as *four neighbor* (Figure 3.7, left). If in addition the top-left and bottom-right (or alternatively, top-right and bottom-left) neighbors are connected, we have a *six-neighbor* connectivity (Figure 3.7, center). If all the neighbors in the grid are connected, we call this the *eight-neighbor* connectivity (Figure 3.7, right).

3.6 Delaunay tessellation

Because the connectivity graph is merely a formal expression of adjacency, different definitions of adjacency will result in different connectivity. For example, we may call adjacent a pair of points x_i, x_j, whose Voronoi cells are adjacent (separated by a common Voronoi edge V_{ij}). The corresponding connectivity graph can be realized by connecting each pair of adjacent points by a minimal geodesic $\Gamma_{ij} : [0, L] \to X$, which we assume to be parameterized by arclength.[7] A traveler walking along Γ_{ij} on the surface starts his journey at $\Gamma_{ij}(0) = x_i$, makes half of his way in the Voronoi cell V_i, then crosses the Voronoi edge V_{ij} at $\Gamma_{ij}(L/2)$, and continues the remaining half of the journey in V_j, until he reaches the destination point $\Gamma_{ij}(L) = x_j$. The obtained connectivity graph is maximal in the sense that no edge can be added to it without intersecting the other edges.

Let x_i, x_j, and x_k be three points whose Voronoi cells are adjacent, i.e., share a Voronoi vertex. The region on the surface enclosed by the paths Γ_{ij}, Γ_{ik}, and Γ_{jk} is called a *geodesic triangle*, which is a generalization of the Euclidean triangle. Because the Voronoi vertex V_{ijk} adjacent to V_i, V_j, and V_k is equidistant from x_i, x_j, and x_k, an open metric ball of radius $r = d_X(x_i, x)$ centered at V_{ijk} is empty of any points of X'. If the sampling is sufficiently dense, the ball is convex. This implies that the closed ball $\overline{B}_r(x)$ entirely includes the geodesic triangle formed by x_i, x_j, and x_k. Moreover, $\overline{B}_r(x)$ is the smallest metric ball enclosing the geodesic triangle and having the vertices x_i, x_j, and x_k on its boundary. Such a ball is called the *circumscribed ball*[8] of the geodesic triangle and is unique if the sampling is sufficiently dense [247].

Cutting the surface along the edges Γ_{ij} splits it into tiles formed by the geodesic triangles (regions enclosed by Γ_{ij}), and if the surface has a boundary, additional tiles between the outmost edges and the boundary. In [248], Leibon and Letscher show that under the sampling density conditions that guarantee the existence of a Voronoi tessellation, these tiles form valid tessellation of the surface. Such a tessellation is called a *Delaunay tessellation* after the Russian mathematician Boris Delaunay (1890–1980),[9] who first described it in 1934 [133]. The cells and the edges of a Delaunay tessellation are called *Delaunay cells* and *Delaunay edges*, respectively.[10]

We have already seen that the circumscribed ball of a Delaunay cell contains no points of X' in its interior. This property is called the *empty circumscribed ball property* and is often used as an alternative, axiomatic definition of the Delaunay tessellation. The radius of the circumscribed ball can be thought of as a measure of *coarseness* of a geodesic triangle; the coarseness of the entire tessellation is measured as the maximal coarseness of its cells [15]. It appears that among all tessellations of X into geodesic triangles having X' as the vertices (and, possibly, the non-triangular cells including the boundary), the Delaunay tessellation has the minimal coarseness.[11] Delaunay tessellation also minimizes the total length of the edges, $\sum L(\Gamma_{ij})$ [247]. In the Euclidean case, additional optimality properties are known. For example, the Delaunay tessellation maximizes the sum of the inscribed circles radii, and the minimal angle of all triangles [15].

3.7 Triangular meshes

Delaunay tessellation associated with a sufficiently dense sampling allows us to decompose the surface into a finite set of continuous two-dimensional cells. However, the definition of a Delaunay cell relies on the notion of geodesic

3.7 Triangular meshes

triangle, which can be neither represented nor processed by a computer. A practical way of constructing a continuous approximation of the surface is by replacing the geodesic triangles with the Euclidean ones, built upon the same vertices, i.e., constructing an approximation of the surface by "gluing" a collection of planar triangular tiles along their edges (see Figure 3.1, bottom). In computational geometry and computer graphics, such an approximation is called a *triangular mesh*.[12] In a sense, if a point cloud is a "zero-dimensional" approximation of a surface, and a connectivity graph is a "one-dimensional" approximation, a triangular mesh can be thought of as "two-dimensional" one.

In discrete geometry, a mesh is usually defined as an abstract structure of the form (I, E, F), consisting of a set of vertices I, edges E, and triangular *faces*

$$F = \{(i,j,k) \in I \times I \times I : (i,j), (i,k), (k,j) \in E\}$$

(this is simply to say that a triangular face is composed of three edges). The faces can be represented as an $N_F \times 3$ matrix T of indices, where the kth row $t_k = (t_k^1, t_k^2, t_k^3)$, $t_k^i \in \{1, ..., N\}$ is the set of vertices constituting the kth triangle and N_F is the number of triangular faces.[13]

The object (I, E, F) is purely topological, as it does not contain any geometric properties of the underlying surface. A geometric realization of the mesh is defined by specifying the coordinates of the vertices in $\mathbb{R}^3$, which can be represented as an $N \times 3$ matrix of coordinates X' with the kth row given by (x_k^1, x_k^2, x_k^3). Together, the matrix of faces T and the matrix of coordinates X' give a complete description of the triangular mesh. This is one of the most common representations of discrete surfaces used in computer graphics.

Example 3.2. A simple example of a triangular mesh is a *tetrahedron* (Figure 3.8). It has four vertices, six edges,[14] and four triangular faces.

It is natural to think of the mesh as of a piecewise-planar approximation of the underlying smooth surface X, defined as the union of all the triangular faces, which we denote by

$$T(X') = \bigcup_{k=1}^{N_F} \mathrm{conv}(x_{t_k^1}, x_{t_k^2}, x_{t_k^3})$$

(here, $\mathrm{conv}(x_{t_k^1}, x_{t_k^2}, x_{t_k^3})$ is the *convex hull* the vertices of the kth triangle, containing all the points belonging to this triangle). In the following, we will use the term *mesh* referring to this piecewise planar approximation. Any point x on the mesh $T(X')$ can be represented providing the triangle index k and the coefficients of the convex combination of the triangle vertices,

$$x = u^1 x_{t_k^1} + u^2 x_{t_k^2} + (1 - u^1 - u^2) x_{t_k^3}, \quad u^i \in [0, 1].$$

The vector (u^1, u^2) is called the *barycentric coordinates* [279] of x and constitutes a local (face-wise) parameterization of the mesh.

56   3 Discrete Geometry

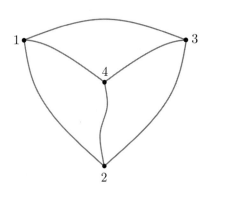
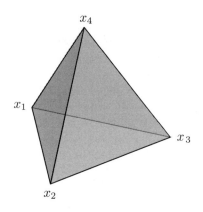

Vertices 1   2   3   4
Edges   (1,2) (1,3) (1,4) (4,2) (4,3) (2,3)
Faces   (2,4,3) (1,4,2) (3,4,1) (2,3,1)

Figure 3.8. A tetrahedron considered from topological (left) and geometric (right) point of view.

Considering a mesh $T(X')$ as a topological space, we can ask whether it is topologically equivalent to the underlying continuous surface. If the neighborhood of each vertex can be continuously mapped into a disk (or to half-disk in case the manifold has boundary vertices), we say that $T(X')$ is a *manifold mesh*. Equivalently, any edge in a manifold mesh belongs to at most two triangles (or a single triangle if the edge is part of the boundary) [193]. Because we assume that all our surfaces are manifolds, we consider only manifold meshes as valid.

Example 3.3 (geometry images). In Chapter 2, we mentioned that some surfaces can be represented by a global parameterization of the form $x(U) = (x^1(U), x^2(U), x^3(U))$, where U denotes the parameterization domain (for simplicity, let us restrict the discussion to parameterization on the unit square $U = [0,1] \times [0,1]$). Sampling of parametric surfaces is convenient because it can be done in the parameterization domain. We can sample parametric surfaces by sampling the parameterization domain on a uniform Cartesian grid $U' = \{(i\Delta u^1, j\Delta u^2)\}$, where $\Delta u^1 = 1/M$ and $\Delta u^2 = 1/N$ denote the grid sampling step (Figure 3.9, left). Applying the parameterization on U', we obtain a point cloud $X' = (x^1(U'), x^2(U'), x^3(U'))$, which can be stored as three $N \times M$ matrices. Such a representation can be thought of as a three-channel *geometry image*, a term coined by Hugues Hoppe and his co-authors [192]. Surfaces reconstructed by range acquisition devices such as structured and coded light three-dimensional scanners are often readily representable as geometry images [62]. A particular case of parameterization is the so-called *Monge form*, given by $(u^1, u^2, x^3(u^1, u^2))$ and representing a surface that can

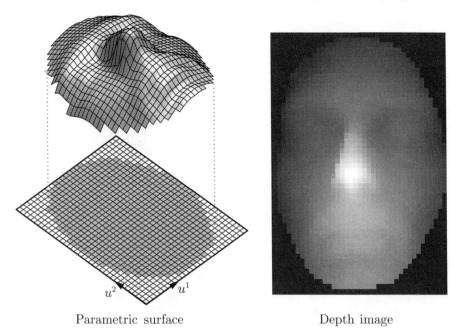

Parametric surface   Depth image

Figure 3.9. Scanned facial surface of one of the authors, given in parametric representation (left). Representation of the same surface as a depth image (right).

be realized as the graph of a function $x^3(u^1, u^2)$. For a Monge surface, the representation is even simpler as we only have to store the matrix of the $x^3(U')$ values. Such a single-channel geometry image is often called a *depth image* or *height field* (Figure 3.9, right).

The connectivity can be derived from the connectivity of the Cartesian grid in the parameterization domain, like in Example 3.1 (see Figure 3.9, left). The set of edges produced in this way defines a triangular mesh (Figure 3.9, right). However, not every connectivity pattern results in a manifold mesh. For example, eight-neighbor connectivity produces a mesh where some edges are shared by four triangles, whereas a six-neighbor connectivity produces a valid manifold mesh.

3.8* Local feature size and curvature-dependent sampling

In Example 3.3, we have seen two ways to triangulate a surface: a valid manifold mesh and an invalid non-manifold mesh with self-intersecting faces. The reader may come to the conclusion that being a manifold is enough for

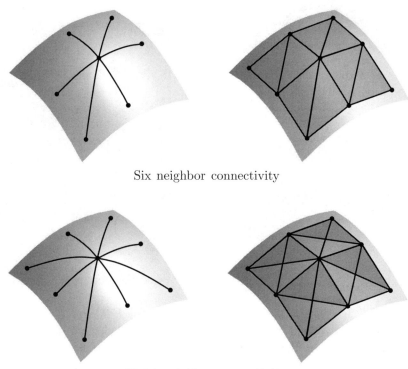

Figure 3.10. Representation of a geometry image as a triangular mesh. Top row: six-neighbor connectivity produces a manifold mesh. Bottom row: eight-neighbor connectivity produces an non-manifold mesh with self-intersecting faces.

the mesh to be valid. However, the following example shows that this is not enough:

Example 3.4 (self-intersecting manifold mesh). Consider a manifold mesh obtained by triangulating a planar patch (e.g., using six-neighbor connectivity). A different intrinsically equivalent embedding of the planar patch can be obtained by rolling it as shown in Figure 3.11 (left) and keeping the triangulation. In this case, though we have a manifold mesh, it still contains self-intersections (Figure 3.11, middle).

Example 3.4 accentuates the difference between two different notions of mesh "validity." Topological validity implies that the mesh is topologically equivalent to the underlying surface (in other words, is a manifold mesh). Validity in the geometric sense means that the realization of the mesh in $\mathbb{R}^3$ does not contain self-intersecting faces.

Why is not topological validity sufficient for a mesh to be also geometrically valid? The problem is easy to illustrate recalling the step we made when

3.8* Local feature size and curvature-dependent sampling

Figure 3.11. A manifold mesh representing a smooth surface (left) can be self-intersecting, depending on the embedding (middle). Making the sampling denser allows us to overcome the problem (right).

passing from Delaunay tessellation to a triangular mesh, replacing geodesic triangles by Euclidean ones. Whereas the definition of the former is intrinsic, the latter is extrinsic, i.e., depends on the specific embedding. In the case shown in Example 3.4, depending on the embedding of the surface in $\mathbb{R}^3$, the Euclidean distance between some points may be much smaller than the geodesic ones. Therefore, replacing the geodesic triangles by Euclidean ones does not guarantee that the resulting triangular mesh will not contain self-intersections.

An additional condition that guarantees a geometrically valid Delaunay triangulation has to do with the way the surface is embedded into the Euclidean space. First, we want our mesh to be topologically equivalent (homeomorphic) to the underlying smooth surface. Secondly, we want the embedding to be free of self-intersections. It can be shown that a smooth compact surface X embedded into $\mathbb{R}^3$ has an open *envelope* V_X (an open set containing the surface X), such that every point u in V_X is continuously mappable to a unique point x on X, realizing the distance from u to X [156] (we leave the proof as Problem 3.9). This implies that if our triangular mesh $T(X')$ is completely contained in V_X, then X and $T(X')$ can be continuously mapped one to the other, and particularly, this guarantees that our triangular mesh is valid.

The question is how to make sure that this condition is satisfied. The definition of the envelope V_X assumes that every point in it has a unique nearest point on X, otherwise the map between V_X and X is not well-defined. In other words, all the problematic points are those points in $\mathbb{R}^3$ that are equidistant from two or more points on X. The closure of the set of all such points is called the *medial axis* of X. Medial axis is sometimes referred to as *medial axis transform* (MAT), *symmetric axis*, a term attributed to Blum [39, 40, 41], who introduced this notion in the context of shape description and recognition of biological applications, and *skeleton* [225]. The latter term is used predominately in image processing, and the reason for this name is clear from a two-dimensional example shown in Figure 3.12.

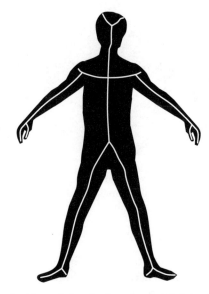

Figure 3.12. Example of medial axis transform (skeleton) of a two-dimensional shape used in image processing. Black indicates the locus of points equidistant from the boundary of the shape. The resemblance of the medial axis to the human body "bones" explains the term *skeleton* used in image processing.

The distance from a point x on X to the medial axis is called the *local feature size* and denoted here by $\rho(x)$ [7]. Local feature size is related to the maximum curvature radius of the surface,

$$\rho(x) \leq \frac{1}{\max\{\kappa_1(x), \kappa_2(x)\}}, \tag{3.3}$$

where κ_1 and κ_2 are the principal curvatures of X at the point x. Unlike the Gaussian curvature, the maximum curvature radius is an extrinsic quantity. A global bound on the local feature size is called the *reach* of the surface X, defined as the distance of X to its medial axis [156].

Based on the result of Amenta and Bern [7], Leibon and Letscher [248] showed that if the surface is sampled sufficiently densely, such that for every $x \in X$, an open ball of radius $\frac{1}{4}\rho_X(x)$ contains a point of X', it is guaranteed that the triangular mesh $T(X')$ formed by Delaunay tessellation does not intersect the medial axis. Consequently, $T(X')$ is completely contained in V_X, which implies a valid triangulation. This condition becomes more intuitive if we interpret it as a curvature-adaptive sampling: at point where the surface is more curved, the sampling is denser. We can see that the condition on sampling density heals the problem we observed in Example 3.4: if the sampling density is increased as shown in Figure 3.11 (right), the triangulation becomes valid.

3.9* Approximation quality

Using a triangular mesh as an approximation of our surface, we can approximate various geometric quantities. As an example of such quantities, we will consider here the area and the normals to the surface, which play an important role when measuring extrinsic similarity of discretized shapes. Intrinsic geometric invariants of shapes, for which we will need to compute geodesic distances on triangular meshes, will be discussed in the next chapters.

The area of X can be approximated as the sum of the areas of all the triangular faces of $T(X')$,

$$\mu(X) \approx \sum_{k=1}^{N_F} \frac{1}{2} \|(x_{t_k^2} - x_{t_k^1}) \wedge (x_{t_k^3} - x_{t_k^1})\|_2,$$

using a simple formula for triangle area (see Problem 3.12). The normals to the surface are approximated as the normals to the triangles of $T(X')$. For triangle k, the normal can be computed as the cross-product,

$$n_k = (x_{t_k^3} - x_{t_k^1}) \wedge (x_{t_k^2} - x_{t_k^1}).$$

We implicitly assume that the triangle vertices are numbered consistently such that all the normals are pointing inwards or outwards. Along the edges, the average of the normal vectors in the two adjacent triangles is used. At vertices, the normal is approximated as the average of the normals of the triangles sharing the same vertex [284].

An important question the reader should ask at this point is how good an approximation of geometric quantities is possible with a given mesh, and in particular, is it enough to have a valid mesh to approximate well the underlying surface? The first and the most obvious requirement is that the piecewise planar approximation $T(X')$ is "close" to the continuous surface X. A simple way to quantify this proximity is by regarding X and $T(X')$ as subsets of $\mathbb{R}^3$ and measuring the minimum size of envelopes (closed neighborhoods) that have to be created around X and $T(X')$, such that X is completely contained in the envelope around $T(X')$ and vice versa. More formally, we can define,

$$d_{H,\mathbb{R}^3}(X, T(X')) = \max \left\{ \sup_{x \in X} d_{\mathbb{R}^3}(x, T(X')), \sup_{x' \in T(X')} d_{\mathbb{R}^3}(x', X) \right\},$$

called the *Hausdorff distance* after the German mathematician Felix Hausdorff (1868–1942), who introduced it in 1914 [200].

Though the Hausdorff distance allows us to measure how "close" $T(X')$ and X are to each other, it is too crude to express the quality of approximation of many geometric quantities. In order to accentuate this problem, let us analyze the following example shown by Hermann Schwartz (1843–1921) in 1890 [346] and named in his honor[15]:

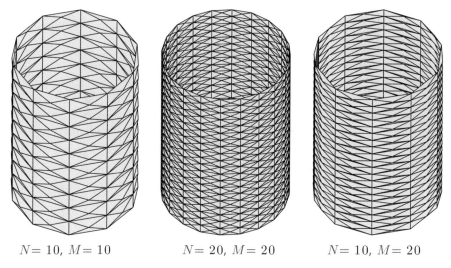

$N = 10, M = 10$ $N = 20, M = 20$ $N = 10, M = 20$

Figure 3.13. Schwartz lantern is an example of triangular mesh that can be made arbitrarily close to a cylinder, yet its area and normals will be arbitrarily different from those of a cylinder. Shown (left to right) are examples with $N = M = 10$, $N = M = 20$, and $N = 10, M = 20$.

Example 3.5 (Schwartz lantern). The Schwartz lantern is a triangular mesh approximating a cylindrical surface, constructed in the following way: a planar rectangular patch of size $2\pi \times 1$ is divided into $N \times M$ rectangles. The patch is then rolled to form a prism. Each of the $N \times M$ rectangular faces is further replaced by four triangular faces, by adding a vertex lying on the cylinder As the result, a shape shown in Figure 3.13 is obtained. Setting $M = N$, the area of the Schwartz lantern will approach 2π as N grows to infinity. Yet, if we set $M = N^3$, the area will become infinite in the limit $N \to \infty$. The approximated normals in this case tend to be orthogonal to the normals of the continuous surface [284].

Schwartz's example shows that while a triangular mesh can be made infinitesimally close in the sense of the Hausdorff distance to the continuous one (taking N and M to infinity, $d_{H,\mathbb{R}^3}(X, T(X'))$ vanishes), their areas and normals can differ arbitrarily. This gives a negative answer to our question: in general, without additional assumption, a valid triangular mesh does not necessarily approximate well a continuous surface, and the approximation quality depends on the properties of the triangulation itself.

The reason for a pathological behavior in the Schwartz lantern is that the triangles tend to become infinitely "thin." Morvan and Thibert [284] showed that if the triangulation is sufficiently *fat* (i.e., does not have pathologically elongated triangles like in the Schwartz lantern), a bound exists for the area approximation error. A similar result exists for the normals. Amenta and Bern

[7] showed that if the triangles are sufficiently small and fat such that they can be inscribed in a ball with a radius sufficiently small compared with the local feature size, the maximum angle between the actual and the approximate normals can be bounded.[16]

Suggested reading

A presentation of geometry images and their applications in computer graphics can be found in [192, 335, 258]. Farthest point sampling strategy is introduced for images in [151] and extended for surfaces in [312, 313]. In the latter two papers, curvature-adaptive sampling schemes are also studied. For a systematic overview of Voronoi tessellations (mainly Euclidean), the reader is referred to [14, 15]. Voronoi tessellations of parametric surfaces are discussed in [235, 236]. Centroidal Voronoi tessellations and a variety of their applications are explored in [140]. For proofs and discussions of properties of Delaunay tessellations of general Riemannian manifolds, the reader is referred to [247, 248]. Computation of Voronoi tessellations of surfaces using the fast marching method is described in [226, 227]. A good introduction to discrete geometry of polyhedral meshes can be found in [271] and on the *Discrete differential geometry* site ddg.cs.columbia.edu. Estimation of differential geometric properties on discretized surfaces is discussed in [272]. Approximation properties of triangular surfaces are discussed in [166, 7, 284, 101]. For a study of properties of the medial axis transform, refer to [400, 228].

Software

Both the VTK and ITK toolboxes provide basic surface manipulation routines in C++. An excellent tool for mesh simplification is Michael Garland's QSlim C++ library [168]. A MATLAB interface to QSlim is provided in TOSCA. Farthest point sub-sampling for triangular meshes in MATLAB is also available as a part of TOSCA. Remeshing and surface extraction routines are also available in the Afront C++ library, based on [343, 342, 338]. One of the fastest codes for construction of Delaunay tessellations, Voronoi diagrams, and convex hulls in $\mathbb{R}^n$ is Qhull. An interface to Qhull is available in MATLAB through the functions qhull, convhulln, delaunayn, and voronoin.

Problems

3.1. Show that if in a compact metric space there exists an $r/3$-covering containing n points, then an r-separated set cannot contain more than n points.

3.2. Show that in a compact metric space, a maximal r-separated set is an r-covering.

3.3. Prove that the farthest point strategy produces an r-separated r-covering.

3.4. Prove that a convex set is homeomorphic to a disk.

3.5. Prove that centroidal Voronoi tessellation minimizes the variance of the representation error.

3.6. Show an example of a geodesic triangle, whose circumscribing ball is non-unique due to insufficient sampling density.

3.7. Show an example when the Delaunay tessellation of a surface does not exist due to insufficient sampling density.

3.8. Show an example when the Delaunay tessellation is not unique.

3.9. Show that given a smooth compact surface X embedded into $\mathbb{R}^3$, there exists an open set U_X such that $X \subseteq U_X$, and a continuous map $\xi : U_X \to X$, such that for all $u \in U_X$, the point $\xi(u)$ is the orthogonal projection of u onto X and it is unique. Hint: see [156].

3.10. Show that $|\rho(x) - \rho(x')| \le d_{\mathbb{R}^3}(x, x')$ for all $x, x' \in X$.

3.11. Show the relation between the local feature size and the maximum curvature radius in equation (3.3).

3.12. Prove that the area of a triangle with vertices $x_1, x_2, x_3 \in \mathbb{R}^3$ can be expressed as $\frac{1}{2}\|(x_2 - x_1) \wedge (x_3 - x_1)\|_2$.

3.13. Validate the results of Example 3.5 by a formal proof.

Notes

[1] The closedness of X is required in order to have all the infima and suprema realized.

[2] Voronoi regions are named after the Russian mathematician Georgy Voronoi. For the sake of historical justice, we should note that the concept of Voronoi regions emerged independently in different fields of science. The earliest mentioning in René Descartes' *Principia Philosophiae* dates back to the seventeenth century. Voronoi regions were formally described by the German mathematician Gustav Lejeune Dirichlet in his 1850 paper [138], almost sixty years before Voronoi's renowned work [390]. As a tribute to Dirichlet, Voronoi regions are sometimes called *Dirichlet* or *Voronoi-Dirichlet* regions. Other synonyms are *Thiessen polytopes* or *polygons* [378] in meteorology and geography, *Wigner-Seitz zones* in chemistry and physics, and *domains of action* in crystallography.

[3] If the points of X' are not in general position, there might be more than three Voronoi regions adjacent to a Voronoi vertex.

[4] If the sampling is sufficiently dense, the existence and uniqueness of the minimizer is guaranteed by the convexity of V_i.

[5] The intrinsic centroid, defined in terms of the intrinsic metric d_X on the surface, should be distinguished from its extrinsic counterpart, defined in terms of the Euclidean metric in $\mathbb{R}^3$. In Chapter 6, we are going to encounter the extrinsic centroid when discussing moment signatures.

[6] As a historical remark, we should mention that Lloyd and Max were not coauthors. The algorithm was first proposed by Lloyd in 1957 [255, 256]. Unfortunately, Lloyd's paper was not published in a wide-audience journal and remained unnoticed, until Max, not acquainted with Lloyd's work, rediscovered the algorithm in 1960 [263]. As a tribute to both researchers, the algorithm is usually referred to as Lloyd-Max quantization algorithm.

[7] A graph created in this way is *dual* to the Voronoi tessellation.

[8] A related notion is that of a *minimum bounding ball*, which is the smallest metric ball containing the geodesic triangle (without the demand that all vertices lie on the boundary of the ball). The simplest example where the two notions do not coincide is an obtuse triangle in the plane, whose minimum bounding ball has the hypotenuse as diameter and does not pass through the opposite vertex.

[9] It is widely believed that the name Delaunay suggests his French origins. In reality, Boris Delaunay got his surname from an Irish ancestor called Deloney, who was among the mercenaries left in Russia after the Napoleonic invasion of 1812. The closest transliteration from Russian should be "Delone," yet, as Delaunay published his works in French, he preferred to transliterate his name *à la française*.

[10] The term *Delaunay geodesic triangulation* (or simply *Delaunay triangulation*) is also frequently used in the discrete geometry referring to Delaunay tessellation, especially in the Euclidean case. However, we reserve the term *triangulation* to denote a different concept.

[11] Coarseness can also be defined in terms of the minimum bounding ball. In the Euclidean case, d'Azevedo and Simpson [124] showed that this alternatively defined measure of coarseness is also minimized by the Delaunay tessellation. To the best of our knowledge, no analogous property has been studied in the general case.

[12] Approximation of a surface by a triangular mesh constitutes a particular case of a more general *polyhedral approximation*, where the surface is represented as a collection of polygons glued along their edges.

[13] Note that the adjacency matrix E can be obtained from T.

[14] Being more rigorous, the tetrahedron has twelve edges, if we count the symmetric ones as well (i.e, if we consider, for example, $(1,2)$ and $(2,1)$ as separate edges).

[15] Schwartz showed his lantern as an example of erroneous definition of the area of a curved surface (the paper entitled in French *Sur une définition erronée de l'aire d'une surface courbe* [346]). This paper did not address the approximation quality in terms of the Hausdorff distance, which had not been defined yet at that time.

[16] A related bound expressed in terms of curvature appears in [284].

4

Shortest Paths and Fast Marching Methods

> As regards obstacles, the shortest distance between two points can be a curve.
>
> B. BRECHT

Dealing with discrete representation of surfaces in the previous chapter, our main concerns were finding a sampling approximating the surface as a set of points and constructing a polyhedral approximation, which conveys the extrinsic geometry of the surface with sufficient accuracy. However, because a surface is also characterized by an intrinsic geometry, such a discretization is incomplete until the length and the metric structures are also discretized. The ability to approximate the intrinsic geometry of a surface is crucial in our exploration of the non-rigid world. It is also important for producing the sampling itself. Recall that in our description of the farthest point sampling, we tacitly assumed the availability of the intrinsic metric, which measures the length of the minimal geodesic connecting two points on the surface. In this chapter, we explore numerical tools that will allow us to compute the intrinsic metric and the underlying shortest paths. We start our discussion with graphs and then extend it to triangular meshes.

4.1 The shortest path problem

Imagine you plan a railroad trip from Paris to Vienna, and because there is no direct connection between the two cities, you may choose between different routes shown in Figure 4.1. Your travel agent informs you that the railroad operator charges according to mileage. Which route is the most convenient? The simplest way to answer this question is to go over all possible paths between Paris and Vienna (seven in our case, if we exclude loops, i.e., paths passing through the same point more than once) and calculate their lengths:

1. Paris–Brussels–Prague–Vienna ($183 + 566 + 194 = 943$ miles);
2. Paris–Brussels–Munich–Vienna ($183 + 504 + 285 = 972$ miles);
3. Paris–Brussels–Bern–Munich–Vienna ($183 + 407 + 271 + 285 = 1146$ miles);
4. Paris–Bern–Brussels–Prague–Vienna ($346 + 407 + 566 + 194 = 1513$ miles);

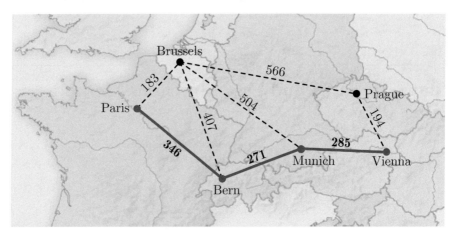

Figure 4.1. Visualization of the shortest path problem in graphs: a traveler has to choose the best route from Paris to Vienna. The route Paris–Bern–Munich–Vienna (solid line) appears to be the shortest.

5. Paris–Bern–Brussels–Munich–Vienna ($346 + 407 + 504 + 285 = 1542$ miles);
6. Paris–Bern–Munich–Brussels–Prague–Vienna ($346+271+504+566+194 = 1881$ miles); and
7. Paris–Bern–Munich–Vienna ($346 + 271 + 285 = 902$ miles).

Clearly, the preferred route is the last one. Note that the answer was obtained by exhaustively searching all possible routes, which is not a feasible algorithm, as its complexity grows exponentially as the number of cities increases. Indeed, if it were used by GPS navigators in our cars, we could wait years before getting the best route to the location of interest. This example serves as an illustration to a fundamental graph theoretic problem known as the *shortest path problem*, which arises in a variety of applications including networking, traffic control, and navigation to name just a few. Our problem of computing the distances between points on a surface approximated as a graph also falls into this category.

Formally, we consider an undirected graph (X, E) endowed with a length function $L : E \to \mathbb{R}$. Recall that when the graph represents a surface, the length is simply given by the Euclidean distance $L(x, x') = \|x - x'\|_2$ for every two adjacent points $(x, x') \in E$. A *path* between two points is a sequence $\Gamma(x, x') = \{x_k\}_{k=1}^{K}$ with $x_1 = x$, $x_K = x'$, and $(x_k, x_{k+1}) \in E$. The length of a path is given by the sum of the lengths of its edges,

$$L(\Gamma) = \sum_{k=1}^{K-1} L(x_k, x_{k+1}).$$

As we have previously seen, such a length structure induces the intrinsic metric

$$d_L(x, x') = \min_{\Gamma} L(\Gamma(x, x')).$$

The shortest path problem consists essentially of computing d_L. Depending on the application, we can distinguish between three problems: computation of $d_L(x_1, x_2)$ for a given pair of points $x_1, x_2 \in X$; computation of a *distance function* or *map* $d(x) = d_L(x_0, x)$ from a given source $x_0 \in X$ to every $x \in X$; and the computation of a *distance matrix* $d_{ij} = d_L(x_i, x_j)$ of all pair-wise distances.

4.2 Dijkstra's shortest path algorithm

Observe that given a shortest path $\Gamma(x_0, x)$ from x_0 to x and some point x' on it, the two sub-paths $\Gamma(x_0, x')$ and $\Gamma(x', x)$ constitute shortest sub-paths from x_0 to x' and from x' to x, respectively. Indeed, if for example there existed a shorter path $\tilde{\Gamma}(x_0, x')$ from x_0 to x, then by concatenating it with $\Gamma(x', x)$, we could construct a new path $\tilde{\Gamma}(x_0, x) = \tilde{\Gamma}(x_0, x') \cup \Gamma(x', x)$ from x_0 to x' with the length

$$\begin{aligned} L(\tilde{\Gamma}(x_0, x)) &= L(\tilde{\Gamma}(x_0, x')) + L(\Gamma(x', x)) \\ &< L(\Gamma(x_0, x')) + L(\Gamma(x', x)) = L(\Gamma(x_0, x)). \end{aligned}$$

This, however, would be a clear contradiction to the fact that $\Gamma(x_0, x)$ has minimum length. This apparently obvious fact is a particular instance of the *Bellman principle of optimality*, which guides a large family of methods known as *dynamic programming*. A direct consequence of the Bellman principle is that for every $x \in X$, there exists some point $x' \in \mathcal{N}(x_0)$ such that

$$d_L(x_0, x) = L(x_0, x') + d_L(x', x),$$

that is, given the second-last point x' in the shortest path between x_0 and x, it can be decomposed into two shortest sub-paths: the one between x_0 and x', plus the edge (x', x). This relation does not give us, however, any indication of how to select such an x' among all the points contained in $\mathcal{N}(x)$. Yet, because $d_L(x_0, x)$ is required to be the shortest distance, we need to select the one that minimizes $L(x_0, x') + d_L(x', x)$, i.e.,

$$d_L(x_0, x) = \min_{x' \in \mathcal{N}(x)} \{L(x_0, x') + d_L(x', x)\}. \qquad (4.1)$$

The latter means that the problem of computing $d_L(x_0, x)$ can be reduced to a smaller sub-problem of computing $d_L(x_0, x')$. Equation (4.1) can be applied recurrently until $d_L(x_0, x_0) = 0$ is reached. Such a recursive relation is called a *dynamic programming functional equation*.

It is probable that similar reasoning guided the Dutch computer scientist Edger W. Dijkstra when in 1959 he proposed his celebrated algorithm for the

70   4 Shortest Paths and Fast Marching Methods

---

    input   : undirected graph (X, E), length function $L : E \to \mathbb{R}$,
                           source point $x_0 \in X$.
    output   : distance map $d : X \to \mathbb{R}$ from the source point.

Initialization
1 **foreach** *point $x \in X$* **do**
2      $d(x) \longleftarrow \infty$
3 **end**
4 $d(x_0) \longleftarrow 0$
5 $Q \longleftarrow X$

Iteration
6 **while** $Q \neq \emptyset$ **do**
      Get the point with smallest value of $d(x)$
7      $x \longleftarrow \arg\min_{x \in Q} d(x)$
      Update all points x' adjacent to x
8   **foreach** $x' \in \mathcal{N}(x) \cap Q$ **do**
9          $d(x') \longleftarrow \min\{d(x'), d(x) + L(x, x')\}$
10   **end**
      Remove x from the unprocessed set
11    $Q \longleftarrow Q \setminus \{x\}$
12 **end**

Algorithm 4.1. Dijkstra's shortest path algorithm.

---

shortest path problem [137]. Essentially, Dijkstra's algorithm (Algorithm 4.1) is a dynamic programming *successive approximation* procedure based on (4.1). It works by keeping for each point x the length $d(x)$ of the shortest path found so far between x_0 and x (i.e., an approximation to the exact distance $d_L(x_0, x)$). Initially, this value is zero for the source point ($d(x_0) = 0$) and infinity for all other points. The algorithm then processes all the points in X in an attempt to improve the approximation $d(x)$. Each point is processed exactly once. Initially, all points are marked as unprocessed and placed into the set Q. Starting with x_0, the next point to be processed each time is the one whose $d(x)$ value is the smallest over all unprocessed points. The processing of $x \in X$ consists of updating the distance approximation of points $x' \in \mathcal{N}(x)$ adjacent to x. The shortest path from x_0 to x can be extended to a path from x_0 to x' by adding the edge (x, x') at the end. The new path length $d(x) + L(x, x')$ is taken as the new approximation of $d(x')$ unless the current value of $d(x')$ is smaller (Step 9 in Algorithm 4.1). Once processed, the point is removed from Q. When all points are processed (i.e., when Q is empty), the algorithm stops and returns $d(x)$, which contain the exact values of $d_L(x_0, x)$.

Example 4.1 (Dijkstra's algorithm). Running Dijkstra's algorithm on our Europe touring example with $x_0 =$ Paris would produce the following sequence of events visualized in Figure 4.2:

1. *Paris* is processed by updating the approximations of *Brussels* and *Bern* to 183 and 346.
2. The minimum unprocessed point is *Brussels* ($d = 183$). The approximations of *Munich* and *Prague* are be updated to $183 + 504 = 687$ and $183 + 566 = 749$, respectively, whereas the approximation of *Bern* remains unchanged, as $346 < 407 + 183$.
3. The minimum unprocessed point is *Bern* ($d = 346$). The only unprocessed adjacent point is *Munich*, which is updated to $346 + 271 = 617$.
4. The minimum unprocessed point is *Munich* ($d = 617$). The only unprocessed adjacent point is *Vienna*, which is updated to $617 + 285 = 902$.
5. The minimum unprocessed point is *Prague* ($d = 749$). The only unprocessed adjacent point is *Vienna*, which is not updated, as $902 < 749 + 194$.
6. The only remaining unprocessed point is *Vienna*, yet it has not unprocessed adjacent points, so no update is performed.

Updating a single point requires finding the minimum of d over all elements of Q (Step 7). As $N = |X|$ points are updated, this results in $\mathcal{O}(N^2)$ time complexity. However, because in our case each point is connected only to its local neighbors so that the graph has $N_E = \mathcal{O}(N)$ edges, extraction of the minimum value can be implemented more efficiently using a binary or a Fibonacci heap, reducing the time complexity to $\mathcal{O}(N \log N)$.

Dijkstra's algorithm computes the distance from a single point in X to all other points (one-to-all). If we need to compute the distances between all pairs of points in X, we have to run the algorithm N times, each time selecting a different point in X as the source.

4.3 Fast marching methods

Dijkstra's algorithm is, beyond any doubt, one of the best choices for finding shortest path lengths in graphs. However, if we try to use it for measuring distances on surfaces, we are likely to encounter serious discrepancies between the measured distances and the expected ones. To understand why this happens, consider a unit square patch in the plane evenly sampled on a regular Cartesian grid and approximated as a graph with four-neighbor connectivity (Figure 4.3). We expect the shortest path between the lower left and the upper right corners of the square to be simply the straight line of length $\sqrt{2}$ connecting them. Yet, Dijkstra's algorithm measures a length of 2, no matter how we refine the grid. We say that the distance measured by Dijkstra's algorithm in *inconsistent* with the Euclidean distance. There is no flaw in the algorithm itself; in fact, every graph-based shortest path algorithm will yield the same result.

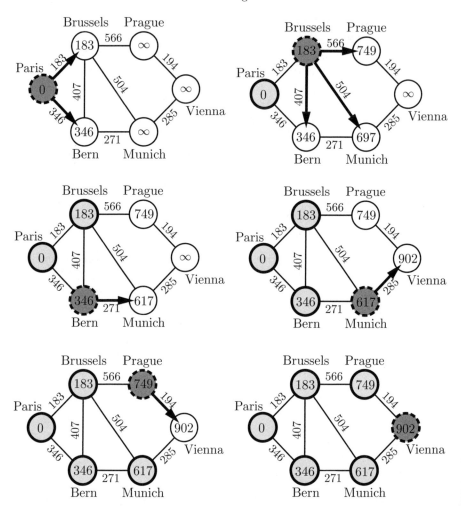

Figure 4.2. Running Dijkstra's algorithm on the European tour example. The algorithm successively improves the approximation of the shortest distance to Paris (iterations are presented from left to right, top to bottom order). White circles indicate unprocessed vertices, dashed circles stand for vertices being processed, and bold circles indicate processed vertices.

The inconsistency is due to the fact that we are allowed to move in the graph only parallel to the axes. This way, there exist many shortest paths between the two corners of the square (two of them are depicted as solid lines in Figure 4.3) all having the same length of 2 and giving no apparent reason to prefer one over the other. Formally, the intrinsic metric induced by the four-neighbor connectivity is called the L_1 or *Manhattan distance*

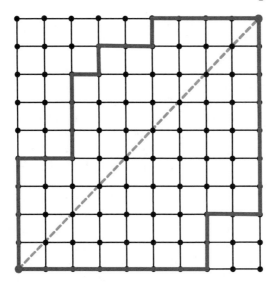

Figure 4.3. Shortest paths measured by Dijkstra's algorithm (solid bold lines) do not converge to the true shortest path (dashed diagonal), no matter how much the grid is refined.

$$d(x_1, x_2) = \|x_1 - x_2\|_1 = \sum_i |x_1^i - x_2^i|$$

(in analogy to the way a taxi driver measures distance in Manhattan's orthogonal grid of streets). This metric is clearly inconsistent with the Euclidean (L_2) distance

$$d(x_1, x_2) = \|x_1 - x_2\|_2 = \sqrt{\sum_i (x_1^i - x_2^i)^2}.$$

We could replace the four-neighbor connectivity by the eight-neighbor one; this would definitely resolve the inconsistency in cases where the path of interest is a diagonal directed at 45°, yet the problem would still remain, for example for a diagonal path of 22.5°.

The inconsistency of the metric, usually termed as *metrication error*, appears to be an inherent problem of representing the surface as a graph.[1] However, it could be resolved by letting the paths pass "between" the graph edges. Practically, this implies transition to a polyhedral representation of the surface, which allows the paths to traverse the faces of the polyhedron without being restricted to the edges. This, of course, requires a new algorithm for computing the shortest path lengths. Here enters a family of algorithms called the *fast marching methods* introduced independently by Sethian [348] and Tsitsiklis [383] in 1995. Sethian presented a method for computing consistent distance maps on regular orthogonal grids, which was later extended to general triangular meshes by R. K. and Sethian [226]. R. K. and Sethian also proposed the geometric interpretation of fast marching methods. Because our objects of interest are non-regular meshes, we focus on the latter algorithm, assuming the surface is given as a triangular mesh.

In order to understand how fast marching works, we start with an illustration. Imagine an imprudent visitor that leaves unextinguished fire at some location in a natural reserve. The flame quickly becomes a forest fire, which expands outwards with constant velocity of $1\ m/sec$. Trees reached by the fire are consumed so the fire never propagates backward. When the firemen arrive, they find the flames so fierce that every effort to extinguish them is in vain. The best they can do is to record the fire front position at different points in time; using these measurements, the firemen can predict when the fire arrives to unburnt regions of the forest and order evacuation of the people. The firemen soon discover that the time of arrival of the fire front to a point in the forest is related to the shortest distance from that point to the source (i.e., the point where the fire started). It appears that the fire traverses the route having the smallest propagation time (and hence, the shortest length). In optics and acoustics this fact is known as *Fermat's principle* or, in a more general form, the *least action principle*. In plain language, Fermat's principle states that light traveling between two points always chooses the quickest path. Snell's law of refraction follows directly from this principle.

We associate the distance function $d(x) = d_X(x_0, x)$ with the time of arrival of the front to a point x. Making an infinitesimal step in a direction $v \in T_x X$, the value of d changes by

$$d(x+v) - d(x) = D_v d(x) + \mathcal{O}(\|v\|^2), \tag{4.2}$$

where $D_v d$ is the directional derivative we have already encountered in Chapter 2. $D_v d$ is a scalar measuring the change in the value of d as we make a differential step in the direction v. Among the different directions a vector v may have, there is some $v_1 \in T_x X$ such that $d_X(x, x_0) = d_X(x+v_1, x_0)$. Such a direction, characterized by $D_{v_1} d(x) = 0$, is said to be *tangential* to the front at the point x. On the other hand, if we make a step in the perpendicular direction v_2, the value of d will change the most. Such a direction of the steepest change of $d : X \to \mathbb{R}$ is usually referred to as the *intrinsic gradient* of d at the point x, denoted by $\nabla_X d(x)$. Formally, the intrinsic gradient is a map $\nabla_X d : T_x X \to T_x X$, satisfying $\langle \nabla_X d(x), v \rangle = D_v d(x)$ for any $v \in T_x X$.

Thinking of d as a function of $x \in \mathbb{R}^3$, we may also define its *extrinsic gradient* as a map $\nabla_x d : \mathbb{R}^3 \to \mathbb{R}^3$ satisfying

$$\langle \nabla_x d(x), dx \rangle_{\mathbb{R}^3} = \left. \frac{d}{dt} d(x + t\, dx) \right|_{t=0} \tag{4.3}$$

for any $dx \in \mathbb{R}^3$. In the standard Euclidean basis, $\nabla_x d$ can be expressed as the vector of partial derivatives,

$$\nabla_x d = \left(\frac{\partial d}{\partial x^1}, \frac{\partial d}{\partial x^2}, \frac{\partial d}{\partial x^3} \right)^{\mathrm{T}}, \tag{4.4}$$

which is usually referred to simply as "the gradient of d" in multivariate calculus. In the coordinate of the parameterization $x : U \to X$, the intrinsic gradient can be obtained by projecting its extrinsic counterpart on the tangent space of X,

$$\nabla_X d = J(J^{\mathrm{T}} J)^{-1} J^{\mathrm{T}} \nabla_x d, \tag{4.5}$$

where $J = (x_1, x_2)$ is the 3×2 Jacobian matrix, whose columns span the tangent space of X at the point x. We leave the formal proof to the reader as Problem 4.8.

It appears that for a distance function, the direction of the steepest change of d coincides with the direction of the front propagation,[2] which is also the direction of the shortest path connecting x_0 with x. Formally, if $\Gamma : [0, L] \to X$ is a minimal geodesic such that $\Gamma(0) = x_0$ and $\Gamma(t) = x$, then

$$\dot{\Gamma}(t) = \frac{d}{dt} \Gamma(t) = -\nabla_X d(\Gamma(t)). \tag{4.6}$$

Since Γ is a geodesic, $\|\dot{\Gamma}\|_2 = 1$ (a formal proof is left to the reader as Problem 4.1). We can therefore conclude that a distance map d has to satisfy

$$\|\nabla_X d(x)\|_2 = 1 \tag{4.7}$$

with the boundary condition $d(x_0) = 0$. This partial differential equation is called the *eikonal equation* (from Greek εικων for "image"), which points to its optical origin. Note that along geodesics, the eikonal equation reduces to the ordinary differential equation (4.6), called the *characteristic equation* [106, 225]. Computation of the distance map $d(x)$ requires solution of the eikonal equation,[3] whose *characteristics* (curves on which the partial differential equation becomes an ordinary differential equation) are minimal geodesics.

Fast marching (Algorithm 4.2) is a numerical procedure that solves the eikonal equation by simulating wavefront propagation. Technically, it is a dynamic programming successive approximation method, very similar to Dijkstra's algorithm, and can be considered a continuous version of the latter. Fast marching keeps for each point x on the mesh the time of arrival $d(x)$ of the wave front originating in x_0. Because of the equivalence of path length and arrival time, we are going to switch freely between these two terms. The initial approximation of $d(x)$ is, like in Dijkstra's algorithm, zero at x_0 and infinity elsewhere. The algorithm classifies the points of the mesh into three categories, which we chose to label with *black*, *red* and *green* colors, suggested by the forest fire example. Black points are points where the arrival time has been computed and is not going to change in the future. Green points are unprocessed points, for which the arrival time has not been computed yet

76   4 Shortest Paths and Fast Marching Methods

---

input   : non-obtuse triangular mesh (X, T), source point x_0.
output   : distance map $d : X \to \mathbb{R}$ from the source point.

Initialization
1 **foreach** *point* $x \in X$ **do** $d(x) \longleftarrow \infty$
2 $d(x_0) \longleftarrow 0$
3 $B \longleftarrow \{x_0\}$
4 $R \longleftarrow \mathcal{N}(x_0)$
5 $G \longleftarrow X \setminus (B \cup R)$

Iteration
6 **while** $B \neq X$ **do**
      Get the point with smallest value of $d(x)$
7     $x_1 \longleftarrow \arg\min_{x \in R} d(x)$
      Update all triangles that share x
8   **foreach** *triangle* $(x_1, x_2, x_3) \in \{(x_1, x_2, x_3) \in T : x_2 \in X, x_3 \in B^c\}$
      do
9        $R \longleftarrow R \cup \{x_3\}$
10   Update (x_1, x_2, x_3)
11   **end**
      Remove x_1 from the unprocessed set
12    $R \longleftarrow R \setminus \{x_1\}$
13    $B \longleftarrow B \cup \{x_1\}$
14 **end**

Algorithm 4.2. Fast marching algorithm.

(corresponding with live trees). Red points are those belonging to the propagating wave front, which can be considered an interface between the black and the green regions of the mesh. In our forest fire example, red points correspond with trees that are currently in flames. Initially, only the source x_0 is marked as black and all points adjacent to it are marked as red. The remaining points are marked as green. Like Dijkstra's algorithm, at each iteration we process the red point with the smallest value of $d(x)$ by updating the approximation of all the non-black points in triangles sharing it. The red point is then tagged as black and the updated adjacent points are tagged as red. The process continues until all points become black.

The numerical heart of the fast marching method and its main difference from Dijkstra's algorithm is hidden in the *update* procedure (Step 10 in Algorithm 4.2). Recall that in Dijkstra's algorithm the path was restricted to the graph edges, and a graph vertex was updated each time from an adjacent vertex. In fast marching, because the path can pass through the triangular faces of the mesh, a vertex has to be updated from a triangle, requiring two supporting vertices. We assume that the update step is applied to a triangle (x_1, x_2, x_3), where x_1 is the red point with the smallest arrival time $d_1 = d(x_1)$, x_2 is a point for which some arrival time approximation $d_2 = d(x_2)$ is available, and x_3 is the red or green point, whose arrival time approximation $d_3 = d(x_3)$ is being updated. For simplicity, we further assume without loss of generality that the triangle lies in the plane with $x_3 = 0$, and leave to the reader to derive the update equations in the general case.

In essence, given that the front reaches x_1 at time d_1 and x_2 at time d_2, the update step has to estimate the time when the front arrives to x_3. Let us assume that the front is planar,[4] i.e., propagating from some planar source described by the equation $n^T x + p = 0$. The unit vector n determines the propagation direction, whereas the scalar p determines the source origin. Clearly, x_1 and x_2 must be distant d_1 and d_2 from the plane, respectively. Using the point-to-plane distance, we obtain the following set of equations

$$d_1 = n^T x_1 + p;$$
$$d_2 = n^T x_2 + p.$$

In matrix notation, the later system of equations can be expressed as

$$V^T n + p \cdot 1_{2\times 1} = d,$$

where V is a 2×2 matrix whose columns are x_1, and x_2, $1_{2\times 1} = (1,1)^T$, and $d = (d_1, d_2)^T$. Our goal is to solve for the wavefront parameters n and p, and compute d_3 as the distance of x_3 from the plane,

$$d_3 = n^T x_3 + p = p. \tag{4.8}$$

Assuming the triangle (x_1, x_2, x_3) is non-degenerate, V is full-rank and we can solve for n, obtaining

$$n = (V^T)^{-1}(d - p \cdot 1_{2\times 1}) = V^{-T}(d - p \cdot 1_{2\times 1}).$$

Apparently, we have only two equations with three unknowns, but in reality the wavefront has only two degrees of freedom: direction and origin. The "missing" degree of freedom hides in the fact that n is a unit vector. Enforcing $\|n\| = 1$ yields

$$1 = n^T n = (d - p \cdot 1_{2\times 1})^T V^{-1} V^{-T} (d - p \cdot 1_{2\times 1})$$
$$= (d - p \cdot 1_{2\times 1})^T (V^T V)^{-1} (d - p \cdot 1_{2\times 1})$$
$$= p^2 \cdot 1_{2\times 1}^T Q 1_{2\times 1} - 2p \cdot 1_{2\times 1}^T Q d + d^T Q d,$$

where $Q = (V^T V)^{-1}$. Since $d_3 = p$, we conclude that d_3 is given as a solution to the quadratic equation

$$d_3^2 \cdot 1_{2\times 1}^T Q 1_{2\times 1} - 2 d_3 \cdot 1_{2\times 1}^T Q d + d^T Q d - 1 = 0. \tag{4.9}$$

Note that equation (4.9) has two solutions, stemming from the fact that both n and $-n$ satisfy the condition $\|n\| = 1$. The smallest solution corresponds with n forming an acute angle with x_1 and x_2 (Figure 4.4, left). In this case,

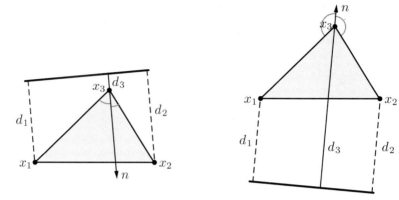

Figure 4.4. Fast marching updates the triangle (x_1, x_2, x_3) by estimating the planar wavefront direction n and origin p based on d_1 at x_1 and d_2 at x_2, and propagating it further to x_3. d_3 has two possible solutions: the one shown on the left is inconsistent, as $d_3 < d_1, d_2$. The solution on the right is consistent, as $d_3 > d_1, d_2$. Geometrically, in order to be consistent, the update direction n has to form obtuse angles with the triangle edges (x_3, x_1) and (x_3, x_2).

the front arrives to x_3 *before* it arrives to x_1 and x_2. This, of course, contradicts the construction of the fast marching algorithm, which assumes that the updated distance d_3 is larger than the supporting distances d_1 and d_2. We term such a solution *inconsistent* and discard it. As shown in Figure 4.4 (right), a consistent solution has to form an obtuse angle with both triangle edges (x_3, x_1), (x_3, x_2), which can be expressed simply as $V^T n < 0$. The reader is invited to check (Problem 4.4) that this condition is satisfied by the largest solution for d_3.

It appears that in order to make our distance map convergent to the true solution of the eikonal equation as the mesh is refined, it is also required that an increase of d_1 or d_2 increases d_3. This *monotonicity condition* can be stated as

$$\nabla_d d_3 = \left(\frac{\partial d_3}{\partial d_1}, \frac{\partial d_3}{\partial d_2} \right)^T > 0,$$

where the inequality is interpreted coordinate-wise. In order to obtain an expression for $\nabla_d d_3$, we differentiate equation (4.9) with respect to $d = (d_1, d_2)^T$,

$$0 = d_3 \cdot \nabla_d d_3 \cdot 1_{2\times 1}^T Q 1_{2\times 1} - \nabla_d d_3 \cdot 1_{2\times 1}^T Qd - d_3 \cdot Q 1_{2\times 1} + Q 1_{2\times 1},$$

from where

$$\nabla_d d_3 = \frac{Q(d - p \cdot 1_{2\times 1})}{1_{2\times 1}^T Q(d - p \cdot 1_{2\times 1})}.$$

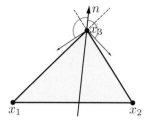 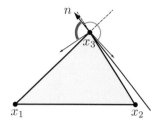

Figure 4.5. The monotonicity condition $QV^T n < 0$ requires the wavefront propagation direction n to form obtuse angles with the normals to the triangle edges (x_3, x_1) and (x_3, x_2). Geometrically, this can be interpreted as demanding n to lie within the triangle (left). Right: update direction violating the monotonicity condition lies outside the triangle.

Substituting $n = V^{-T}(d - p \cdot 1_{2 \times 1})$, we can write

$$\nabla_d d_3 = \frac{QV^T n}{1_{2 \times 1}^T QV^T n}.$$

Observe that the monotonicity condition $\nabla_d d_3 > 0$ is satisfied when either $QV^T n > 0$, or $QV^T n < 0$, that is, both coordinates of $QV^T n$ have the same sign. However, because the consistency of the solution requires $V^T n$ to be negative, and Q is positive semi-definite, $QV^T n$ cannot have both coordinates positive. We therefore conclude that the solution has to satisfy $QV^T n < 0$.

From the relation

$$QV^T V = (V^T V)^{-1} V^T V = I,$$

it follows that the rows of QV^T are orthogonal to the triangle edges (x_3, x_1), (x_3, x_2). The monotonicity condition has therefore the following geometric interpretation: the wavefront propagation direction n must lie within the triangle (Figure 4.5). Adding the consistency condition, n must also form obtuse angles with the triangle edges. To satisfy this demand for any update coming from within the triangle, the angle $\triangleleft x_1 x_3 x_2$ must be *acute*. When the triangle is *obtuse*, even if n lies inside the triangle, the front may arrive to x_3 before reaching x_1 or x_2 (Figure 4.6). To cure this problem, R. K. and Sethian proposed to split an obtuse angle into two acute ones by connecting the vertex x_3 to another point on the mesh [226]. The splitting is performed as a pre-processing stage.

If the wavefront propagation direction happens to violate the monotonicity condition, we can force it to lie within the triangle. In such cases, n will coincide with the direction of one of the edges (x_3, x_1) or (x_3, x_2), and the update will assume the simple Dijkstra-like form

$$d_3 = \min\{d_1 + \|x_1\|, d_2 + \|x_2\|\}.$$

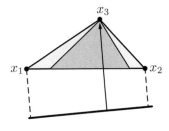 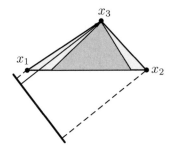

Figure 4.6. When updating an obtuse triangle, the front may come only from a limited angular section (shaded) in order to obey $d_3 > d_1, d_2$ (left). When the front comes outside this section (right), it first arrives to x_1 and x_3 and only then to x_2. Consequently, x_3 is supported by a single point, x_1 only, which does not allow to recover the actual front direction.

As in Dijkstra's algorithm, if the computed d_3 is larger than the current approximation of $d(x_3)$, x_3 is not updated. The entire update step can be summarized as in Algorithm 4.3.

The described update procedure constructs a linear approximation of $d(x)$ over the triangle (x_1, x_2, x_3). This construction can be viewed as fitting a tilted plane to the given values of d_1 and d_2 anchored to the corresponding triangle vertices. The tilt (the gradient) of the plane is set to 1 (45°) in order to satisfy the eikonal equation; the remaining two degrees of freedom are orientation and origin, which correspond with n and p in our notation. By a sequence of updates applied to different triangles, fast marching gradually constructs a piecewise planar approximation to the true $d(x)$. James Sethian compared the

---

input : non-obtuse triangle with the vertices x_1, x_2, x_3, and the corresponding arrival times d_1, d_2, d_3
output : updated d_3

1 Solve the quadratic equation

$$p = \frac{1_{2\times 1}^T Qd + \sqrt{(1_{2\times 1}^T Qd)^2 - 1_{2\times 1}^T Q 1_{2\times 1} \cdot (d^T Qd - 1)}}{1_{2\times 1}^T Q 1_{2\times 1}}.$$

where $V = (x_1 - x_3, x_2 - x_3)$, and $d = (d_1, d_2)^T$.
2 Compute the front propagation direction $n = V^{-T}(d - p \cdot 1_{2\times 1})$
3 if $(V^T V)^{-1} V^T n < 0$ then
4 $d_3 \longleftarrow \min\{d_3, p\}$
5 else
6 $d_3 \longleftarrow \min\{d_3, d_1 + \|x_1\|, d_2 + \|x_2\|\}$
7 end

Algorithm 4.3. Fast marching update step.

Figure 4.7. Distance map measured on a triangular mesh. Equidistant contours from the source located at the right hand are shown.

way such an approximation is constructed with erecting scaffolding around a building: a worker stands on one of the boards, puts a board above his head, and then moves to another board at the same level and puts a board one level up. Once all the boards are placed at a given level, the worker climbs up to the next level and repeats the process. The scaffolding is built from the ground up; each level must be completed before the next is begun.

4.4 Fast marching on parametric surfaces

The fast marching method we have been discussing so far works for general triangular meshes. Spira and R. K. showed that when the surface is given in parametric form, the algorithm can be formulated entirely in the parameterization domain [367]. Let us be given a surface $x : U \to \mathbb{R}^3$ parameterized over $U \subset \mathbb{R}^2$; our goal is to compute the distance map $d : U \to \mathbb{R}$ from some source $u_0 \in U$ (note that d is now a function of the parameterization coordinates u^1, u^2). Using the chain rule, we obtain

$$\frac{\partial d}{\partial u^i} = \frac{\partial d}{\partial x^1}\frac{\partial x^1}{\partial u^i} + \frac{\partial d}{\partial x^2}\frac{\partial x^2}{\partial u^i} + \frac{\partial d}{\partial x^3}\frac{\partial x^3}{\partial u^i}$$

for $i = 1, 2$, or in vector notation,

$$\nabla_u d = J^{\mathrm{T}} \nabla_x d. \tag{4.10}$$

Projecting the extrinsic gradient $\nabla_x d$ on the tangent space, we can express the intrinsic gradient $\nabla_X d$ in terms of the parameterization coordinates,

$$\nabla_X d = J(J^\mathrm{T} J)^{-1} J^\mathrm{T} \nabla_x d = JG^{-1} \nabla_u d, \quad (4.11)$$

where $G = J^\mathrm{T} J$ is the first fundamental form matrix we have encountered in Chapter 2. Substituting the latter expression into the eikonal equation, we get

$$1 = \nabla_X^\mathrm{T} d \nabla_X d = \nabla_u^\mathrm{T} d \, G^{-\mathrm{T}} J^\mathrm{T} J G^{-1} \nabla_u d = \nabla_u^\mathrm{T} d \, G^{-1} \nabla_u d.$$

where $G = J^\mathrm{T} J$ is the first fundamental form matrix we have encountered in Chapter 2. The latter is the eikonal equation expressed in the parameterization coordinates. Note that the metric participates in the expression, accounting for the fact that the distance d is measured on the surface.

The fast marching algorithm aims at computing an approximation to $d(u^1, u^2)$ on a numerical grid in the parameterization domain. For simplicity, let us assume that U is discretized on a Cartesian grid with unit step in each direction. A grid point (u^1, u^2) is connected to neighboring grid points $(u^1 + m^1, u^2 + m^2)$ according to some planar *connectivity pattern*, of which the simplest is the *four-neighbor* connectivity $(m^1, m^2) = (\pm 1, 0), (0, \pm 1)$. Another possible grid connectivity is based on eight neighbors $(m^1, m^2) = (\pm 1, 0), (0, \pm 1), (\pm 1, \pm 1)$. Each vertex in the grid becomes supported by a set of triangles (four and eight in the case of the former two connectivity patterns), formed upon the neighboring vertices. The distance map d is computed by traversing the numerical grid in the fast marching order and updating grid points from the neighboring triangles using the previously described update step.

Let $u_3 = (u^1, u^2)$ be a grid point being updated from the triangle formed by $u_1 = (u^1 + m_1^1, u^2 + m_1^2)$ and $u_2 = (u^1 + m_2^1, u^2 + m_2^2)$. We denote by x_1, x_2 and x_3 the corresponding points on the surface, and assume without loss of generality that $x_3 = 0$. Neglecting second-order terms, we can write

$$x_1 \approx m_1^1 r_1 + m_1^2 r_2 = J m_1$$
$$x_2 \approx m_2^1 r_1 + m_2^2 r_2 = J m_2,$$

or in matrix form, $V \approx JM$, where $V = (x_1, x_2)$ and $M = (m_1, m_2)$. We can now define

$$E = V^\mathrm{T} V \approx M^\mathrm{T} G M, \quad (4.12)$$

a 2×2 matrix, whose elements are the dot products $e_{ij} = \langle x_i, x_j \rangle$. It is remarkable that the triangle geometry required for the fast marching update step is contained entirely in that matrix. Indeed, the diagonal elements e_{11} and e_{22} are the squared lengths of the triangles edges $x_1 x_3$ and $x_2 x_3$, respectively, whereas $e_{12}/\sqrt{e_{11} e_{22}}$ is the cosine of the angle at x_3. The knowledge of the metric coefficients and the connectivity pattern of the grid is sufficient to compute a distance map on the parametric surface. This is a very handy property in some applications, for example, where surface acquisition techniques do not provide the surface itself, but rather its gradients [77].

As a final remark on this method, recall that the fast marching update scheme was valid for acute triangles only. On parametric surfaces, this condition can be stated as $e_{12} > 0$. In case where a triangle is obtuse ($e_{12} < 0$), Spira and R. K. proposed to split it into two acute triangles by adding a connection to another "virtual" non-adjacent neighbor on the grid [367]. The virtual connection is selected as the one producing two acute angles and having the shortest length in the parameterization domain. Adding such virtual connections to grid points can be done in $\mathcal{O}(N)$ at the grid initialization stage.

4.5 Marching even faster

One of the core components of all Dijkstra-type distance computation algorithms, including fast marching methods, is the heap capable of extracting the vertex with the smallest value of d in logarithmic time.[5] Selecting the next vertex to be updated according to minimum distance ensures that the grid points are visited in an order simulating the propagation of a wavefront. This fact conceals one of the major drawbacks of fast marching: the grid traversal order depends on the data and cannot be known *a priori*. Moreover, this order is not well-structured, making problematic an efficient use of memory systems. Such a strident lack of structure calls for searching alternative traversal orders.

In his classic paper [121], Per-Erik Danielsson studied the computation of distance maps from arbitrary sources in the plane. He observed that as the Euclidean geodesics are straight lines, the characteristics of the eikonal equation fall into one of the four plane quadrants and can be therefore covered by a sequence of four directed scans, where in each scan a grid point is updated from the previously updated "causal" neighbors in the scan order (Figure 4.8). Each point is updated four times, implying linear complexity in the grid size, $\mathcal{O}(N)$. Because the order in which the grid points are visited is known in advance and is independent of the data, Danielsson's raster scan algorithm is characterized by regular access to the memory and can benefit significantly from the *caching* mechanism supported by most modern processor architectures.

Equipping the Danielsson's raster scan algorithm with the fast marching update scheme gives rise to a family of distance computation algorithms, where the Dijkstra-type wavefront propagation is replaced by sweeping the grid in four alternating directions. One of the earliest mentioning of this technique dates back to Dupuis and Oliensis' studies on shape from shading from 1994 [142]. A raster-scan algorithm for solving the eikonal equation

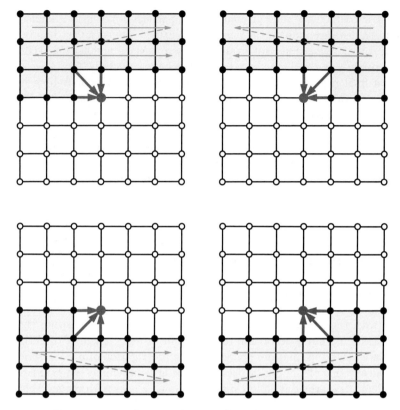

Figure 4.8. Raster scan grid traversal used in Danielsson's algorithm. Four directed raster scans are sufficient to cover all the geodesic directions in the plane.

on weighted Euclidean domains was also studied by Zhao, who introduced the name *fast sweeping* [411].

Because the raster scan algorithm operates on a regular Cartesian grid, it is an attractive alternative to fast marching for computation of distance maps on parametric surfaces [61, 60]. However, it is important to realize that unlike the Euclidean case where the geodesics are straight lines and thus can be covered by four directed raster scans, on a general surface geodesics are usually curved. This implies that four raster scans may cover only a part of an eikonal equation characteristic; in order to obtain a consistent distance map, the scans have to be repeated several times. This fact is visualized in Figure 4.9, where the raster scan algorithm is used to compute the distance map in a "maze" with complicated spiral-shaped geodesics. Six iterations of alternating raster scans are required in order to cover it completely. We can therefore conclude that the complexity of the raster scan algorithm is *data-dependent*. Nevertheless, it appears that the maximum number of iterations required to produce a consistent distance map on a parametric surface

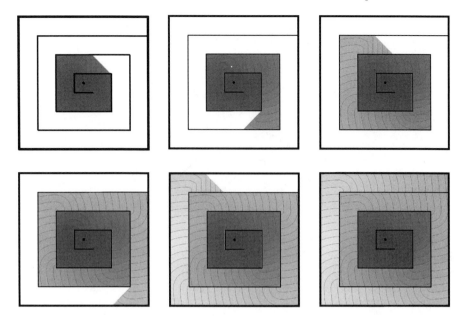

Figure 4.9. Distance map computation on the "maze" surface using the raster scan algorithm after an increasing number of iterations (left-to-right, top-down). Note that a single repetition of the four directed raster scans is insufficient to cover the complicated spiral-shaped characteristic.

can be bounded assuming some regularity of the metric coefficients and the second-order derivatives of the parameterization [61, 60]. The bound does not depend on the grid size, leaving the theoretical complexity of the algorithm $\mathcal{O}(N)$. Yet, the number of iterations depends on the properties of both the surface itself and its parameterization. This means that some parameterizations of the same surface may be less advantageous than others from the point of view of the raster scan algorithm. For example, with the trivial parameterization of the plane $(x^1, x^2, x^3) = (u^1, u^2, 0)$, the geodesics are straight lines requiring one iteration. Using a more bizarre parameterization, say, $(x^1, x^2, x^3) = (u^1 \cos u^2, u^1 \sin u^2, 0)$, several iterations are needed.

4.6 Parallel distance computation

In addition to the better structured access to memory, the raster scan algorithm has another important advantage over traditional fast marching: unlike fast marching methods, which are inherently *sequential*, raster scan can benefit from *parallelization*. To demonstrate the parallelism, let us consider for example the right-down scan (upper left in Figure 4.8) starting from the top leftmost grid point d_{11} (we denote by d_{ij} the value of the distance map d

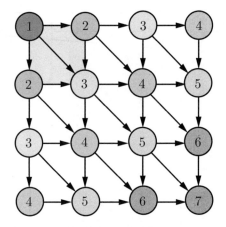

Figure 4.10. Dependency graph of the right-down raster scan. Grid points lying on the lines $i + j = \text{const}$ can be updated concurrently.

at the point (i, j) on the grid). After d_{11} has been updated, the point d_{12} is updated. Note that the point d_{21} can also be updated in parallel to d_{12}, as it depends only on d_{11}. Next, points d_{31}, d_{22}, and d_{13} can be updated concurrently, and so on. Figure 4.10 visualizes the fact that grid vertices lying on the lines $i + j = k$ depend only on the vertices lying on the lines $i + j = k - 1$ and $i + j = k - 2$ and can be therefore updated in parallel. Similar parallelization can be used for the other three raster scans.

Whereas for sufficiently long lines the speedup due to parallelization is significant, points located along shorter lines benefit less from concurrent computation. Also, the fact that the number of update operations performed in parallel is not constant may complicate efficient implementation. A way to overcome this difficulty was found by Ofir Weber [61, 60], who proposed to rotate the directions of all scans by 45°, as depicted in Figure 4.11. Let us examine again the right-down scan from the previous example, now rotating the scan direction by 45° counter-clockwise (Figure 4.11, upper left). Because a grid point d_{ij} is updated from $d_{i-1,j-1}, d_{i,j-1}$, and $d_{i+1,j-1}$, all points in the j-th column of the grid can be processed simultaneously once the $(j - 1)$-st column has been updated (Figure 4.12). In other words, depending on the scan direction, all grid vertices belonging to the same row or column can be updated concurrently. On k processors,[6] such a parallelization allows one to speed up the distance map computation by k which is referred to as "embarrassingly parallel" in the parallel computing jargon.

Practically, the structured memory access pattern of the raster scan algorithm combined with parallel execution gives a dramatic improvement in the execution time compared with that of conventional fast marching. An implementation on standard 32-bit Intel architecture with SSE extensions[7] shortens the execution time by one order of magnitude. Implementation on a

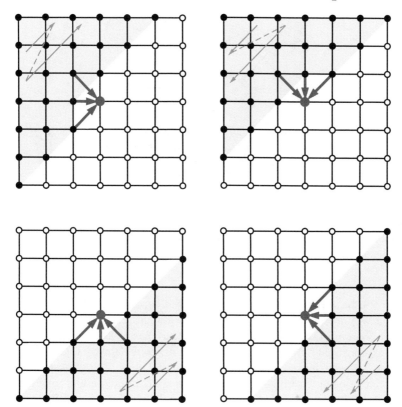

Figure 4.11. Raster scan grid traversals from Figure 4.8 rotated by 45° counterclockwise.

graphics processing unit[8] exhibits an improvement by another order of magnitude. To give an impression, a single iteration of the raster scan algorithm on a grid consisting of nine million vertices takes below 40 milliseconds, enabling real-time application of many computer graphics and computational geometric algorithms based on geodesic distances [61, 60].

4.7* Minimal geodesics

Thus far, we have focused our attention on computing the geodesic distances on a surface. However, it is often important to compute paths realizing those distances, that is, minimal geodesics. As an illustration, imagine that NASA has commissioned us the development of a navigation system of an autonomous exploration vehicle that is supposed to land on the surface of planet Venus. Because of the densely clouded Venusian atmosphere, little sunlight reaches the ground, which precludes the use of solar cells. Hence, the

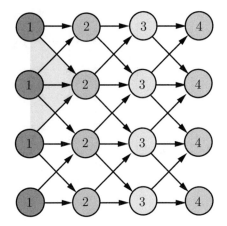

Figure 4.12. Dependency graph of the right-down raster scan rotated by 45°. Grid points belonging to the same column can be updated concurrently.

vehicle will have to use batteries with limited capacity. Because a similar expedition will not be repeated in the nearest few decades, the scientists expect to get maximum information from it. For this reason, the navigation system of the exploration vehicle has to be smart: it should always follow the shortest path between source and destination points, thus taking most advantage of the battery life. Astronomers provide us with radar measurements of the ground topography in the vicinity of the landing point; the optimal path of the rover can be described as the shortest path on that surface.

Computation of shortest paths is important in many applications such as the above navigation problem (though, usually in more down-to-earth scenarios). We have already mentioned that the characteristics of the eikonal equation are shortest paths on the surface (the proof is left as an exercise in Problem 4.3). This means that given the distance map d computed from some source point x_0 on the surface, a curve Γ satisfying the ordinary differential equation

$$\dot{\Gamma} = -\nabla_X d$$

(the characteristic equation) with the initial condition $\Gamma(0) = x_1$ is a minimal geodesic passing between x_0 and x_1. In order to find the geodesic, we have to solve (or *integrate*) the characteristic equation. This is usually done using a numerical ODE solver, sometimes referred to as *backtracking* [226]. Starting at x_1, we construct a curve passing on the surface and following the direction of $-\nabla_X d$, until $d = 0$ is reached at x_0. Intuitively, this process can be thought of as descending "downhill" on the distance map. The selection of the negative gradient direction guarantees the fastest decrease of the distance. In practice, when the surface is represented as a triangular mesh and the distance map is computed using the fast marching method, d is piecewise-linear. This means that $\nabla_X d$ is constant on any triangle of the mesh and changes only when we transit from one triangle to another. As a consequence, the backtracked geodesic is constructed of linear segments, lying in the planes of the mesh faces. There exist higher-order methods for backtracking geodesics on the surface, e.g., Huen's (or modified Euler) numerical integration method, which approximates Γ as a second-order polynomial. However, special care has to be taken when using such methods, as the distance map and, consequently, the geodesics, are not $\mathcal{C}^1$. The continuity of the gradient is violated at points called *shocks*, where two wavefronts meet. At such points, it is common to switch back to the first-order backtracking scheme.

When the surface is given in parametric form, the computation of geodesics can be performed entirely in the parameterization domain. Let us denote by γ the parameterization of a geodesic Γ. Substituting $\nabla_u d$ from equation (4.10), we obtain

$$\nabla_u d = J^\mathrm{T} \nabla_x d = -J^\mathrm{T} \dot{\Gamma} = -J^\mathrm{T} J \dot{\gamma} = -G \dot{\gamma}.$$

We can now express the characteristic equation in terms of γ, as

$$\dot{\gamma} = -G^{-1} \nabla_u d, \qquad (4.13)$$

which can be interpreted as a *scaled gradient descent*. Again, backtracking techniques can be used to construct γ in the parameterization domain. Backtracking of minimal geodesics is intimately connected to the family of *gradient descent* optimization techniques that we will encounter in the next chapter.

Suggested reading

For a mathematically oriented reader, we recommend Bellman's book *Dynamic Programming* [23], which shows in depth the foundations of dynamic programming and presents a different outlook on the calculus of variations from the perspective of the principle of optimality. A detailed overview of fast marching methods and their applications can be found in James Sethian's book *Level Set Methods and Fast Marching Methods* [349] and R. K.'s book *Numerical Geometry of Images* [225]. The paper [268] by Mémoli and Sapiro introduces an interesting fast-marching type algorithm for computation of geodesic distances directly on point clouds and implicit surfaces. A fast marching method for parametric surfaces is described in [367]. A family of raster-scan based fast sweeping methods is discussed in [217, 382, 411]. Parallel algorithms for distance computation and their implementation on SIMD processor architectures and graphics hardware is presented in [61, 60]. For a reader interested in numerical solution of partial differential equations, the paper [350] by Sethian and Vladimirsky presents a number of numerical schemes for solving the eikonal equation as well as the more general class of static Hamilton-Jacobi equations. A fast marching algorithm based on an alternative spherical wavefront model is discussed in [299]. For a completely different, non-fast marching type algorithm for the computation of distance maps on meshes, known as the Mitchell-Mount-Papadimitriou algorithm (MMP for short), the reader is referred to [276]. A fast approximation to the MMP algorithm is studied in [371], which to the best of our knowledge is the fastest sequential algorithm for geodesic distance computation available in the public domain.

Software

An implementation of the fast marching method on triangular meshes is available as part of TOSCA. SSE2 and GPU implementations of the parallel algorithm for parametric surfaces are available as binaries accompanying [60].

Problems

4.1. Show that a geodesic satisfies $\|\dot{\Gamma}\|_2 = 1$.

4.2. Prove that the intrinsic gradient $\nabla_X d$ can be obtained by projecting the extrinsic gradient $\nabla_x d$ onto the tangent plane.

4.3. Prove that a distance function obeys the eikonal equation. Show that the characteristics of the eikonal equation are minimal geodesics.

4.4. Show that the largest solution of the quadratic equation (4.9) is associated with the normal direction n, which satisfies the consistency condition $V^T n < 0$.

4.5.✓ In practical implementation, arithmetics has limited precision. Analyze the sensitivity of the fast marching update scheme to additive truncation and round-off noise. Hint: evaluate the derivative of d_3 with respect to d_1 and d_2.

4.6. In our discussion, the fast marching update scheme was presented in a special system of coordinates, where the triangle was supposed to lie in the plane. Formulate the fast marching update scheme for a triangle in an arbitrary system of coordinates in $\mathbb{R}^3$.

4.7.✓ Devise an alternative update scheme for the fast marching algorithm based on a *circular* wavefront approximation. Prove that such a scheme does not always give a consistent update. What can be said about its numerical stability?

4.8. There exist algorithms, like the Mitchell-Mount-Papadimitriou algorithm [276, 371], capable of computing the exact geodesic distances on a triangular mesh. Assuming that the mesh is only a first-order approximation of some underlying continuous surface, do such algorithms have better accuracy compared with fast marching that produces a first-order approximation of the distance map?

4.9. One of the main limitations of the raster scan algorithm is the fact that it works only with a single-chart parametric surface. However, sometimes it is impractical or impossible to represent a surface using a single chart. Extend the raster scan algorithm to the multiple-chart case.

Notes

[1] In our regular example, we oversimplified the metrication problem in graphs. In reality, there exist conditions on sampling and connectivity that guarantee convergence to the true metric on the surface [47]. However, if the sampling of the surface is a part of the acquisition process (e.g., the surface is sampled on a regular grid), such conditions are usually violated.

[2] Formally, a wave traveling between two points must traverse a path whose length is stationary with respect to variations of the path. According to calculus of variations, this path is exactly the characteristic of the eikonal equation.

[3] The eikonal equation differs from the majority of other familiar differential equations in the fact that $d(x)$ does not have to be (and usually is not) everywhere differentiable, i.e., $\nabla_X d(x)$ may not exist at some $x \in X$. The introduction of the concept of *viscosity solution* by Pierre-Louis Lions and Michael Crandall [120] in the early 1980s permitted to address this problem formally. When computing the distance function, we actually are looking for a viscosity solution of the eikonal equation.

[4] This assumption is accurate for points far away from the source; for points nearby, a somewhat better assumption is a spherical model of the source. Such a scheme was proposed by Klein and Novotni in [299] but appears to have numerical stability and consistency problems [61, 60].

[5] Several alternatives to heap sorting have been proposed, among which a notable example is the *untidy* priority queue mentioned by Tsitsiklis [383] and later by Yatziv et al. [404]. Untidy priority queue is based on quantization of the distance values used for selecting the next vertex to be updated, allowing one to extract the minimum in $\mathcal{O}(1)$ with a slight reduction of the algorithm accuracy.

[6] Assuming the grid size $M \times N$ is such that $M, N \geq k$.

[7] SSE, standing for *Streaming SIMD Extensions*, is a single instruction multiple data (SIMD) instruction set introduced by Intel in 1999 in their Pentium III processors, superseding the MMX extensions. SSE allows performing single-precision floating point operations on four operands simultaneously.

[8] A *graphics processing unit (GPU)* is a dedicated hardware rendering device for a personal computer, workstation, or game console. Modern GPUs are very efficient at manipulating and displaying computer graphics, and their highly parallel structure makes them by far more efficient than conventional CPUs for a range of complex algorithms. Since the introduction of programmable shading capabilities, a trend of general-purpose computation on GPUs is becoming increasingly popular [304]. Usually, GPU's fragment processors are exploited for performing arithmetic operations, and texture memory is used for storing the data.

5

Numerical Optimization

> Nature and Nature's Laws lay hid in night;
> God said, "Let Newton be!" – and all was light.
>
> A. POPE, *Epitaph for Newton*

In problems we have encountered thus far, the reader can recognize a repeating motif: how to find a solution that is "the best" in some sense. For example, the centroidal Voronoi tessellation could be viewed as finding a sampling with the minimum mean squared error, or the measurement of geodesic distances as finding a path with the shortest length. Such problems are generically called *optimization problems* and play a fundamental role in our applications. In this chapter, we will explore numerical algorithms that will allow us to solve problems of this kind.

5.1 Local versus global optimization

The "bread and butter" of any optimization problem is a function f, called the *objective* or the *cost*. This function measures how "good" a particular solution is. In general, we deal with functions of the form $f : \mathbb{X} \to \mathbb{R}$, where $\mathbb{X}$ is some vector space (in particular, we will encounter the Euclidean space $\mathbb{R}^N$ and the space of real matrices, $\mathbb{R}^{N \times M}$). A generic *unconstrained optimization problem* can be formulated as

$$\min_{x \in \mathbb{X}} f(x).$$

A solution $x^* = \mathrm{argmin}_{x \in \mathbb{X}} f(x)$ is called a *global minimizer* and the value $f(x^*)$ a *global minimum* of f.

The question discussed in this chapter is how to practically find the minimum of $f(x)$. Obviously, the naïve suggestion to perform an exhaustive search in the space of all the possible solutions would fail to work – even in case when the dimension of $\mathbb{X}$ is low, this space is very large. In the practical problems we deal with, the dimensionality of $\mathbb{X}$ can be as high as a few thousands. It turns out that finding a global minimum of a general function is as hard a problem as trying to find a needle in a haystack. However, under certain assumptions on the function, it is possible to find the minimum by analyzing the local behavior of the function, and thus the problem can be significantly simplified.

As an illustration, consider an objective function of two variables whose graph can be visualized as a landscape with valleys and hills. A traveler staying at the bottom of a valley and seeing mountain walls rising up around him will identify his location as a minimum. Yet, unable to see beyond the valley, he may not be aware of other deeper valleys a few miles away – therefore, we say that the traveler location is a *local minimum*. The traveler example is a pictorial interpretation of *local optimization*. At the other end, a satellite equipped with a radar will be able to measure precisely the depth of each valley and find the global minimum, or, in our terminology, perform *global optimization*.

5.2 Optimality conditions

How does the traveler verify that he has arrived at the bottom of the valley? In the one-dimensional case, we know from basic calculus that a local minimum of a twice-differentiable function $f : \mathbb{R} \to \mathbb{R}$, is obtained at the point x^* where the following optimality conditions hold:

(O1) $f'(x^*) = 0$;
(O2) $f''(x^*) > 0$.

Condition (O1) on the first-order derivative is necessary for the point x^* to be an extremum, but it does not guarantee that it is a minimum. Condition (O2) is a sufficient condition, which ensures that locally around x^* the function is increasing, i.e., that x^* is a *local minimizer*. This can be seen by writing a second-order Taylor expansion of f around x^*,

$$f(x^* + dx) = f(x^*) + f'(x^*)dx + \frac{1}{2}f''(x^*)dx^2 + \mathcal{O}(|dx|^3), \qquad (5.1)$$

and observing that in order for such an approximation to be a convex parabola with respect to dx with the minimum at x^*, conditions (O1) and (O2) must hold.

In order to generalize these conditions to the multidimensional case, we first need to extend the definition of derivative to multivariate functions. For this purpose, let us consider the value of a C^2 function $f : \mathbb{X} \to \mathbb{R}$ at a point x in $\mathbb{X}$ and see how it changes when we displace the point by some small value dx. We can linearize the change of the function in the following way

$$f(x + dx) = f(x) + \langle \nabla_x f(x), dx \rangle_\mathbb{X} + \mathcal{O}(\|dx\|^2),$$

where $\langle \cdot, \cdot \rangle_\mathbb{X}$ denotes an inner product on $\mathbb{X}$. The second-order term can be neglected assuming dx to be infinitesimal. In the one-dimensional case, the inner product reduces to a simple multiplication $\nabla_x f(x)dx$; rearranging the terms as

5.2 Optimality conditions

$$\nabla_x f(x) = \frac{f(x+dx) - f(x)}{dx},$$

we obtain the standard definition of a derivative of f at x by taking dx to zero.

The function $\nabla_x f(x) : \mathbb{X} \to \mathbb{X}$ is the *gradient* of f we have already encountered in Chapter 4 and in this context can be thought of just as a generalization of the notion of derivative. In the Euclidean space, we can represent $\nabla_x f(x)$ in the standard basis $\{e_1, \ldots, e_n\}$, where e_i is a vector containing one at the i-th position and zeros elsewhere, in the following way: we take the step $dx = \epsilon e_i$, resulting in

$$\begin{aligned} f(x + \epsilon e^i) &= f(x) + \langle \nabla f(x), \epsilon e_i \rangle_{\mathbb{X}} + \mathcal{O}(\epsilon^2) \\ &= f(x) + \nabla f^i(x)\epsilon + \mathcal{O}(\epsilon^2). \end{aligned}$$

Taking $\epsilon \to 0$ we have the i-th coordinate of the gradient as $(\nabla f(x))^i = \frac{\partial f(x)}{\partial x^i}$, which leads to the familiar expression,

$$\nabla f(x) = \left(\frac{\partial f(x)}{\partial x^1}, \ldots, \frac{\partial f(x)}{\partial x^n} \right)^{\mathrm{T}},$$

often used as a definition of the gradient.

Example 5.1 (gradient of matrix functions). We exemplify the gradient computation for functions defined on the space $\mathbb{R}^{N \times M}$ of real matrices, equipped with the standard inner product $\langle A, B \rangle_{\mathbb{R}^{N \times M}} = \mathrm{trace}(A^{\mathrm{T}} B) = \mathrm{trace}(BA^{\mathrm{T}})$ (note that matrix multiplication is commutative under the trace operator; see Problem 5.2).

1. $f(X) = \mathrm{trace}(AX)$, where A is an $M \times N$ matrix. Expanding the function around X and neglecting $\mathcal{O}(\|dX\|^2)$ terms, we have

$$\begin{aligned} f(X + dX) &= \mathrm{trace}(A(X + dX)) = \mathrm{trace}(AX) + \mathrm{trace}(AdX) \\ &= f(X) + \mathrm{trace}(dX^{\mathrm{T}} A^{\mathrm{T}}) = f(X) + \langle A^{\mathrm{T}}, dX \rangle_{\mathbb{R}^{N \times M}}. \end{aligned}$$

Identifying the last term with the definition of the gradient, we obtain $\nabla f(X) = A^{\mathrm{T}}$.

2. $f(X) = \mathrm{trace}(X^{\mathrm{T}} BX)$, where B is an $N \times N$ matrix. In a similar way,

$$\begin{aligned} f(X + dX) = \mathrm{trace}((X + dX)^{\mathrm{T}} B(X + dX)) = \mathrm{trace}(X^{\mathrm{T}} BX) + \\ + \mathrm{trace}(dX^{\mathrm{T}} BX) + \mathrm{trace}(X^{\mathrm{T}} BdX) + \mathrm{trace}(dX^{\mathrm{T}} BdX). \end{aligned}$$

Neglecting the second-order term, $\mathrm{trace}(dX^{\mathrm{T}} BdX)$, we have

$$\begin{aligned} f(X + dX) &= f(X) + \mathrm{trace}(dX^{\mathrm{T}} BX) + \mathrm{trace}(X^{\mathrm{T}} BdX) \\ &= f(X) + \mathrm{trace}(dX^{\mathrm{T}} (B + B^{\mathrm{T}})X) \\ &= f(X) + \langle (B + B^{\mathrm{T}})X, dX \rangle_{\mathbb{R}^{N \times M}}, \end{aligned}$$

from which $\nabla f(X) = (B + B^{\mathrm{T}})X$.

The second-order derivative can be obtained in a similar way by linearizing the gradient,

$$\nabla f(x+dx) = \nabla f(x) + H_x(dx) + \mathcal{O}(\|dx\|^2). \tag{5.2}$$

The map $H : \mathbb{X} \times \mathbb{X} \to \mathbb{X}$ is called the *Hessian* of f (after the German mathematician Ludwig Otto Hesse (1811-1871)) and is denoted by $\nabla^2_{xx} f(x)$ or simply as $\nabla^2 f(x)$. Similarly to the gradient, the Hessian can be expressed in the standard basis as an $n \times n$ matrix of second-order partial derivatives,

$$\nabla^2 f(x) = \left(\frac{\partial^2 f(x)}{\partial x^i \partial x^j} \right),$$

which allows us to rewrite equation (5.2) as

$$\nabla f(x+dx) = \nabla f(x) + \nabla^2 f(x)dx + \mathcal{O}(\|dx\|^2).$$

Note that hereinafter we tacitly assume x to be a column vector. Because we assume f to be a C^2 function, the mixed derivatives satisfy $\frac{\partial^2}{\partial x^i \partial x^j} f(x) = \frac{\partial^2}{\partial x^j \partial x^i} f(x)$, i.e., the Hessian is symmetric.

Equipped with the gradient and the Hessian, we are now ready to generalize the one-dimensional optimality conditions to multivariate functions. Approximating f around x^* by a quadratic function, we rewrite equation (5.1) as

$$f(x^* + dx) = f(x^*) + \nabla f(x^*)^\mathrm{T} dx + dx^\mathrm{T} \nabla^2 f(x^*) dx + \mathcal{O}(\|dx\|^3).$$

First, it is clear that the first-order condition (O1) trivially generalizes as $\nabla f(x^*) = 0$ or alternatively, as $\|\nabla f(x^*)\| = 0$. Second, in order for the function value to be increasing around x^*, the term $dx^\mathrm{T} \nabla^2 f(x^*) dx$ must be positive for all $dx \in \mathbb{X}$. This is equivalent to saying that the Hessian at x^* is positive-definite and is guaranteed by requiring that all the eigenvalues of $\nabla^2 f(x^*)$ are positive (we leave this simple exercise in algebra to the reader as Problem 5.4). We arrive at the following conditions:

(O'1) $\|\nabla f(x^*)\| = 0$;
(O'2) $\nabla^2 f(x^*) \succ 0$,

where the symbol "$\succ 0$" is a short notation for a positive-definite matrix.

As in the one-dimensional case, the optimality conditions (O'1–O'2) guarantee a local minimum, which is not necessarily a global one. Particular exceptions of this rule are *convex functions*. Formally, we say that a function $f : A \subseteq \mathbb{X} \to \mathbb{R}$ defined on a subset A of $\mathbb{X}$ is convex if A is a convex set and for any $x_1, x_2 \in A$ and $\lambda \in [0, 1]$

$$f(\lambda x_1 + (1-\lambda)x_2) \leq \lambda f(x_1) + (1-\lambda)f(x_2),$$

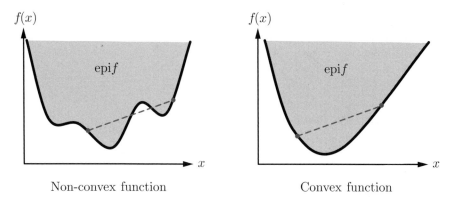

Figure 5.1. One-dimensional example of non-convex (left) and a convex (right) functions. The shaded area represents the epigraph of the function. The dashed line shows a chord connecting two points on the graph of the function.

and *strictly convex* if the above inequality is strong (a subset A of a vector space is called convex if for all $x_1, x_2 \in A$ and $\lambda \in [0, 1]$, the vector $\lambda x_1 + (1 - \lambda)x_2$ is also in A).

In the one-dimensional case (see Figure 5.1), this definition can be easily visualized geometrically: the graph of a convex function is always below the chord connecting any two points on it. This property can be formulated as $f(x) \geq f(x_0) + \nabla f(x_0)^{\mathrm{T}}(x - x_0)$, for all $x, x_0 \in \mathbb{X}$. Obviously, the definition of a convex function is very similar to the definition of a convex set. A convex function can be alternatively defined as a function whose *epigraph* (set of points lying on or above the graph of the function) is a convex set, which generalizes the one-dimensional intuition (see Problem 5.6).

From this geometric interpretation, it is clear why a convex function cannot have more than one local minimum: indeed, if we had two local minima, going from one to another would require the function graph to first go up and then once more down, thus violating the convexity property. Consequently, for a convex function, a local minimum is necessarily the global one[1] (a rigorous proof is left as Problem 5.8). This property makes convex functions especially favorable for optimization: using local optimization methods, we can guarantee reaching the global minimum.

5.3 Unconstrained optimization algorithms

The traveler from our example, now equipped with the optimality conditions, knows where to stop, yet now he has to figure out how to move in order to get to the minimum. To make the situation even more dramatic, imagine that our traveler has no topographic map of the area and the visibility conditions are poor. Having no possibility to plan the entire route from his current location

---
input : objective function f.
output : minimizer x^* of $f(x)$.
initialization: some initial $x^{(0)}$ and $k = 0$.

1 **repeat**
2   Determine a *descent direction* $d^{(k)}$.
3   Choose a *step size* $\alpha^{(k)}$.
4   Update $x^{(k+1)} \longleftarrow x^{(k)} + \alpha^{(k)} d^{(k)}$.
5   $k \longleftarrow k+1$.
6 **until** *convergence*
7 $x^* \approx x^{(k)}$.

Algorithm 5.1. Generic minimization algorithm.

to the bottom of the valley (the local minimum), the traveler has to make small steps that lead him to lower altitude, based only on local information. For example, he can make a step in the direction in which the slope of the mountain inclines downward. Then, he finds a "downward" direction in the new point and makes another step and so on until the destination is reached. Unconstrained minimization algorithms follow essentially the same idea, which can be formalized as the following iterative procedure:

The algorithm starts with some *initialization*, $x^{(0)}$, which can be derived from some *a priori* information about the optimal solution or a random vector. It then chooses a vector $d^{(0)}$ and a scalar $\alpha^{(0)}$ such that $f(x^{(0)}) > f(x^{(0)} + \alpha^{(0)} d^{(0)})$; d is called a *descent direction* and α a *step size*. The current point $x^{(0)}$ is replaced with $x^{(1)} = x^{(0)} + \alpha^{(0)} d^{(0)}$ and the process is repeated iteratively. An optimization algorithm is said to *converge* if it produces a convergent sequence of points $x^{(0)}, x^{(1)}, \ldots \to x^*$ (called a *minimizing sequence*), such that x^* is a local minimizer of the objective function f. When f is convex, x^* is also its global minimizer; otherwise we say that the algorithm is *locally convergent*.

Optimization algorithms differ in three basic components: Step 3 (the choice of the descent direction), Step 4 (the choice of the step size), and Step 5 (the *stopping criterion*). We do not include Step 1 (initialization) into this list, because it is problem- rather than algorithm-specific. An ideal way to stop the optimization is when $|f(x^{(k)}) - f(x^*)| = 0$ or $\|x^{(k)} - x^*\| = 0$. However, because we do not know x^* in advance, this criterion is unusable. The first-order optimality condition tells us that $\nabla f = 0$ in the minimum, and this condition can be used to stop the algorithm. In practice, due to the use of finite-precision arithmetic, it is unlikely that the gradient will vanish completely. For this reason, practical optimization algorithms are usually stopped when $\|\nabla f\| \leq \epsilon_g$, for some tolerance ϵ_g. It is also common to stop the algorithm when the relative change of the function value $(f(x^{(k)}) - f(x^{(k+1)}))/f(x^{(k)})$ drops below some tolerance threshold ϵ_f, or when the step size $\|x^{(k+1)} - x^{(k)}\|$ becomes sufficiently small.[2] Combinations of one or more of these stopping criteria are often used.

Provided that d is a descent direction, it is guaranteed that a sufficiently small step in this direction will decrease the value of f. That is, if we define the one-dimensional function $f_d(\alpha) = f(x + \alpha d)$, it is guaranteed that $f_d(\alpha) < f_d(0)$ for a sufficiently small α. Step size selection determines where along the ray $\{x + \alpha d : \alpha \geq 0\}$ the next iterate will be. The simplest choice is the *constant step size*, $\alpha^{(k)} = \alpha_0$. Though widely used, such a strategy is problematic. Indeed, d is a decrease direction only in the proximity of x; too large a step size may increase the function resulting in oscillatory behavior of $f(x^{(k)})$ and often preventing convergence. A possible remedy is reducing the step size, however, in this case a minimization algorithm will suffer from slow convergence. For this reason, a far better strategy is to chose $\alpha^{(k)}$ adaptively by searching for suitable values along the ray $\{x + \alpha d : \alpha \geq 0\}$. Such a procedure is usually referred to as *line search*.[3]

The best way of choosing the step size is by using *exact line search*, in which α is chosen to minimize f along the ray $\{x + \alpha d : \alpha \geq 0\}$:

$$\alpha = \arg\min_{\alpha \geq 0} f(x + \alpha d). \tag{5.3}$$

There exists a variety of numerical procedures for solving the above one-dimensional minimization problem [30, 46]. In some special cases, the solution may have an analytic form. Exact line search is used when its complexity is significantly lower than the complexity of computing d itself.

Many times, the exact minimizer of $f(x + \alpha d)$ is not necessary or too expensive to find. Line search methods that find a step size that reduces f "sufficiently" are called *inexact*. One of the most popular versions of such line searches is known as the *Armijo rule* or *backtracking line search* (Algorithm 5.2).

Backtracking line search starts with some initial step size α_0 and then gradually reduces it by the factor β until the condition $f(x + \alpha d) \leq f(x) + \sigma \alpha \nabla f(x)^T d$ is satisfied. The geometric meaning of this condition is visualized in Figure 5.2: on one hand, for a sufficiently small α, the one-dimensional

---

input   : descent direction d, objective function $f(x)$ and its gradient $\nabla f(x)$, parameters $\sigma \in (0, 0.5)$ and $\beta \in (0, 1)$.
output   : step size α.
initialization: initial step size α_0.

1   $\alpha \longleftarrow \alpha_0$
2   **while** $f(x + \alpha d) > f(x) + \sigma \alpha \nabla f(x)^T d$ **do**
3      $\alpha \longleftarrow \beta \alpha$.
4   **end**

Algorithm 5.2. Backtracking line search (Armijo rule).

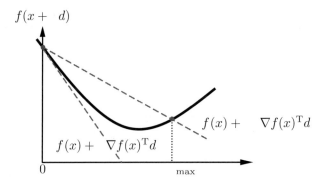

Figure 5.2. Visualization of the Armijo rule applied to a one-dimensional function $f(x+\alpha d)$. The lower dashed line shows the linear extrapolation of f, whereas the upper dashed line has a slope smaller by a factor of σ. Armijo rule accepts any value of α, for which the function lies between the two lines, i.e., $0 \leq \alpha \leq \alpha_{\max}$.

function $f(x+\alpha d)$ behaves very similarly to its first-order Taylor approximation $f(x)+\alpha \nabla f(x)^\mathrm{T} d$, hence f can be decreased by $\alpha \nabla f(x)^\mathrm{T} d$. On the other hand, for larger values of the step size that are desired for faster convergence, $f(x+\alpha d)$ may lie above the line $f(x)+\alpha \nabla f(x)^\mathrm{T} d$ and sometimes a too large α may even increase the value of the function. The Armijo rule provides a reasonable trade-off between the two situations by accepting a decrease of factor σ of that suggested by first-order extrapolation. Typically, σ ranges between 0.01 to 0.3 and β between 0.1 (fast, yet crude search) and 0.8 (more accurate, yet slower search).

Both exact and inexact types of line search guarantee a decrease of f at every iteration of a minimization algorithm, which produces a monotonically non-increasing sequence $f(x^{(0)}) \geq f(x^{(1)}) \geq \cdots \geq f(x^{(k)}) \geq \cdots$ of function values. For this reason, algorithms that use line search are often termed *safeguarded*.

5.4 The quest for a descent direction

Thus far, we have assumed that the decrease direction d was given. We will now explore several ways to find it. Observe the first-order Taylor approximation of $f(x+d)$ around x,

$$f(x+d) \approx f(x) + \nabla f(x)^\mathrm{T} d.$$

The term $\nabla f(x)^\mathrm{T} d$ is called the *directional derivative* of f at x in the direction d. It describes the approximate change in f for a small step in the direction d. Observe that $f(x+d) < f(x)$ if the directional derivative $\nabla f(x)^\mathrm{T} d$ is negative. In other words, a descent direction must form an acute angle with the negative gradient.

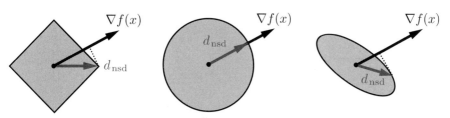

Figure 5.3. Normalized steepest descent directions in sense of the L_1 (left), L_2 (middle), and a general Q-norm (right).

Intuitively, to make the descent the steepest, it is desirable to make $\nabla f(x)^T d$ as negative as possible. Because the term $\nabla f(x)^T d$ is linear, it is unbounded below and thus we can make it as negative as we like by selecting a very large d, provided that d is a descent direction. This, of course, does not make much sense, as the above approximation holds only locally. To make the problem well-defined, we will constrain d to have unit length. The direction we are looking for is

$$d_{nsd} = \arg \min_{d: \|d\|=1} \nabla f(x)^T d. \tag{5.4}$$

Such a d is often called a *normalized steepest descent direction* [46]. Note that d_{nsd} is not unique and depends on the choice of the norm in $\|d\| = 1$.

Geometrically, d_{nsd} can be thought of as the direction in the unit ball $\|d\| = 1$ that has the largest extension in the direction $-\nabla f(x)$; different choices of the norm $\|\cdot\|$ will result in different descent directions (see Figure 5.3). For example, if we choose the L_1 norm, the steepest descent direction is the standard basis vector e^i, onto which the projection of $-\nabla f(x)$ is the largest:

$$d_{nsd} = -\text{sign}\left(\frac{\partial f(x)}{\partial x^i}\right) e^i, \tag{5.5}$$

where $i = \arg\max_{i=1,\ldots,n} |\frac{\partial f(x)}{\partial x^i}|$. Note that a step in the L_1 steepest descent direction will update a single coordinate of x at a time. For this reason, the L_1 norm steepest descent algorithm is often called a *coordinate descent* algorithm.

Another by far more common choice is the standard L_2 norm, which yields simply the negative gradient direction,

$$d_{nsd} = -\frac{\nabla f(x)}{\|\nabla f(x)\|_2}, \tag{5.6}$$

as the descent direction. The resulting algorithm is called *normalized gradient descent*.[4]

Example 5.2 (condition number). Let us derive the gradient descent algorithm with exact line search on the following quadratic function, borrowed from Boyd and Vandenberghe's book *Convex Optimization* [46]:

$$f(x) = \frac{1}{2}((x^1)^2 + \lambda(x^2)^2)$$

(here x^i stands for the i-th coordinate of the vector x, and $(x^i)^2$ is squared value). Because the eigenvalues of the Hessian of $f(x)$ are $\lambda_1 = 1, \lambda_2 = \lambda$, the function is convex for $\lambda > 0$ and its minimum is achieved at $x^* = 0$ with the optimal value 0. The non-normalized gradient descent direction is given by $d = -\nabla f(x) = -(x^1, \lambda x^2)^T$. Exact line search finds the step size α that minimizes the one-dimensional function,

$$f(\alpha) = f(x + \alpha d) = \frac{1}{2}\left((x^1 + \alpha d^1)^2 + \lambda(x^2 + \alpha d^2)^2\right).$$

α can be expressed analytically by imposing

$$\begin{aligned} 0 = f'(\alpha) &= (x^1 + \alpha d^1)d^1 + \lambda(x^2 + \alpha d^2)d^2 \\ &= -(x^1)^2 - \lambda^2(x^2)^2 + \alpha((x^1)^2 + \lambda^3(x^2)^2), \end{aligned}$$

from where

$$\alpha = \frac{(x^1)^2 + \lambda^2(x^2)^2}{(x^1)^2 + \lambda^3(x^2)^2}.$$

We run the gradient descent algorithm starting at $x^{(0)} = (\lambda, 1)^T$. In this case, the first descent direction will be $d^{(0)} = -\lambda(1,1)^T$ with the step size $\alpha^{(0)} = 2/(\lambda+1)$. Consequently, the next point will be

$$x^{(1)} = (\lambda, 1)^T - \frac{2\lambda}{\lambda+1}(1,1)^T = \left(\lambda\left(\frac{\lambda-1}{\lambda+1}\right), \frac{\lambda-1}{\lambda+1}\right)^T.$$

Continuing in the same manner, we obtain

$$x^{(k)} = \left(\lambda\left(\frac{\lambda-1}{\lambda+1}\right)^k, \left(-\frac{\lambda-1}{\lambda+1}\right)^k\right)^T,$$

which yields

$$f(x^{(k)}) = \frac{1}{2}\lambda(\lambda+1)\left(\frac{\lambda-1}{\lambda+1}\right)^{2k} = \left(\frac{\lambda-1}{\lambda+1}\right)^{2k} f(x^{(0)}) = \left(\frac{\lambda-1}{\lambda+1}\right)^2 f(x^{(k-1)}).$$

Note that the *suboptimality* $f(x^{(k)}) - f(x^*)$ of the solution drops by the factor $(\lambda-1)^2/(\lambda+1)^2$ after each gradient descent iteration. In other words, λ influences the *convergence rate* of the algorithm: for $\lambda = 1$, it converges in a single iteration; for $\lambda \sim 1$, the convergence is fast, whereas for $\lambda \gg 1$ or $\lambda \ll 1$, the convergence is extremely slow. Two examples of convergence are shown in Figure 5.4.

5.4 The quest for a descent direction

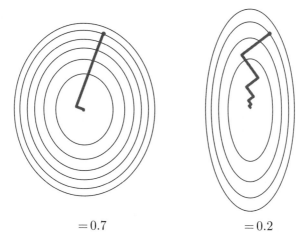

$= 0.7 \qquad\qquad = 0.2$

Figure 5.4. Convergence trajectory of the gradient descent algorithm on the function $f(x^1, x^2) = (x^1)^2 + \lambda(x^2)^2$ for $\lambda = 0.7$ (left) and $\lambda = 0.2$ (right).

Example 5.2 shows that on a simple quadratic function, for which the Hessian is a constant 2×2 matrix with the eigenvalues 1 and λ, we have

$$f(x^{(k+1)}) - f(x^*) \leq \gamma(f(x^{(k)}) - f(x^*)),$$

where the ratio

$$\gamma = \left(\frac{\lambda - 1}{\lambda + 1}\right)^2 < 1$$

is called the *convergence rate*. In simple terms, the solution suboptimality $f(x^{(k)}) - f(x^*)$ converges to zero at least as fast as a geometric series. This is usually referred to as *linear convergence*, as $f(x^{(k)}) - f(x^*)$ falls below a line on a log-linear plot of suboptimality versus iteration number. Setting some tolerance ϵ on the solution suboptimality, $f(x^{(k)}) - f(x^*) \leq \epsilon$ is achieved after at most

$$k_{\max} = -\frac{\log(f(x^{(0)}) - f(x^*))/\epsilon)}{\log \gamma}$$

iterations. Note that whereas the numerator of the above expression depends on the initialization of the algorithm, the denominator is a function of γ only.

This simple convergence analysis can be generalized in the following way: For a C^2 function f with the Hessian $\nabla^2 f(x)$ having its eigenvalues in the range $[\lambda_{\min}, \lambda_{\max}]$, the convergence rate of the gradient descent algorithm is

$$\gamma = \left(\frac{\lambda_{\max} - \lambda_{\min}}{\lambda_{\max} + \lambda_{\min}}\right)^2 = \left(\frac{\xi - 1}{\xi + 1}\right)^2,$$

where the ratio $\xi = \lambda_{\max}/\lambda_{\min}$ is referred to as the *condition number* of the Hessian matrix.[5] Because in the proximity of some point x a C^2 function can be expressed in terms of a quadratic form,

$$f(x+d) \approx f(x) + \nabla f(x)^\mathrm{T} d + \frac{1}{2} d^\mathrm{T} \nabla^2 f(x) d,$$

the condition number of $\nabla^2 f(x)$ describes how isotropic are the level lines of $f(x)$ close to x. In our mountain example, $\xi \approx 1$ corresponds with a valley, in which the slope in every direction is approximately the same. On the other hand, $\xi \ll 1$ or $\xi \gg 1$ means that along some direction the valley is much flatter than along some other.

The Hessian condition number governs the convergence speed of the gradient descent algorithm. For example, for $\xi = 10^6$, about 5000 iterations are required to decrease the solution suboptimality by 1%, meaning that the convergence is extremely slow. Problems with a large ξ are usually called *ill-conditioned*.

5.5 Preconditioning

Example 5.2 demonstrates that the gradient descent algorithm is not suitable for solving ill-conditioned minimization problems. However, recall that we have the flexibility to use a different norm in (5.4). Particularly, consider the Q-norm $\|x\|_Q = (x^\mathrm{T} Q x)^{1/2} = \|Q^{1/2} x\|_2$, where Q is a positive-definite matrix.[6] This change of the norm can be considered as a change of coordinates. The Q-norm steepest descent for $f(x)$ with respect to x is equivalent to the gradient descent for $h(y) = f(Q^{-1/2} y)$ with respect to y, where $y = Q^{1/2} x$. The normalized gradient descent direction at the point $x = Q^{-1/2} y$ is given by the chain rule,

$$d_{\mathrm{nsd}} = - \left(\nabla f(x)^\mathrm{T} Q^{-1} \nabla f(x)\right)^{-1/2} Q^{-1} \nabla f(x), \quad (5.7)$$

(see Problems 5.11 and 5.12).

The change of the norm has a tremendous effect on the convergence rate. Observe that in the new coordinate system, the Hessian of $h(y)$ at a point $y = Q^{1/2} x$ is given by

$$\nabla^2 h(y) = \nabla^2 g(Q^{1/2} x) = Q^{-1/2} \nabla^2 f(x) Q^{-1/2}.$$

If in the initial problem the Hessian is ill-conditioned, and the transformation of coordinates by Q improves its condition number, the Q-norm steepest descent will converge faster than does the regular gradient descent. The matrix Q that improves the condition number of a problem is called a *preconditioner*. In general, Q should be chosen such that after the transformation the Hessian is well-conditioned, at least in the proximity of the minimum. Thus, if an approximation $\tilde{H} \approx \nabla^2 f(x^*)$ of the Hessian at the solution point is available, the choice $Q = \tilde{H}$ results in $\nabla^2 h(y^*) = \tilde{H}^{-1/2}\nabla^2 f(x^*)\tilde{H}^{-1/2} \approx I$. However, it should be stressed that there is no universal recipe for finding a good preconditioner; in the majority of cases, it remains more of an art.

5.6 Let Newton be!

We have seen that the choice of a Q-norm with $Q \approx \nabla^2 f(x^*)$ in the steepest descent algorithm significantly improves its convergence, yet, clearly, except in some special cases, the Hessian at the solution as well as the solution itself are not known in advance. However, if f is sufficiently smooth, $\nabla^2 f(x) \approx \nabla^2 f(x^*)$ in the proximity of x^*, we can select $Q = \nabla^2 f(x)$ (different at each iteration) obtaining,

$$d = -(\nabla^2 f(x))^{-1}\nabla f(x), \qquad (5.8)$$

called the *Newton direction*. When the Hessian is positive definite, d forms an acute angle with $\nabla f(x)$, and thus is a descent direction. A steepest descent algorithm using the Newton direction is called the *Newton algorithm* and a single step in the Newton direction is called the *Newton step*.[7] When line search is used in the Newton algorithm, the method is usually referred to as the *damped* or *safeguarded* Newton method.

Newton's step can be considered a gradient descent iteration in a "canonical" system of coordinates obtained by scaling x by the approximate inverse Hessian matrix. It appears that such a scaling is the best possible preconditioning in the sense that any affine change of coordinates in $f(x)$ will not influence the convergence of the method (Problem 5.13). The Newton step can be interpreted as the minimizer of a second-order Taylor approximation $f(x+d) \approx f(x) + \nabla f(x)^T d + d^T \nabla^2 f(x)d$, which is a quadratic function of d and is minimized by $d = (\nabla^2 f(x))^{-1}\nabla f(x)$. Because a twice differentiable function is locally quadratic, $x+d$ should be a very good estimate of x^* in the proximity of the solution. In the trivial case where the function is quadratic, the Newton algorithm converges in a single iteration. Because of the fact that the Newton direction is derived from a second-order Taylor approximation of the function and uses second-order derivatives, the Newton method is often termed as a *second-order* optimization algorithm.

Yet another way to interpret the Newton step is by considering the solution of the first-order (necessary) optimality condition $\nabla f(x^*) = 0$. Linearizing this equation around x with respect to d, we obtain

$$\nabla f(x+d) \approx \nabla f(x) + \nabla^2 f(x)d = 0,$$

which is solved by selecting d to be the Newton direction. In other words, the Newton step is the displacement that must be added to x in order to satisfy the linearized optimality condition.

The Newton algorithm exhibits very rapid convergence compared with gradient descent. As soon as the iterations bring $x^{(k)}$ sufficiently close to the solution x^* (practically, when $\|\nabla f(x^{(k)})\|_2$ is sufficiently small), the method starts converging *quadratically*. Roughly, this means that the number of correct digits in the solution suboptimality, $f(x^{(k)}) - f(x^*)$, doubles after each iteration. This stage is also characterized by a rapidly dropping gradient norm.

Another typical behavior is the fact that the Hessian usually does not change significantly in the proximity of the optimum. It is possible to reduce the number of Hessian evaluations and inversions by keeping the Hessian from previous iterations and updating it only every few iterations. Such an acceleration of the Newton method is referred to as *frozen Hessian* [285].

In practice, the Newton system is rarely solved by inverting the Hessian matrix. A far more common way is to express the Hessian as a product $\nabla^2 f = LL^T$, where L is a lower triangular matrix. Such an expression is called *Cholesky decomposition* or *factorization*, published posthumously and honoring its inventor, the French military engineer Andre-Louis Cholesky [24]. Cholesky decomposition exists for any positive-semidefinite matrix.[8] Using this approach, the Newton system $LL^T d = -\nabla f$ is solved by two stages: first, the system $Ly = -\nabla f$ is solved for y (a step called *forward substitution*). Second, the system $L^T d = y$ is solved for d. This step is referred to as *backward substitution*. Because L is a triangular matrix, both systems are solved efficiently, with $\mathcal{O}(N^2)$ operations (Algorithm 5.3). Cholesky decomposition requires $\frac{1}{6}N^3 + \mathcal{O}(N^2)$ multiplication operations and the same amount of addition operations, which is usually more efficient than straightforward matrix inversion (though the theoretical complexity is still $\mathcal{O}(N^3)$) [177].

When the objective function is non-convex, its Hessian may contain negative or zero eigenvalues, which means that the Newton direction may cease to be a descent direction or $\nabla^2 f(x)$ may not be invertible. In order to cope with this problem, the Hessian is often *modified* by adding some positive-definite matrix, e.g., $\tilde{H} = \nabla^2 f(x) + \epsilon I$, which makes its eigenvalues positive. Cholesky decomposition can still be used with the modified Hessian.

5.7 Truncated Newton

The quadratic convergence of the Newton method is a clear advantage over steepest descent algorithms. However, despite this advantage, the practical

---

input : Hessian $\nabla^2 f(x)$, gradient $\nabla f(x)$.
output: Newton direction d.

1 Find Cholesky decomposition of the Hessian, $\nabla^2 f(x) = LL^T$.
2 Forward substitution: solve $Ly = -\nabla f$ for y.
3 Backward substitution: solve $L^T d = y$, obtaining the Newton direction d.

Algorithm 5.3. Newton system solution using Cholesky factorization.

use of the Newton method is limited. First, computation of the Hessian matrix is required at each iteration, which may be computationally intensive. Second, the case where ∇f is not Lipschitz continuous may raise numerical problems, as the Hessian may have unbounded spectrum (when f is not twice differentiable, the Hessian does not exist and the Newton algorithm is not applicable). Yet, the major difficulty stems from the fact that at each iteration, we have to solve the linear system $\nabla^2 f(x)d = -\nabla f(x)$). Except some special cases where the Hessian has a nice structure, this stage requires $\mathcal{O}(N^3)$ operations. This makes the Newton algorithm applicable in general only to small to medium-scale problems, in which the number of variables does not exceed a few hundreds.

Fortunately, there exist various methods that were designed to achieve rapid convergence without computing or inverting the Hessian matrix. A family of algorithm called *truncated* or *inexact* Newton methods computes the Newton direction approximately, requiring only that

$$\|\nabla^2 f(x)d + \nabla f(x)\|_2 \leq \eta \cdot \|\nabla f(x)\|_2 \tag{5.9}$$

instead of solving the exact system $\nabla^2 f(x)d = -\nabla f(x)$. Approximate solution of the Newton system is carried out using an iterative method for the solution of linear equations,[9] which is stopped as soon as (5.9) is satisfied. It is common to refer to the sequence of Newton steps as the *outer iterations*, whereas the iterations used to solve the linear system at each outer iteration are called the *inner iterations*.

5.8 Quasi-Newton algorithms

Another class of approximate Newton algorithms replaces the exact Newton step by

$$d^{(k)} = -(H^{(k)})^{-1}\nabla f(x^{(k)}), \tag{5.10}$$

where $H^{(k)}$ is an approximate Hessian, which is usually initialized to $H^{(0)} = I$ and then gradually built using the gradients from some or all previous iterations, $\{\nabla f(x^{(0)}), \ldots, \nabla f(x^{(k)})\}$. Such algorithms are usually referred to as *quasi-Newton*.

In order to understand how to obtain the new approximate Hessian $H^{(k+1)}$ from the previous approximation $H^{(k)}$, recall how a derivative of a one-dimensional function $h(t)$ can be replaced by a finite-difference

$$h'(t) \approx \frac{h(t) - h(t')}{t - t'}$$

or, alternatively, $h'(t)(t - t') \approx h(t) - h(t')$. This intuition can be generalized to the multi-variate case, leading to

$$\nabla^2 f(x^{(k+1)})(x^{(k+1)} - x^{(k)}) \approx \nabla f(x^{(k+1)}) - \nabla f(x^{(k)}).$$

Instead of using the exact Hessian, this condition can be enforced on some approximate matrix $H^{(k+1)} \approx \nabla^2 f(x^{(k+1)})$,

$$H^{(k+1)}(x^{(k+1)} - x^{(k)}) \approx \nabla f(x^{(k+1)}) - \nabla f(x^{(k)}).$$

This system provides only n equations to determine the $n \times n$ matrix $H^{(k+1)}$. In 1965, Broyden [84] proposed the following update formula

$$H^{(k+1)} = H^{(k)} + \frac{y - H^{(k)}s}{s^T s} s^T,$$

where $y = \nabla f(x^{(k+1)}) - \nabla f(x^{(k)})$ and $s = x^{(k+1)} - x^{(k)}$. It can be shown that $H^{(k+1)}$ defined in such a way leads to the least change of $H^{(k)}$ (see Problem 5.14).

A somewhat more frequently used update was proposed independently by Broyden [85], Fletcher [158], Goldfarb [175], and Shanno [352]:

$$H^{(k+1)} = H^{(k)} - \frac{H^{(k)} s (H^{(k)} s)^T}{s^T H^{(k)} s} + \frac{y y^T}{y^T s}.$$

Using $H^{(k)}$ instead of the true Hessian matrix $\nabla^2 f(x^{(k)})$ in the Newton step gives rise to a quasi-Newton algorithm named the Broyden-Fletcher-Goldfarb-Shanno or *BFGS* method. The approximate Hessian remains positive definite as long as $y^T s > 0$.

Of course, we still face the problem of inverting $H^{(k)}$. A remedy is found in the matrix inversion lemma from linear algebra, which allows us to express explicitly the inverse of the sum of a symmetric positive-definite matrix with a k-rank matrix (rank 1 in case of Broyden's update and rank 2 in case of the BFGS). This allows us to update directly an approximation of the inverse Hessian $B^{(k)} \approx (H^{(k)})^{-1}$. The BFGS formula rewritten in terms of $B^{(k)}$ assumes the form of

$$B^{(k+1)} = \left(I - \frac{s y^T}{y^T s}\right) B^{(k)} \left(I - \frac{s y^T}{y^T s}\right) + \frac{s s^T}{y^T s}. \tag{5.11}$$

Like the steepest descent algorithms, the BFGS quasi-Newton method uses only first-order derivatives and line search. Though BFGS requires slightly more calculations per iteration and some additional storage compared with steepest descent methods, it greatly outperforms the latter in convergence speed. To date, the BFGS quasi-Newton method is considered one of the best choices for medium and large-scale problems among the variety of existing unconstrained minimization algorithms.

5.9 Non-convex optimization

All the optimization algorithms we have discussed so far are suitable for finding local minima and have global convergence only if the objective function

5.9 Non-convex optimization

is convex. In case of a non-convex objective, unfortunately, there is no magic recipe that can always guarantee global convergence. Therefore, local optimization algorithm cannot be applied blindly in such cases.

There exist several heuristic methods for minimization of non-convex functions. Assume that our objective function has one global and several local minima. The simplest way to prevent local convergence is by starting with an initialization $x^{(0)}$, sufficiently close to the global minimum. Often, non-convex functions exhibit patterns referred to as *basins of attraction* around their minima. Starting the optimization in such a "basin" will eventually "attract" the solution to a local minimum. However, in many optimization problems (including the ones we are focusing on), some *a priori* knowledge about the solution is available to choose the initialization in the correct basin of attraction, thus usually preventing local convergence.

When a good initialization is not readily available, in some problems it is still possible to find one using *multiresolution* approaches. The key idea is starting with the solution of a coarse-resolution version of the problem, which contains significantly less variables and approximates in some sense the original problem,[10] proceeding to higher-resolution levels with the obtained solution used as the initialization, until arriving to the full-resolution problem. We will meet such approaches in Chapter 7 in our discussion of multidimensional scaling methods.

In some cases, the *iterative majorization* approach can be used. Instead of minimizing the function f, we minimize a convex *majorizing function* $h : \mathbb{X} \times \mathbb{X} \to \mathbb{R}$. A majorizing function is a function of two variables, $h(x,q)$, which touches f at the point q, i.e., $h(q,q) = f(q)$ and is above f at all the rest of the points, i.e., $f(x) \leq h(x,q)$ for all $x \in \mathbb{X}$. The idea of iterative majorization consists of sequentially updating the point q at which the two functions touch each other with a point decreasing the value of the majorizing function. Formally, the procedure is as shown in Algorithm 5.4.

Step 3 of Algorithm 5.4 can be carried out using local optimization. The algorithm produces a sequence of points $x^{(1)}, x^{(2)}, \ldots$ on which the value of f decreases. Though we cannot guarantee in general that such a sequence

---

input : objective function f, majorizing function h.
output : minimizer x^* of $f(x)$.
initialization: some initial $q^{(0)}$ and $k = 0$.

1 **repeat**
2   Find $x^{(k+1)}$ for which $h(x^{(k+1)}, q^{(k)}) \leq h(q^{(k)}, q^{(k)})$
3   Update $q^{(k+1)} = x^{(k)}$.
4   $k \longleftarrow k + 1$
5 **until** *convergence*
6 $x^* = q^{(k)}$.

Algorithm 5.4. Iterative majorization algorithm.

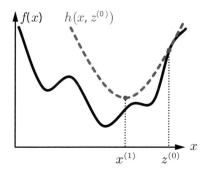
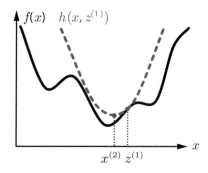

Figure 5.5. Illustration of two steps of the iterative majorization algorithm.

converges to the global minimum of f, it is reported that in many non-convex problems, the majorization approach helps prevent convergence to a local minimum. Another advantage is that the majorizing function is usually much simpler than the objective (for example, a quadratic function is a popular choice of h) and, as a result, easier to minimize. We will touch upon the majorization method when discussing the multidimensional scaling problem in Chapter 7.

5.10 Constrained optimization

Let us return once more to our mountain traveler and put him in an unpleasant situation: walking in the mountain, he suddenly comes across a lake. Unable to swim, the traveler is constrained to walk around the lake and cannot get into the water. Taking this illustration to the domain of optimization theory brings us to *constrained optimization problems*, formulated in general form as

$$\min_{x \in \mathbb{X}} f(x) \quad \text{s.t.} \quad \begin{cases} g_k(x) \leq 0, \ k = 1, \ldots, K \\ h_l(x) = 0, \ l = 1, \ldots, L. \end{cases} \quad (5.12)$$

(here "s.t." is read as "subject to"). The functions $g_k : \mathbb{X} \to \mathbb{R}$ are called *inequality constraints* and $h_l : \mathbb{X} \to \mathbb{R}$ *equality constraints* (for notation convenience, the constraints are often written as vector functions, $g : \mathbb{X} \to \mathbb{R}^K$ and $h : \mathbb{X} \to \mathbb{R}^L$, respectively; $g(x) \geq 0$ and $h(x) = 0$ are interpreted element-wise). For simplicity, we assume that f, g_k, and h_k are $\mathcal{C}^2$. The constraints define the subset of $\mathbb{X}$ to which the solution must belong, referred to as the *feasible set* (in our illustration, the zone where the traveler can walk). A point x satisfying all the constraints is said to be a *feasible solution* to (5.12), and *infeasible* otherwise. If x is feasible and $g_i(x) = 0$, the inequality constraint $g_i(x) \leq 0$ is said to be *active* at x. If, on the other hand, $g_i(x) < 0$, the constraint is *inactive*. If the gradients $\nabla h_l(x)$ of the equality constraints and the gradients $\nabla g_k(x)$ of the active inequality constraints are linearly-independent at x, the point x is called *regular*. Note that the minimizer of $f(x)$ without the constraints is not necessarily the solution of the constrained problem (5.12), as it is not necessarily a feasible solution. Thus, the traveler from our example, while descending down the valley, may find that he cannot go any further without having to get into the water. The minimum in this case is located on the boundary of the feasible set.

The main idea of solving constrained problems is to arrange the objective and the constraints into a single function

$$\mathcal{L}(x, \lambda, \mu) = f(x) + \sum_{k=1}^{K} \lambda^k g_k(x) + \sum_{l=1}^{L} \mu^l h_l(x)$$
$$= f(x) + g(x)^{\mathrm{T}} \lambda + h(x)^{\mathrm{T}} \mu, \qquad (5.13)$$

which can be minimized like in a usual unconstrained problem. The function $\mathcal{L}$ is called the *Lagrangian* and the vectors $\lambda = (\lambda^1, \ldots, \lambda^K)$ and $\mu = (\mu^1, \ldots, \mu^L)$ the *Lagrange multipliers*. Informally, we can think of this approach as of adding penalty for constraint violation to the objective. The Lagrange multipliers define the trade-off between the importance of the objective and the constraint violation. A set of conditions known as *Karush-Kuhn-Tucker conditions* (or KKT for short; sometimes the name *Kuhn-Tucker* is also used) guarantees that the constrained problem (5.12) and the unconstrained problem (5.13) are equivalent [234]. Formally, they state the following: If x^* is a regular point and a local minimum of the constrained problem (5.12), then

(KKT1) $g_k(x^*) \leq 0$ for all $k = 1, \ldots, K$ and $h_l(x^*) = 0$ for all $l = 1, \ldots, L$;
(KKT2) $g(x^*)^{\mathrm{T}} \lambda^* = 0$ and all the elements of λ^* are non-negative;[11]
(KKT3) there exist unique Lagrange multipliers λ^* and μ^* such that

$$\nabla \mathcal{L}(x^*, \lambda^*, \mu^*) = \nabla f(x^*) + \nabla g(x^*)^{\mathrm{T}} \lambda^* + \nabla h(x^*)^{\mathrm{T}} \mu^* = 0.$$

Note that in general, the KKT conditions are necessary but not sufficient. If the objective f and the constraints g_k, h_l ($k = 1, \ldots, K$, $l = 1, \ldots, L$) are convex, then the KKT conditions are also sufficient, i.e., if (KKT1–KKT3) hold, then x^* is a solution of the constrained problem (5.12).

Example 5.3 (geometric interpretation of KKT conditions). In order to get more geometric intuition about the KKT conditions, let us consider a simpler minimization problem with only one equality constraint $h(x) = 0$. Condition (KKT3) has the following geometric interpretation: the gradient of the objective function, ∇f, has to be parallel to the gradient of the constraint, ∇h (Figure 5.6). To understand its necessity, assume that at some point x' satisfying $h(x') = 0$, the two gradients do not line up. This means that the projection of $\nabla f(x')$ on the tangent of the curve $h(x) = 0$ at x' is non-zero.

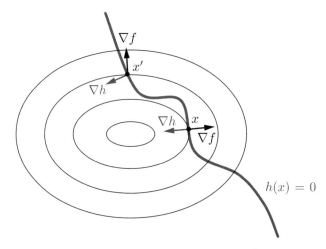

Figure 5.6. Geometric interpretation of Karush-Kuhn-Tucker conditions: the gradients of the objective f and the constraint h must line up at the constrained minimum.

As consequence, a small displacement along that curve will decrease f while still obeying the constraint, implying that x' is not optimal.

5.11 Penalty and barrier methods

As we have mentioned, we can think of the Lagrangian as a way to convert the constrained problem into an unconstrained one. A family of constrained optimization algorithms called *penalty methods* is similar in its spirit to this idea, defining the *penalty aggregate*

$$F_p(x) = f(x) + \sum_{k=1}^{K} \varphi_p(g_k(x)) + \sum_{l=1}^{L} \psi_p(h_l(x)),$$

which includes the objective and penalty on constraint violation. Here $\varphi_p, \psi_p : \mathbb{R} \to \mathbb{R}$ is a parametric family of penalty functions defined as shown in Table 5.11.

5.11 Penalty and barrier methods

Table 5.1. Penalties for equality and inequality constraints.

Inequality constraints penalty	Equality constraints penalty
$\varphi_p(t) = \frac{1}{p}\varphi(pt)$	$\psi_p(t) = \frac{1}{p}\psi(pt)$
$\varphi'_p(t) = \varphi'(pt)$	$\psi'_p(t) = \psi(pt)$
$\lim_{p \to \infty} \varphi_p(t) = \begin{cases} 0 & t \leq 0 \\ \infty & t > 0 \end{cases}$	$\lim_{p \to \infty} \psi_p(t) = \begin{cases} 0 & t = 0 \\ \infty & \text{else} \end{cases}$

For example, $\psi(t) = t^2$ is a popular choice for the equality constraint penalty, and $\varphi(t) = \exp(t) - 1$ is often used as the penalty for inequality constraints (see Figure 5.7). When p grows to infinity, we approach the ideal penalty function, which gives infinite penalty for constraint violation and zero otherwise. Yet, the function becomes non-smooth, therefore, we must start with a small p and then increase it gradually. In this way, we create a sequence of problems, which are solved using unconstrained optimization algorithms, and the solution is used as the initialization for the next problem with a larger p. Formally, the penalty method proceeds as shown in Algorithm 5.5.

The stopping condition is usually a combination of a condition on the constraint violation ($g(x) \leq \epsilon_c$, $|h(x)| \leq \epsilon_c$) and a tolerance on the objective function change. It is also common to bound the values of p, such that the algorithm is stopped when $p > p_{\max}$. Typically, $4 \leq \beta \leq 10$ is used in Step 4.

Another class of algorithms very similar to the penalty method is called the *barrier methods*, whose only difference is that the constraints cannot be violated.[12] Barrier methods can be used in problems with inequality constraints only, and the barrier functions $\phi_p : \mathbb{R} \to \mathbb{R}$ are defined similarly to the inequality constraint penalty, with the difference that $\phi_p(t) \to \infty$ as $t \to 0^-$ for all $p > 0$ and undefined for $t > 0$. Commonly used barrier functions

---

input : Penalty aggregate F_p, parameter β.
output : approximate solution x^* of constrained problem (5.12).
initialization: a small p and some initial $x_p^{(0)}$.

1 **repeat**
2   Find $x_p^* = \operatorname{argmin}_{x \in \mathbb{X}} F_p(x)$ by unconstrained optimization initialized with x_p^0.
3   Set $p' = \beta p$.
4   Set $x_{p'}^{(0)} = x_p^*$.
5   Update $p \longleftarrow p'$.
6 **until** *convergence*
7 $x^* = x_p^*$.

Algorithm 5.5. Penalty method.

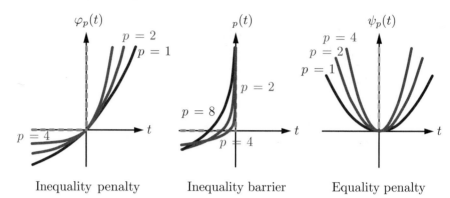

Figure 5.7. Left: inequality constraint penalty $\varphi(t) = \exp(t) - 1$; center: barrier $\phi(t) = -\log(-t)$; right: equality constraint penalty $\psi(t) = t^2$. Shown in dashed line is the asymptotic penalty/barrier for $p \to \infty$.

are $\phi(t) = -t^{-1}$ or $\phi(t) = -\log(-t)$, shown in Figure 5.7. The barrier method should be used when it is absolutely important that the constraints are never violated; it is more tricky than the penalty method, which is arguably preferred in general. The algorithm must be initialized with a feasible solution and during all the optimization stages must always remain in the feasible set. This usually requires certain modification in the line search, in order to ensure that the selected step does not cause constraint violation.

5.12* Augmented Lagrangian method

The main disadvantage of the penalty and barrier methods is the fact that for finite values of p, the penalty aggregate gives only an approximate way to solve the constrained problem. Typically, large values of p are required to achieve sufficient accuracy. At the same time, large values of p are hard to handle, as $|\varphi'_p|$ and $|\varphi''_p|$ become very large.

Recall that the meaning of Lagrangian according to the KKT conditions is that the solution of the constrained problem (5.12) can be found by unconstrained minimization of $\mathcal{L}(x, \lambda^*, \mu^*)$, once the optimal Lagrange multipliers λ^* and μ^* are known. In practice, we do not know λ^* and μ^* in advance, however, we can search for them together with x^*.

This observation gives rise to a method introduced by Hestenes [204] and Powell [319] and referred to as *augmented Lagrangian* (AL). Instead of penalty functions $\varphi_p(t)$ and $\psi_p(t)$ used for inequality and equality constraints in the penalty method, two-parametric functions of the form $\varphi_p(t; \lambda)$ and $\psi_p(t; \mu)$ satisfying

5.12* Augmented Lagrangian method

---

input   : Augmented Lagrangian function F_p, parameter β.
output   : approximate solution x^* of constrained problem (5.12).
initialization: $k = 0$, a small p, some initial $x^{(0)}, \lambda^{(0)}$ and $\mu^{(0)}$.

1 **repeat**
2   Find $x_p^* = \mathrm{argmin}_{x \in \mathbb{X}} F_p(x, \lambda^{(k)}, \mu^{(k)})$ by unconstrained optimization initialized with $x^{(k)}$.
3   Update the Lagrange multipliers: $\mu_i^{(k+1)} = \varphi_p'(g_i(x_p^*); \mu_i^{(k)})$, $\lambda_l^{(k+1)} = \psi_p'(h_l(x_p^*); \lambda_l^{(k)}) > \beta \lambda_l^{(k)}$.
4     $x^{(k+1)} = x_p^*$.
5   Set $p \longleftarrow \beta p$.
6   Set $k \longleftarrow k+1$.
7 **until** *convergence*
8 $x^* = x_p^*$.

Algorithm 5.6. Augmented Lagrangian method.

$$\psi_p'(0; \lambda) = \lambda,$$
$$\varphi_p'(0; \mu) = \mu,$$

are used. For example, such functions can be defined as $\varphi(t; \mu) = \mu \varphi_p(t)$ and $\psi_p(t; \lambda) = \psi_p(t) + \lambda t$. The penalty aggregate is replaced by

$$F_p(x, \lambda, \mu) = f(x) + \sum_{k=1}^{K} \varphi_p(g_k(x); \mu_k) + \sum_{l=1}^{L} \psi_p(h_l(x); \lambda_l).$$

Differentiating $F_p(x)$ with respect to x, we have

$$\nabla_x F_p(x, \lambda, \mu) =$$
$$\nabla_x f(x) + \sum_{k=1}^{K} \varphi_p'(g_k(x); \mu_k) \nabla_x g_k(x) + \sum_{l=1}^{L} \psi_p'(h_l(x); \lambda_l) \nabla_x h_l(x),$$

which bears resemblance to the gradient of the Lagrangian $\nabla \mathcal{L}(x)$ in (KKT3). The terms $\varphi_p'(g_k(x); \mu_k)$ and $\psi_p'(h_l(x); \lambda_l)$ can therefore be considered as estimates of the Lagrange multipliers μ_k and λ_l, respectively. The entire algorithm can be summarized as shown in Algorithm 5.6.

The update of the Lagrange multipliers in Step 3 of Algorithm 5.6 is usually damped, in order to avoid significant change of $\lambda^{(k)}$ and $\mu^{(k)}$ over subsequent iterations. The advantage of augmented Lagrangian over the standard penalty method is the fact that once μ^*, λ^* are found, the optimal solution x^* is known after unconstrained minimization of $F_p(x, \lambda^*, \mu^*)$. For this reason, the augmented Lagrangian method usually converges for small values of p.

Suggested reading

A good starting point to get a general picture of numerical optimization is the comprehensive books of Bertsekas [30], Kelley [221], and Nocedal and Wright [298]. Boyd and Vandenberghe [46] dig deeper into the theory and show multiple examples of different applications where optimization problems arise. The most recent state-of-the-art theory and practice of convex optimization is presented in Arkadi Nemirovsky's lecture notes [293, 295]. He also has a review of classic methods in [294], more accessible to the general reader. A good survey of truncated Newton algorithm is presented by Stephen Nash [292]. An excellent book oriented toward specialists in optimization and covering in depth different issues related to constrained optimization problems is Nesterov and Nemirovsky [296]. An overview of augmented Lagrangian methods can be found in [110, 29].

We left beyond the scope of this book a whole class of optimization problems with discrete variables (the graph shortest path problem is one example we encountered in Chapter 4). Such problems are usually referred to as *combinatorial optimization*. For an overview of combinatorial optimization problems and methods, the reader is referred to [116].

Software

The best starting point is the commercial MATLAB optimization toolbox, available as part of MATLAB distribution. TOMLAB is another commercial toolbox for MATLAB. Numerous optimization codes in MATLAB are available in public domain. Kelley's book [221] is accompanied by MATLAB codes. OPT++ is a free C++ library of optimization algorithms from the Sandia Lab, including different versions of Newton and quasi-Newton algorithms. LANCELOT is a free Fortran software package for large-scale constrained and unconstrained problems, implementing, among the rest, the augmented Lagrangian method [111]. For a comprehensive survey of public domain and commercial optimization software in different languages, we refer the reader to the online document maintained by Robert Fourer [165].

Problems

5.1. Describe a function satisfying the optimality condition (O'1) and a weaker version of (O'2): $\nabla^2 f(x^*) \geq 0$. Is it guaranteed that x^* is a local minimum?

5.2.✓ Show that matrix multiplication is commutative under the trace operator, i.e., $\text{trace}(A^T B) = \text{trace}(B A^T)$ for A and B of size $n \times m$.

5.3. Verify the result of Example 5.1 by coordinate-wise differentiation in the standard basis.

5.4. Show that a matrix is positive-semidefinite if and only if its eigenvalues are non-negative.

5.5. Show that an intersection of convex sets is a convex set.

5.6. Show that a function $f : A \subseteq V \to \mathbb{R}$ is convex if and only if its epigraph $\text{epi} f = \{(v, y) \in A \times \mathbb{R} : f(v) \leq y\}$ is a convex set.

5.7. Show that if a function $f : A \subseteq V \to \mathbb{R}$ is convex, its sub-level sets $\{v \in A : f(v) \leq y\}$ are convex sets.

5.8. Show that a local minimum of a convex function is (i) a global minimum, and (ii) it is unique if the function is strictly convex.

5.9. Show that Armijo rule eventually terminates.

5.10. Generalize Example 5.2 to a general positive-definite quadratic function in $\mathbb{R}^n$. Derive the descent direction and the optimal step size.

5.11. Prove the chain rule $\nabla_x f(Ax) = A \nabla_x f(Ax)$.

5.12. Derive the expression (5.7). (Hint: use the result of 5.11.)

5.13. Derive the Newton iteration as the best possible preconditioning.

5.14. Derive the Broyden approximation for the Hessian by solving
$$H^{(k+1)}(x^{(k+1)} - x^{(k)}) = \nabla f(x^{(k+1)}) - \nabla f(x^{(k)}),$$
$$H^{(k+1)} u = H^{(k)} u \quad \text{s.t.} \quad \langle x^{(k+1)} - x^{(k)}, u \rangle = 0.$$

Show that $H^{(k+1)}$ defined in this way is unique.

5.15. Find the Lagrange multipliers in the following constrained problem,
$$\min_{(x^1, x^2)^T \in \mathbb{R}^2} x^1 + x^2 \quad \text{s.t.} \quad \begin{cases} (x^1 - 1)^2 + (x^2)^2 = 1 \\ (x^1 - 2)^2 + (x^2)^2 = 4. \end{cases}$$

Does this result contradict the KKT conditions? Explain.

5.16. Find the solution of the constrained optimization problem
$$\min_{(x^1, x^2)^T \in \mathbb{R}^2} x^1 + x^2 \quad \text{s.t.} \quad (x^1)^2 + (x^2)^2 = 2,$$
using geometric interpretation of KKT conditions only.

Notes

[1] If A does not contain the local minimizer of f, then the global minimum is obtained on the boundary of A.

[2] Typically, the value of ϵ_g ranges between 10^{-3} and 10^{-8}, and the value of ϵ_f is chosen about 10^{-3}

[3] Boyd [46] suggests that the term *ray search* appears to be more appropriate.

[4] Gradient descent assumes Euclidean geometry of the problem. Shun-Ichi Amari suggested that in some problems, the optimization variable x can be thought of as a vector of parameterization coordinates, representing a point on a non-Euclidean manifold [6]. In such a case, the steepest descent direction has to be computed on the manifold and take into consideration its metric, giving rise to what Amari termed *natural gradient* $G^{-1}\nabla f$ (note that we have already encountered a similar construction in our discussion of minimal geodesic computation in Chapter 3). Natural gradient was first proposed for problems in blind separation and equalization of signals. It can be interpreted as a "naturally" scaled gradient descent, thus being intimately connected to the Newton algorithm.

[5] In optimization literature, condition number is usually denoted by κ. We use a different notation to avoid the confusion with the curvature.

[6] The matrix square root can be defined as follows: if $Q = U\Lambda U^T$ is the eigendecomposition of Q, then $Q^{1/2} = U\Lambda^{1/2}U^T$, where $\Lambda^{1/2}$ is a diagonal matrix with elements $\lambda_1^{1/2}, \ldots, \lambda_N^{1/2}$.

[7] Though bearing Sir Isaac's name, Newton in fact did not invent this algorithm. In his work *De analysi per æquationes numero terminorum infinitas* [297], Newton described an iterative method for finding roots of polynomials, which was later extended by Thomas Simpson for general scalar functions. Today, this method is usually referred to as the *Newton-Raphson method*. One can think of the Newton optimization algorithm as using the Newton-Raphson to find the roots of the gradient $\nabla f(x)$, which is equivalent to finding the local minimum of $f(x)$.

[8] If in addition the matrix is positive-definite, its Cholesky decomposition is unique.

[9] Typically, the *conjugate gradient algorithm* [292] is used in truncated Newton methods for approximate solution of the Newton system.

[10] For example, in multidimensional scaling problems we will encounter in the following chapters, our variables will be coordinates of points sampled on a surface. In multiresolution methods, we start with a coarse sampling and refine it on subsequent resolution levels.

[11] From condition (KKT2), it follows that components of λ^* corresponding with inactive constraints vanish, i.e., inactive constraints are redundant and can be removed without changing the solution.

[12] Often, penalty and barrier methods are referred to under the common name penalty/barrier methods (PBM) [414].

6

In the Rigid Kingdom

> ...He had Cinderella sit down, and, putting the slipper to her foot, he found that it went on very easily, fitting her as if it had been made of wax.
>
> C. PERRAULT, *Cinderella*

Imagine a glamorous royal ball hosted by a young Prince in his palace. Among hundreds of elegantly dressed guests, a fair lady comes uninvited. The Prince, struck by her radiant beauty, falls in love from the first sight. But all of a sudden, as the tower clock bell sounds the first stroke of midnight, the mysterious guest slips from the Prince's arms and vanishes into the darkness without a word of goodbye, leaving as the only evidence of her visit a tiny glass slipper. The Prince swears to marry the girl whose petite foot fits into it. He commands all maids in his kingdom to measure the slipper, and after a long search finally finds a poor girl, whose foot fits perfectly. The Prince recognizes his fair guest, declares his love to her, and they marry and live happily ever after.

In this brief synopsis, the reader will certainly recognize the plot of Cinderella.[1] This fairy tale illustrates the problem of *surface similarity*. Speaking in our language, the Prince was looking for a *distance function* that given two surfaces (those of the slipper and the girl's foot) provides a quantitative measure of their similarity. For this discussion, we will assume that as well as the glass slipper, Cinderella's foot is *rigid* and cannot be bent, folded, or deformed in any way; one can only change its location and orientation in space. Formally, we say that our objects are subsets of $\mathbb{R}^3$ with the standard Euclidean metric, in which the isometry group contains only rigid transformations: translation and rotation (reflections are usually excluded because they are not physical transformations). Similarity of two surfaces in such a case is extrinsic and, up to a Euclidean isometry, can be thought of as a measure of their congruence.

In this chapter, we explore tools for comparison of extrinsic geometries in a way invariant to rigid transformations. We will start our discussion in a pursuit after a representation of two surfaces X and Y that is invariant to Euclidean isometries. Next, we will view the similarity problem through the prism of numerical optimization and see how it is related to another problem of *alignment* or *correspondence* of rigid surfaces.

Figure 6.1. Cinderella trying on the slipper in Gustave Doré's engraving.

6.1 Moments of joy, moments of sorrow

Every rigid transformation in $\mathbb{R}^3$ can be described by six parameters: three rotation angles $\theta = (\theta^1, \theta^2, \theta^3)^{\mathrm{T}}$ about the x, y, and z axes, respectively, and three translation coordinates $t = (t^1, t^2, t^3)^{\mathrm{T}}$. Such a transformation repositions a vector x in $\mathbb{R}^3$ to

$$x' = Rx + t = R_1 R_2 R_3 x + t,$$

where

$$R_1 = \begin{pmatrix} 1 & 0 & 0 \\ 0 & \cos\theta^1 & \sin\theta^1 \\ 0 & -\sin\theta^1 & \cos\theta^1 \end{pmatrix}, \; R_2 = \begin{pmatrix} \cos\theta^2 & 0 & \sin\theta^2 \\ 0 & 1 & 0 \\ -\sin\theta^2 & 0 & \cos\theta^2 \end{pmatrix},$$

and

$$R_3 = \begin{pmatrix} \cos\theta^3 & \sin\theta^3 & 0 \\ -\sin\theta^3 & \cos\theta^3 & 0 \\ 0 & 0 & 1 \end{pmatrix}$$

are rotation matrices.[2]

A straightforward approach for getting rid of rigid isometries is to find a Euclidean transformation that brings a surface X to some "canonical" placement in $\mathbb{R}^3$. For example, if we could identify a landmark point s_0 on X, translating the surface by $t = -s_0$ would always bring that point to the origin, resolving the ambiguity in surface position. However, finding landmark points requires additional information about the surface, which is not always available.

Nevertheless, there exist several points that can be found for every three-dimensional surface. One of such points is the *extrinsic centroid* (the terms *center of mass* and *center of gravity* are often used as synonyms),

$$x_0 = \int_X x\, dx, \tag{6.1}$$

which is essentially the "average location" of X (note that unlike its intrinsic counterpart we have encountered in Chapter 3, the extrinsic centroid does not necessarily belong to X). Clearly, translating the surface in such a way that x_0 coincides with the origin resolves the translation ambiguity.

Next, we have to resolve the remaining three degrees of freedom due to rotation. This can be done by finding a direction in which the surface has maximum extent, and aligning it, say, with the the e_1 axis (Figure 6.2, left). Because a direction is described by a unit vector in $\mathbb{R}^3$, this step resolves only two of the three degrees of freedom. The remaining degree of freedom is due to the rotation ambiguity about the e_1 axis. However, we can apply the same idea again by rotating the surface such that the projection on the $e_2 e_3$ plane, which can be illustrated as the footprint of the shadow cast by the surface (Figure 6.2, right), has the maximum extent in the direction of the e_2 axis.

Formally, the first direction we are looking for can be defined as the one that maximizes the *variance* of the projection of X onto it,

$$d_1 = \arg\max_{d_1 : \|d_1\|_2 = 1} \int_X (d_1^T x)^2 dx,$$

where we assume that the surface has already been translated so that $x_0 = 0$. Observe that the integrand $(d_1^T x)^2$ can be written as $d_1^T x x^T d_1$. Because d_1 does not participate in the integration, we can write

$$d_1 = \arg\max_{d_1 : \|d_1\|_2 = 1} d_1^T \left(\int_X x x^T dx \right) d_1 = \arg\max_{d_1 : \|d_1\|_2 = 1} d_1^T \Sigma_X d_1.$$

Σ, is a 3×3 matrix, whose elements

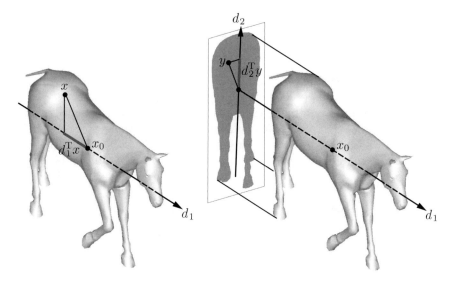

Figure 6.2. The first principal direction d_1 of the surface maximizes the variance of the projection of X onto it (left). Then, the surface is projected onto the plane orthogonal to d_1 (right, grayed) and the second principal direction d_2 is chosen as the maximum variance direction in that plane.

$$\sigma_{ij} = \int_X x^i x^j \, dx \tag{6.2}$$

are usually referred to as the *second-order geometric moments* of the surface,[3] and the direction d_1 maximizing the projection variance is called the first *principal direction*. Observe that the first principal direction, which has to maximize $d_1^T \Sigma d_1$, is nothing but the first eigenvector of Σ corresponding with its maximum eigenvalue. In order to find the second principal direction, we have to project the surface onto the plane orthogonal to d_1 and find the vector d_2 in that plane, which maximizes the variance of the projection. Obviously, d_2 corresponds with the second largest eigenvector of Σ.

Because the matrix Σ is symmetric, it admits unitary diagonalization, that is, $\Sigma = U^T \Lambda U$, where Λ is a diagonal matrix with eigenvalues $\lambda_1 \geq \lambda_2 \geq \lambda_3$ of Σ along the diagonal, and U is a unitary matrix whose columns are the corresponding eigenvectors. We leave as an exercise (Problem 6.1) the proof of the fact that U^T is a rotation matrix aligning d_1 and d_2 with the e_1 and e_2 axes, respectively. Clearly, after such an alignment, the *main* second-order moments σ_{ii} coincide with λ_i, whereas the *mixed* second-order moments (that is, the off-diagonal elements σ_{12}, σ_{13} and σ_{23}) vanish.

Thus far, we have seen that the transformation $(R, t) = (U^T, -U^T x_0)$ resolves the ambiguity of rigid isometries and brings the surface into a "canonical" configuration in the Euclidean space (Figure 6.3). Our goal is now to

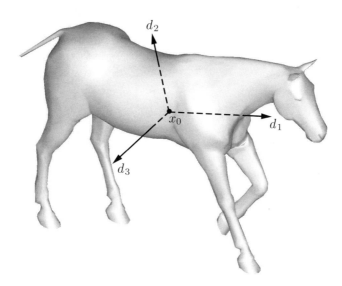

Figure 6.3. The two principal directions d_1, d_2 and a unit vector d_3 orthogonal to them define a natural coordinate system of the surface. Aligning these principal directions with the axes e_i of the standard Euclidean basis resolves the rotation ambiguity.

compare between two surfaces X and Y and quantify their similarity. We observe that the three eigenvalues λ_1, λ_2, and λ_3 of Σ provide some information about the surface extrinsic geometry. Indeed, a shape similar to a sphere is expected to have $\lambda_1 \approx \lambda_2 \approx \lambda_3$, whereas a more elongated surface should definitely have $\lambda_1 \gg \lambda_2$. In other words, the ratios $\lambda_2 : \lambda_1$ and $\lambda_3 : \lambda_1$ describe the shape *eccentricity*, and their magnitude express the shape *scale*.

We do not have to stop at the second-order moments and can define the $(p+q+r)$-th order geometric moment as

$$m_{pqr} = \int_X (x^1)^p (x^2)^q (x^3)^r dx. \tag{6.3}$$

Note that the center of gravity of the surface is a vector of its first-order moments, $x_0 = (m_{100}, m_{010}, m_{001})^{\mathrm{T}}$, whereas the elements of Σ correspond with $\sigma_{11} = m_{200}, \sigma_{22} = m_{020}, \sigma_{33} = m_{002}$ (diagonal elements), and $\sigma_{12} = m_{110}, \sigma_{13} = m_{101}, \sigma_{23} = m_{011}$ (off-diagonal elements). Higher-order moments depend on the surface position and orientation; they should be computed after performing the alignment step that eliminates the first-order and mixed second-order moments. The discretization of the integral in equation (6.3) is left as an exercise to the reader (Problem 6.3).

Intuitively, higher-order geometric moments provide us information about the surface: the more m_{pqr}'s we take, the better we can identify our object. It appears that if all moments of two surfaces coincide, the surfaces are identical. In order to understand this property, let us rewrite the (p, q, r) geometric moment of a surface as

$$m_{pqr}(f) = \int_{\mathbb{R}^3} \psi_{pqr}(x) f(x) dx = \langle \psi_{pqr}, f \rangle, \quad (6.4)$$

where $\psi_{pqr}(x) = (x^1)^p (x^2)^q (x^3)^r$, and $f : \mathbb{R}^3 \to \mathbb{R}$ is a superposition of characteristic functions, taking the value of "infinity" for $x \in X$ and zero elsewhere[4] in $\mathbb{R}^3$. Using these notations, we immediately notice that $\{m_{pqr}\}_{p,q,r=0}^{\infty}$ assume the role of the decomposition coefficients of f in the set of monomial functions $\{\psi_{pqr}\}_{p,q,r=0}^{\infty}$. Because $\{\psi_{pqr}\}$ span the space of all finite energy (more precisely, square integrable or L^2) functions on $\mathbb{R}^3$, the set of coefficients $\{m_{pqr}\}$ is unique for each surface. Indeed, if the functions f and g describing two surfaces X and Y, respectively, differ by some $h = f - g$ with non-zero energy (that is, $\int_{\mathbb{R}^3} h^2(x) dx > 0$), then there must exist some non-zero coefficients $m_{pqr}^h(h)$ such that $h = \sum m_{pqr}(h) \psi_{pqr}$. Consequently,

$$m_{pqr}(f) = \langle \psi_{pqr}, f \rangle = \langle \psi_{pqr}, g + h \rangle = m_{pqr}(g) + m_{pqr}(h) \neq m_{pqr}(f),$$

at least for some values of p, q, and r. This means that the set of all geometric moments constitutes a unique descriptor of a given surface, which is also invariant to rigid isometries if proper alignment is performed. This descriptor is also complete, meaning that, at least theoretically, one can recover[5] the surface from $\{m_{pqr}\}_{p,q,r=0}^{\infty}$.

Generally, all moments are needed to uniquely identify a surface. If we are given only a truncated set $\{m_{pqr}\}_{p,q,r=0}^{P}$ of moments up to the P-th order, there exist an infinitely large number of surfaces differing only in moments above the P-th order. However, it appears that this variety of surfaces becomes more and more similar to our surface as we increase P. In other words, even a finite set of high-order moments serves as a "fingerprint" or "signature" that identifies a sufficiently narrow class of surfaces. Ideally, we would like to be able to say that surfaces with bounded "frequencies" can be uniquely described by a finite set of moments.[6] Unfortunately, in the case of geometric moments, it is difficult to express the geometric properties of such surfaces. For this reason, geometric moments are not the best choice for measuring similarity of shapes. Other types of moments having a more clear "frequency" interpretation such as the spherical harmonics [188] or the Legendre moments [376] are usually preferred.

Using a finite set of moments, either geometric or other, we can quantify the similarity of two surfaces X and Y by applying some norm to the difference between their finite moment signatures $\{m_{pqr}(X)\}$ and $\{m_{pqr}(Y)\}$, for example,

$$d_{\text{MOM}}(X,Y) = \sum_{p,q,r=0}^{P} (m_{pqr}(X) - m_{pqr}(Y))^2. \tag{6.5}$$

Said differently, d_{MOM} is a *distance function* that measures the dissimilarity between two surfaces (hereinafter, we use the term "distance" in a broad sense, not necessarily implying that d_{MOM} is a metric). Provided that X and Y are aligned prior to computing d_{MOM}, this distance function is invariant to rigid isometries. Surfaces having small distance between them are supposed to be nearly congruent (extrinsically similar), and conversely, nearly congruent surfaces result in a small d_{MOM}.

However, it is important to mention that the moment signature distance d_{MOM} has several flaws. First, recall that the continuous surfaces X and Y that we have been using freely are never available; all we have are samplings of the surfaces. It appears that the computation of moments is sensitive to the sampling, or more precisely, to sampling non-uniformity. Second, a relatively dense sampling is required in order to obtain reliable results. Third, computation of high-order geometric moments is sensitive to acquisition noise and inaccuracies due to the use of finite-precision arithmetics (see Problem 6.5). These shortcomings may limit the applicability of surface comparison methods based on moment signatures. Yet, a more serious disadvantage of d_{MOM} is that we cannot use it as a criterion of *partial similarity*.

Returning to our fairy tale example, imagine that the Prince imprudently drops the glass slipper, which breaks apart. Using moments signatures, he would never succeed in finding Cinderella, as a part of the slipper obviously has different moments than the does complete one. It is clear that the Prince needs a better distance function that still works even when the surfaces are given only partially. To his help comes a family of the so-called *iterative closest point* algorithms (*ICP* for short), first introduced by Chen and Medioni [99], and then independently by Besl and McKay [31].

6.2 Iterative closest point algorithms

The idea behind the iterative closest point algorithms is simple: given two surfaces, X and Y, find the rigid transformation (R,t), such that the transformed surface $Y' = RY + t$ is as "close" as possible to X. "Closeness" is expressed in terms of some *surface-to-surface distance* $d(RY + t, X)$. More precisely, ICP can be formulated as the minimization problem,

$$d_{\text{ICP}}(X,Y) = \min_{R,t} d(RY + t, X). \tag{6.6}$$

The minimum surface-to-surface distance expresses the extrinsic similarity of X and Y. Because the minimum is searched over all Euclidean transformations, d_{ICP} is clearly invariant to rigid isometries. ICP was first proposed and is currently used mainly for *registration* (alignment) of surfaces. In fact, the optimal rigid transformation (R^*, t^*) is the best alignment between Y and X.

Iterative closest point algorithms differ in the choice of the surface-to-surface distance $d(Y', X)$ and the numerical method for solving the minimization problem. One of the possible candidates for such a distance could be the Hausdorff distance

$$d_{H,\mathbb{R}^3}(Y', X) = \max\left\{\sup_{x \in X} d_{\mathbb{R}^3}(x, Y'), \sup_{y \in Y'} d_{\mathbb{R}^3}(y, X)\right\},$$

which we have already encountered in Chapter 3. However, the Hausdorff distance is rarely used in practice due to its sensitivity to outliers: difference in a single sample can make d_H arbitrarily large. Most commonly, $d(Y', X)$ is expressed as the sum of squared distances between all points on Y' to the surface X,

$$d(Y', X) = \sum_{y \in Y'} d^2(y, X). \tag{6.7}$$

Because Y is discrete, the sum is finite and can be thought of as an L_2 approximation of the Hausdorff distance. Note that in this formulation $d(Y', X)$ is not symmetric, yet this "unaesthetic" lack of symmetry allows ICP to handle partially missing data. Indeed, assume that Y' is congruent to a subset of the surface X. Because every point y on Y' also exists on X, we obtain $d(Y, X) = 0$ and, consequently, $d_{ICP}(X, Y) = 0$. That is, we are able to tell that a part is similar to the whole. If we now take X to be congruent to a part of Y, no matter how we rotate and translate Y, there will always be points on it that have no corresponding points on X and thus $d_{ICP}(X, Y)$ will not vanish. This means that the whole surface is not similar to its part, which in most cases satisfies our intuition.

The variety of choices of the surface-to-surface distance $d(Y', X)$ is now shifted to the choice of the squared *point-to-surface distance* $d^2(y, X)$. The simplest possibility is to find for every $y \in Y'$ the *closest point*[7] x^* on X and define $d^2(y, X)$ as the Euclidean distance to that point,

$$d^2(y, X) = \min_{x \in X} \|x - y\|_2^2 = \|x^* - y\|_2^2. \tag{6.8}$$

This *point-to-point distance* (Figure 6.4, left) was first proposed by Besl and McKay [31] and was probably the origin of the name "iterative closest point" that labeled this family of rigid registration algorithms. Finding the closest point on X for every y on Y establishes a correspondence between the two surfaces. Clearly, every y may have its own closest point, and theoretically, we have to go over all the points of X to find it for a given y.

Observe that the point-to-point distance treats X as a cloud of points. However, in reality X is a surface, and when a point gets sufficiently close to it, X can be approximated locally as a plane. Hence, if X is given as a triangulated mesh, we can choose $d^2(y, X)$ to be the *point-to-plane distance*

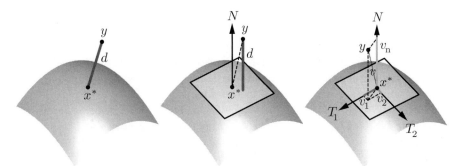

Figure 6.4. The point-to-surface distance approximated using the point-to-point (left), point-to-plane (center), and point-to-quadratic surface (right) distances. v and v_n, v_1, and v_2 denote the vector $y - x^*$ and its projections on N, T_1, and T_2, respectively.

$$d^2(y, X) = \min_{x \in X} \langle N(x), x - y \rangle^2, \qquad (6.9)$$

where $N(x)$ denotes the unit normal vector to the surface X at the point x (Figure 6.4, center). However, now our situation is even worse than before, as the closest point x^* is no more restricted to be one of the samples of the surface X and can be therefore found anywhere on its triangular faces. Obviously, it is impractical to search for the exact closest point. A reasonable compromise is to approximate x^* by the closest sample of X as we did in the point-to-point distance. Such an approximate point-to-plane distance was used by Chen and Medioni [99].

The point-to-plane distance is based essentially on a local first-order approximation of the surface by a plane. We can refine this model by using a second-order approximation, which in addition to the normal vector N also requires the two principal curvatures κ_1, κ_2 and the corresponding principal directions T_1 and T_2 at every point.[8] Pottmann and Hofer [316] showed that the second-order Taylor approximant[9] of the squared point-to-surface distance can be expressed as

$$d^2(y, X) \approx \frac{d}{d - \rho_1} \langle T_1(x^*), y - x^* \rangle^2 + \frac{d}{d - \rho_2} \langle T_2(x^*), y - x^* \rangle^2 \\ + \langle N(x^*), y - x^* \rangle^2, \qquad (6.10)$$

where $\rho_i = 1/\kappa_i$ are the principal curvature radii at the point x^*, and d is the signed distance to the closest point, defined as $d = \|y - x^*\|_2$ when x^* is found at the same side of the surface pointed by the normal, and $d = -\|y - x^*\|_2$ when x^* is located at the other side (Figure 6.4, right).

Observe that for $d \ll \rho$, the first two terms vanish and $d(y, X)$ becomes the point-to-plane distance (6.9). At the other extremity, when $d \gg \rho$, one has

$$d^2(y, X) \approx \langle T_1, y - x^* \rangle^2 + \langle T_2, y - x^* \rangle^2 + \langle N, y - x^* \rangle^2 = \|x^* - y\|_2,$$

which is nothing but the point-to-point distance (6.8). Using wave terminology, the point-to-point distance is a second-order accurate "far field" approximation of the true point-to-surface distance, whereas the point-to-plane distance is second-order accurate in the "near field." This corresponds with our intuition: observed from a distance, X behaves like a point, whereas at short distances, the planar approximation is more accurate.

The Pottmann-Hofer distance (6.10) gives an accurate approximation to the point-to-surface distance for all ranges of d. Its only problem is that for some values of d, this approximation may become negative. To avoid this problem, Pottmann and Hofer proposed the following non-negative quadratic approximant

$$d^2(y, X) \approx \frac{d}{d + \rho_1} \langle T_1(x^*), y - x^* \rangle^2 + \frac{d}{d + \rho_2} \langle T_2(x^*), y - x^* \rangle^2 + \langle N(x^*), y - x^* \rangle^2.$$

In general, it appears to be the best choice for the squared point-to-surface distance; its only disadvantage is the need to compute the principal curvatures and directions on the surface X. When X is contaminated by noise or sparsely sampled, this is not a trivial task.

6.3 Enter numerical optimization

Thus far, we have explored three types of functions measuring the squared distance between a point y and the surface X. Any of these distances can be employed in the ICP algorithm by plugging it into $d(RY + t, X)$ in (6.6). Our next goal is to find such a Euclidean transformation (R, t) that minimizes $d(RY + t, X)$. A straightforward way is to find the correspondence between Y and X, construct the objective function (whose terms $d^2(y', X)$ depend on the correspondence), and find the rigid isometry (R, t) that minimizes this objective function. Once we have the optimal rigid isometry, we apply it to the surface Y hoping that now it is aligned in the best way with X. However, we may discover that the transformation has changed the correspondence and the new objective can be further minimized by another rigid transformation. Therefore, we repeat the entire process again until the surface Y comes to a halt, that is, the optimal rigid transformation is close enough to the identity transformation. Formally, this leads to the iterative procedure shown in Algorithm 6.1.

This is essentially the way the first ICP algorithms worked. Step 3 can be performed using any unconditional minimization method. For the point-to-point distance, there even exists a closed-form solution for the optimal (R, t) [208].

6.3 Enter numerical optimization

---

input : Surfaces X and Y.
output : Optimal alignment (R, t), extrinsic similarity d_{ICP}.
initialization: $Y' = Y$.

1 repeat
2   Find the correspondence $x^*(y) = \arg\min_{x \in X} \|y - x\|_2^2$ for all $y \in Y$.
3   Minimize the error function

$$(R, t) = \arg\min_{R,t} \sum_{y \in Y'} d^2(Ry + t, X)$$

4   Transform $Y' \longleftarrow RY' + t$.
5 until *convergence*

Algorithm 6.1. Iterative closest point algorithm.

The above ICP algorithm is actually a heuristic approach, and little can be said about its convergence. The optimal rigid transformation (R, t) found in Step 3 minimizes the objective based on the correspondence found in Step 2. However, after the transformation is applied, the function $d(RY' + t, X)$ may be different from the one for which the transformation was found. Consequently, it is not guaranteed that this simple ICP algorithm will generate a monotonically decreasing sequence of objective function values and eventually converge.[10] On the other hand, we are already acquainted with various numerical optimization methods that guarantee convergence at least to a local minimum. An attempt to fill this apparent gap by putting the iterative closest point algorithms on this solid numerical ground seems to be imminent.

This was probably the motivation that guided Mitra *et al.* [277], who in 2004 made an important step toward this goal. The authors noted that the quadratic approximant to $d^2(y, X)$ can be written as

$$d^2(y, X) \approx y^{\text{T}} Q(y) y + b(q)^{\text{T}} y + c(y), \qquad (6.11)$$

where $Q(y)$ is a 3×3 symmetric positive definite matrix, $b(y)$ is a 3×1 vector, and $c(y)$ is a scalar. Clearly, this function is valid only locally in the neighborhood of y, implying that Q, b, and c depend on y.

Example 6.1 (quadratic approximation of squared distances). In this example, we show how different squared distances can be brought into the form of (6.11). For the squared point-to-point distance, we can write

$$d^2(y, X) = \|y - x^*\|_2^2 = (y - x^*)^{\text{T}}(y - x^*) = y^{\text{T}} y - 2y^{\text{T}} x^* + x^{*\text{T}} x^*;$$

hence, $Q(y) = I$, $b(y) = -2x^*$, and $c(y) = x^{*\text{T}} x^*$. For the squared point-to-plane distance,

$$d^2(y, X) = \langle N, y - x^* \rangle^2 = (N^T y - N^T x^*)^2$$
$$= (N^T y)^2 - 2N^T y N^T x^* + (N^T x^*)^2$$
$$= y^T (NN^T) y - 2(NN^T x^*)^T y + (N^T x^*)^2;$$

hence, $Q(y) = NN^T$, $b(y) = -2NN^T x^*$, and $c(y) = (N^T x^*)^2$.

We can plug this quadratic form into the ICP objective function (6.7), obtaining

$$d(RY + t, X) = \sum_{y' \in RY+t} d^2(y', X) = \sum_{y' \in RY+t} y'^T Q(y') y' + b(y')^T y' + c(y')$$
$$= \sum_{y \in Y} (Rq + t)^T Q(Ry + t)(Ry + t) + b(Ry + t)^T (Ry + t) + c(Ry + t),$$

which should be minimized with respect to the rigid transformation (R, t).

This function is hard to minimize, as it involves $Q(Ry+t)$, $b(Ry+t)$ and $c(Ry+t)$, whose functional dependence on R and t might be complicated due to the possible changes in the correspondence between Y and X. However, assuming small motion (i.e., the rigid transformation is nearly the identity transformation, $RY + t \approx Y$), we can omit this dependence, writing

$$d(RY + t, X) \approx \sum_{y \in Y} (Ry + t)^T Q(y)(Ry + t) + b(y)^T (Ry + t) + c(y)$$

(the scalar $c(y)$ can be discarded, as it does not depend on R or t). The new objective appears much easier to minimize, as it is quadratic in R and t. Yet, if we use the elements of R as our optimization variables, we have to enforce the orthonormality of R in order to guarantee that it remains a rotation matrix. This makes optimization cumbersome.

An alternative is to use the three rotation angles $\theta = (\theta^1, \theta^2, \theta^3)$ as optimization variables. In this case, the objective function becomes nastier, due to the complicated dependence of R on θ, which involves trigonometric functions,

$$R = \begin{pmatrix} 1 & 0 & 0 \\ 0 & \cos\theta^1 & \sin\theta^1 \\ 0 & -\sin\theta^1 & \cos\theta^1 \end{pmatrix} \begin{pmatrix} \cos\theta^2 & 0 & \sin\theta^2 \\ 0 & 1 & 0 \\ -\sin\theta^2 & 0 & \cos\theta^2 \end{pmatrix} \begin{pmatrix} \cos\theta^3 & \sin\theta^3 & 0 \\ -\sin\theta^3 & \cos\theta^3 & 0 \\ 0 & 0 & 1 \end{pmatrix}.$$

At this point, to our help comes the fact that the small motion assumption implies in particular a small rotation, $\theta \ll 1$. Hence, using the first-order Taylor approximations $\cos\theta \approx 1$ and $\sin\theta \approx \theta$, we can linearize the rotation matrix R as follows:

$$R \approx \begin{pmatrix} 1 & \theta^1 & -\theta^2 \\ -\theta^1 & 1 & \theta^3 \\ \theta^2 & -\theta^3 & 1 \end{pmatrix}. \qquad (6.12)$$

Using the linearized R, our objective becomes quadratic with respect to the six rigid isometry parameters $\theta = (\theta^1, \theta^2, \theta^3)$ and $t = (t^1, t^2, t^3)$, and we can use the Newton method to minimize it.

However, in spite of our small motion assumption, in practice the Newton method may find a large transformation as the minimizer of $d(RY + t, X)$. Because our approximation to $d^2(y, X)$ is valid only locally, it may increase the objective function. In such cases, we should only make a small step in the direction of the transformation. In order to do it in a consistent way, let (R', t') be a small transformation that when applied sequentially η times coincides with the large transformation (R, t). Formally, this can be written as

$$Ry + t = \underbrace{R'(\cdots(R'(R'y + t') + t')\cdots) + t'}_{\eta \text{ times}}$$

$$= R'^{\eta} y + (R'^{\eta-1} + R'^{\eta-2} + \cdots + R' + I)t'.$$

Demanding $R'^{\eta} y = Ry$, one has $R' = R^{1/\eta}$, corresponding with a rotation by θ/η. Multiplying the equation $t = (R'^{\eta-1} + R'^{\eta-2} + \cdots + R' + I)t'$ by $(R' - I)$ from the left, one obtains the "telescopic" matrix polynomial

$$(R' - I)t = (R' - I)(R'^{\eta-1} + R'^{\eta-2} + \cdots + R' + I)t'$$
$$= (R'^{\eta} + R'^{\eta-1} + \cdots + R' - R'^{\eta-1} - R'^{\eta-2} - \cdots - I)t' = (R'^{\eta} - I)t',$$

from where $t' = (R - I)^{-1}(R' - I)t$. This simple relation can be extended to non-integer values of η as well. The step size η has to be chosen sufficiently small to guarantee a decrease of the objective function. This can be done using, for example, the Armijo rule, as was proposed by Mitra et al. This approach results in a significantly more stable ICP algorithm, exhibiting better convergence.

6.4 Rigid correspondence

Note that at each iteration of the ICP algorithm where Y is transformed, the correspondence between X and Y may change and has to be recomputed. Even in the elegant formulation proposed by Mitra et al., this need is inevitable, as the parameters $Q(y), b(y)$, and $c(y)$ in the quadratic form (6.11) depend on y and have to be found again once Y is transformed. The simplest way to solve this problem is by computing the parameters *on demand*, i.e., for every point $y \in Y'$ at every iteration of the algorithm, we have to find the closest point,

$$x^*(y) = \arg\min_{x \in X} \|y - x\|_2^2.$$

This sounds like a potentially expensive algorithm. Indeed, if modern ICP algorithms were implemented this way, they would have been terribly slow.

Fortunately, there exist techniques for avoiding exhaustively searching over all points on the surface. Observe that X subdivides $\mathbb{R}^3$ into a collection of Voronoi cells

$$V(x) = \{y \in \mathbb{R}^3 : \|y - x\|_2 < \|y - x'\|_2 \ \forall x' \neq x\},$$

containing all points in $\mathbb{R}^3$ that are closer to x than to any other point on X. Finding the closest point in X given a query point y can be formulated as determining the Voronoi cell to which y belongs. This observation motivates the techniques that use efficient data structures for fast retrieval of the closest point, without exhaustively searching all points in X. It appears that even when X is given as a cloud of points, its Voronoi cells are convex polyhedra with generally complicated shapes, hardly computable efficiently. However, we can approximate the true Voronoi cells using some simpler shapes. One of such approaches is based on a hierarchical data structure called the k-*dimensional* (or kd) *tree* [25]. Each node of the kd tree corresponds with a partition of the space ($\mathbb{R}^3$ in our case) by a plane perpendicular to one of the axes. For example, the root node splits the space into two regions: $\{x^1 < 0\}$ and $\{x^1 \geq 0\}$. The first region is assigned to the left child, whereas the second region is assigned to the right child. Each child may introduce further splitting, e.g., $\{x^1 \geq 0\}$ is divided into $\{x^1 \geq 0\} \cap \{x^2 < 1\}$ and $\{x^1 \geq 0\} \cap \{x^2 \geq 1\}$, and so on. A leaf represents a (possibly unbounded) box-shaped region in $\mathbb{R}^3$. These boxes approximate the Voronoi cells of X. Using versions of the kd tree allows finding the approximate nearest neighbor of y in X with logarithmic complexity [11]. This significantly alleviates the computational burden of the iterative closest point algorithm.

However, if we use the quadratic form (6.11) as proposed by Mitra *et al.*, the need to recompute the correspondence at each iteration for every $y \in Y'$ still seems somewhat superfluous. Indeed, we never use the correspondence explicitly. All we need is to find the quadratic form parameters Q, b, and c for a given query point y. Because the squared distance function $d^2(y, X)$ is at least C^0, these parameters vary smoothly and therefore, for a sufficiently small region around y, the terms Q, b, and c remain nearly constant. Once again, the idea of hierarchical space partitioning can be exploited here. Leopoldseder et al. [249] proposed an *octree*-like structure that recursively splits the space into eight octants, until the variance of the quadratic form parameters in the created box-shaped cell falls below a small threshold. Once the tree is precomputed for the surface X, it allows the retrieval of $Q(y)$, $b(y)$, and $c(y)$ with logarithmic complexity.

It is worthwhile noting that all surface-to-surface distances we have discussed were based on the knowledge of correspondence between the two surfaces. We may therefore say that finding the rigid correspondence is the principal ingredient of ICP. To emphasize this fact, Rusinkiewicz and Levoy even suggested the backronym *iterative corresponding point* for ICP as a replacement for the original *iterative closest point* [329]. In addition to being in the core of ICP, correspondence between two objects is required in many other applications. We defer this discussion to Chapter 12, where the correspondence problem is explored in the more general non-rigid setting.

As a concluding remark, a few words ought to be said about the initialization of the ICP algorithm. Being a non-convex minimization problem, ICP may converge to a wrong local minimum if initialized incorrectly. While finding a good initialization is considered a largely open problem typically solved *ad hoc*, in the past few years guaranteed globally optimal initialization schemes were proposed. In [170], Gelfand *et al.* address this issue using a *branch and bound* global optimization algorithm to find the initial rigid correspondence. A variant of this approach is proposed in [251] by Li and Hartley.

Suggested reading

A good overview of shape similarity techniques can be found in Veltkamp's papers [387, 388, 373]. Moment-based shape descriptors are reviewed in [409, 320]. The reader is also referred to [150] for an interesting discussion on reconstructing a shape from its moments. The review paper [329] discusses efficient variants of the ICP algorithm and shows their convergence in different scenarios. Convergence is also discussed in [317]. An interesting paper by Ezra *et al.* [154] presents lower and upper bounds on the number of iterations in ICP algorithms. A tighter lower bound as well as a probabilistic upper bound are presented in [10]. In [318], Pottmann *et al.* introduce a "correspondence-less" approach to rigid surface registration based on their quadratic approximation to the squared point-to-surface distance previously discussed in this chapter. Another interesting "correspondence-less" approach is proposed by Charpiat *et al.* [98], where a smooth approximation to the Hausdorff distance is studied.

Software

A C++ implementation of ICP is available in the VTK and ITK toolboxes.

Problems

6.1. Show that the rotation matrix aligning the principal directions with the axes is the diagonalizing matrix of Σ.

6.2. Try to characterize the class of surfaces completely described by a *finite* set of their geometric moments $\{m_{pqr}\}_{p,q,r=0}^{N}$.

6.3. Derive a consistent way to discretize the geometric moment integral.

134   6 In the Rigid Kingdom

6.4. Discuss the use of the three-dimensional Fourier harmonics as a replacement to the geometric moments. How can the translation invariance of Fourier harmonics be helpful?

6.5. In reality, finite-precision arithmetics are used to compute the moments. Assume that the coordinates of the surface points are represented with the absolute error of, say, $\epsilon = 10^{-8}$. What will be the relative error of m_{pqr}? How can this complicate the use of geometric moments?

6.6 (Research question). Suppose the surface is acquired each time from a different known viewing angle, with partial occlusions. Given the signature of the surface moments for each angle, what can be said about the moments of the entire surface? Can it be reconstructed from such partial observations?

6.7. Derive the distance in equation (6.10) and show that it is a second-order approximation to the true point-to-surface distance.

6.8. Prove that the squared point-to-surface distance is not $\mathcal{C}^2$ for query points located on the surface's medial axis.

6.9. Derive the quadratic form parameters Q, b, and c for the second-order point-to-surface distance (as was shown in Example 6.1). Compare them with the point-to-point and point-to-plane distances. What can be said about the convexity of the quadratic form?

6.10. Derive a closed-form solution for the optimal rigid isometry (R, t) minimizing the ICP objective function with the squared point-to-point distance.

Notes

[1] The earliest record of this popular fairy tale originated in China in the mid-ninth century. There, the fair Ye Xian had the smallest foot in the kingdom, a synonym of beauty in the Chinese culture. In the West, the most renowned version of Cinderella belongs, perhaps, to the pen of the French author Charles Perrault (1628–1703) [311].

[2] In order to include reflections, we can multiply x' by a diagonal matrix containing ± 1's along the diagonal.

[3] In statistics, Σ is called the *covariance matrix*, and the process of finding the variance-maximizing orthogonal directions is usually referred to as *principal component analysis* (PCA) or the *Karhunen-Loéve transform* (KLT). PCA allows one to construct a low-dimensional approximation of a multi-dimensional random process that captures its "most significant" part (in the L_2 sense). The invention of principal component analysis is usually attributed to the American statistician and economist Harold Hotelling [210], though similar ideas date back to Pearson [308].

[4] Such functions are called *Dirac's delta functions*. If the reader feels uncomfortable with such a formulation, he or she can think of the surface X as of a thin shell; in this case, f takes some constant value for x belonging to the shell, and 0

otherwise. The constant is selected in such way that $\int_X f(x)dx = 1$. For vanishing shell thickness, a delta function is obtained.

[5] The problem of reconstructing the surface from its moments is often called the *inverse moment* problem, though in the signal processing jargon, the term *synthesis* is more adequate [150]. Because the polynomial basis $\{\psi_{pqr}(x) = (x^1)^p(x^2)^q(x^3)^r\}$ is non-orthogonal, reconstruction of a surface from its geometric moments requires the so-called *biorthonormal* basis, which behaves badly in this case. For this reason, in applications where reconstruction is important, geometric moments are of little use. The preference is given to orthonormal bases that allow direct reconstruction of the surface according to $f = \sum m_{pqr}\psi_{pqr} = \sum \langle \psi_{pqr}, f \rangle \psi_{pqr}$. An example of such type of moments are the Legendre moments, which follow in spirit the Fourier transform [376].

[6] Readers familiar with the Nyquist-Shannon sampling theorem will find such a statement analogous to saying that a band-limited signal can be fully represented by its discrete sampling.

[7] The closest point is sometimes referred to as the *normal footpoint*, as the line segment connecting x^* and y is always perpendicular to the surface.

[8] Some surfaces may have *umbilical points*, where the principal curvature directions are not well-defined. In this case, we may take any two orthogonal vectors T_1, T_2 in the tangent plane.

[9] More precisely, the second-order approximant to d^2 does not always exist. At the points located on the surface's medial axis, d^2 is not C^2. Such points have to be detected and excluded in order not to jeopardize the convergence of ICP algorithms based on quadratic surface approximation.

[10] Under certain conditions, Ezra *et al.* [154] show that ICP converges and present a bound on the number of iterations.

7
Multidimensional Scaling

> The world is complex, dynamic, multidimensional; the paper is static, flat. How are we to represent the rich visual world of experience and measurement on mere flatland?
>
> E. R. TUFTE, *Envisioning Information*

Thus far, in our fairy-tale example, the Prince could find Cinderella by comparing the extrinsic geometries of the glass slipper and the feet of all the ladies in his kingdom using rigid similarity methods. Now, assume that instead of dropping a glass slipper, Cinderella has lost a silk glove while escaping from the ball. Trying to naïvely approach the glove fitting problem with rigid similarity tools, the Prince soon finds that, because the glove is a non-rigid object, comparison based on the extrinsic geometry does not work anymore.

Let us leave the desperate Prince for a while and recall what we said in Chapter 2: in order to compare non-rigid shapes, we should look at their intrinsic geometries, which are invariant to isometric deformations. In other words, considering shapes as metric spaces, we need to compare the spaces X and Y with the geodesic metrics d_X and d_Y, respectively. Such a comparison appears to be by far a more complicated task than is the comparison of extrinsic geometry, for the following reason. The relative simplicity of the extrinsic similarity problem we had when discussing rigid shapes in Chapter 6 was due to the fact that the shapes were considered subsets of a common metric space ($\mathbb{R}^3$ with the Euclidean metric). Hence, we could measure their similarity using the Hausdorff distance, which led to the ICP algorithms. In the case of intrinsic similarity, the situation is more difficult: we now have two *different* metric spaces (X, d_X) and (Y, d_Y), which cannot be compared using the Hausdorff distance.

In this and the next few following chapters, we try to build a bridge between the two approaches. We will see how to represent the intrinsic geometries of the shapes in a common metric space where they can be compared using rigid similarity algorithms. This will lead us to a class of computationally tractable methods for measuring intrinsic similarity of non-rigid shapes.

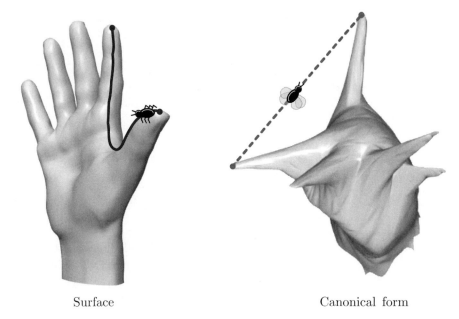

Surface   Canonical form

Figure 7.1. Illustration of the isometric embedding problem. Left: original shape; right: canonical form.

7.1 Isometric embedding problem

Let us come back for a moment to the example we have already used in Chapter 2 to illustrate the concept of intrinsic geometry. Assume that our shape is inhabited by an insect, which always chooses the shortest path to crawl between any two points. Now, imagine that there is another shape embedded into $\mathbb{R}^m$, whose points correspond with those of the original shape, while the Euclidean distances between the points are equal to the original geodesic ones. On this new shape, there lives another winged insect, which flies along straight lines between the points.

The lengths of the paths are the isometry-invariant description of our object. Because the distances traveled by both insects are equal, the descriptions produced by them are the same. However, for our application, the second insect's point of view is preferable, as his world is Euclidean. The advantage stems from the smaller number of degrees of freedom that influence our description: whereas the first insect would not feel any isometric deformation of the shape, the only way we can fool the second one is by applying rigid transformations, which are limited to rotations, translations, or reflections.

Formalizing the above intuition, given a shape (X, d_X), we would like to find a map $f : (X, d_X) \to (\mathbb{R}^m, d_{\mathbb{R}^m})$, such that

$$d_X(x, x') = d_{\mathbb{R}^m}(f(x), f(x')),$$

7.1 Isometric embedding problem

for all $x, x' \in X$. Such an f is an isometric embedding, and the space $\mathbb{R}^m$ is referred to as the *embedding space* in this context. The image $f(X)$, which we call the *canonical form* of X, can be used as an extrinsic representation of the intrinsic geometry of X (see example in Figure 7.1). Note that we regard $f(X)$ as a metric space with the restricted Euclidean metric $d_{\mathbb{R}^m}|_{f(X)}$. Up to isometries in $\mathbb{R}^m$, it defines an equivalence class of all the shapes that are indistinguishable from the point of view of intrinsic geometry. In simple words, two isometric shapes have identical canonical forms, possibly differing by an isometry of $\mathbb{R}^m$.

In a sense, this embedding allows us to undo the non-uniqueness of the way the metric structure of X is realized in $\mathbb{R}^m$ (all the possible bendings), thus reducing its vast number of degrees of freedom. Consequently, considering the canonical forms instead of the shapes themselves, we translate the non-rigid shape similarity problem into a much simpler problem of rigid similarity, with which we already know how to deal. This simple idea, proposed by Asi Elad and R. K. [147, 149], allows us to define the similarity between two shapes as an extrinsic distance between their canonical forms, measured by means of ICP or the moments method as shown in Algorithm 7.1. We call this distance the *canonical form distance* and denote it by d_{CF}.

Though originally formulated with the particular choice of $\mathbb{R}^3$ (i.e., $m = 3$) as the embedding space, the canonical forms approach can be generalized to any embedding space, as we will see in Chapter 9. Here, we stick to the Euclidean embedding, but assume m to be arbitrary. Thus far, the canonical forms method seems an ideal recipe for our problem of non-rigid shape comparison. However, there is still a question whether a shape X is isometrically embeddable into $\mathbb{R}^m$.

Unfortunately, the answer is usually negative. As the simplest case, consider the problem of embedding a sphere into $\mathbb{R}^2$. This problem arose in cartography centuries ago. One of the fundamental problems in map-making is creating a planar map of the Earth, which reproduces, in the best way, the distances between geographic objects. That is, equipped with a simple ruler, we can measure distances on the map, which represent geodesic distances on the Earth (Figure 7.2). Every cartographer knows that it is impossible to create a map of the Earth that preserves all the geodesic distances.[1] This, as a matter of fact, is a consequence of the *theorema egregium*: because the Gaussian curvature of the sphere is positive, whereas the plane has zero curvature, these two surfaces cannot be isometric.

---

input : shapes (X, d_X) and (Y, d_Y).
output: canonical forms distance $d_{\text{CF}}(X, Y)$.

1 Find the isometric embedding f and g of X and Y into $\mathbb{R}^m$.
2 Compute $d_{\text{CF}}(X, Y)$ as $d_{\text{MOM}}(f(X), g(Y))$ or $d_{\text{ICP}}(f(X), g(Y))$.

Algorithm 7.1. Idealized canonical forms distance computation.

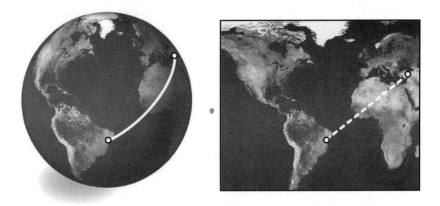

Figure 7.2. The problem of isometric embedding arising in cartography: the spherical surface of the Earth (shown is the upper hemisphere, left) is to be mapped into the plane so that it preserves the geodesic distances (right). A consequence of *theorema egregium* is that such a map does not exist, and a distortion of the distance is inevitable.

Yet, maybe by increasing the embedding space dimension, i.e., trying to embed the sphere into $\mathbb{R}^3, \mathbb{R}^4, \mathbb{R}^5$, and so on, we could succeed in finding an isometric embedding? Even this appears to be impossible. The following example, shown by Nathan Linial [253], demonstrates that even a very simple discrete metric space consisting of only four points cannot be isometrically embedded into a Euclidean space of any finite dimension.

Example 7.1 (Linial's example). Consider four points $x_1, \ldots, x_4$, sampled on the sphere of radius $R = \frac{2}{\pi}$ as shown in Figure 7.3 (one point at the north pole and three points along the equator). The distances between the points are given by the following matrix

$$D_X = \begin{pmatrix} 0 & 1 & 2 & 1 \\ 1 & 0 & 1 & 1 \\ 2 & 1 & 0 & 1 \\ 1 & 1 & 1 & 0 \end{pmatrix}.$$

We denote the embedded points by $z_1, \ldots, z_4$ and assume that the embedding is distortionless, that is, $(D_X)_{ij} = d_{\mathbb{R}^m}(z_i, z_j)$ for $i, j = 1, \ldots, 4$. Let us consider first the triangle with vertices z_1, z_2, z_3, with edges of lengths $d_{\mathbb{R}^m}(z_1, z_2) = d_{\mathbb{R}^m}(z_2, z_3) = 1$ and $d_{\mathbb{R}^m}(z_1, z_3) = 2$. Because $d_{\mathbb{R}^m}(z_1, z_3) = d_{\mathbb{R}^m}(z_1, z_2) + d_{\mathbb{R}^m}(z_2, z_3)$, the triangle is flat, i.e., the points z_1, z_2, z_3 are collinear. Applying the same reasoning, we conclude that the points z_1, z_4, z_3 are collinear, which implies that $z_2 = z_4$ and consequently, $d_{\mathbb{R}^m}(z_2, z_4) = 0$, contradicting the assumption that $d_{\mathbb{R}^m}(z_2, z_4) = d_X(x_2, x_4) = 1$. Because we

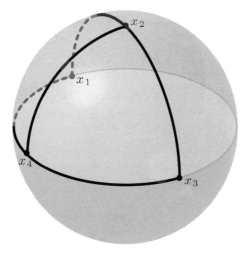

Figure 7.3. Linial's example of a metric space obtained by sampling the sphere at four points, which cannot be embedded into a Euclidean space of any finite dimension.

have not assumed any particular m, we conclude that the given structure cannot be isometrically embedded into a Euclidean space of any finite dimension. Moreover, if this is the case for such a simple object as a sphere, our conclusion is that a general shape cannot be isometrically embedded into $\mathbb{R}^m$.

It is important to emphasize that this result by no means contradicts the Nash embedding theorem. Nash guarantees that any Riemannian structure can be realized as a length metric induced by a Euclidean metric, whereas we are trying to realize it using the restricted Euclidean metric.

Although the embedding error makes it impossible to find a truly isometric embedding, we could try constructing an approximate representation of the shape X, looking for a *minimum-distortion embedding*, i.e., such f that distorts d_X the least, in the sense of some criterion. In Chapter 2, we defined the distortion, which reads in our problem as

$$\operatorname{dis} f = \sup_{x,x' \in X} |d_X(x,x') - d_{\mathbb{R}^m}(f(x), f(x'))|.$$

Adopting this criterion, we can measure how the distances on the original shape differ from those in the embedding space in the sense of the L_∞-norm. In practice, it is useful to replace the L_∞ criterion by an L_p analog,

$$\sigma_p = \int_{X \times X} |d_{\mathbb{R}^m}(f(x), f(x')) - d_X(x,x')|^p \, da \times da, \qquad (7.1)$$

where da denotes the area element on X. The L_∞ criterion can be obtained as the limit of $(\sigma_p)^{1/p}$ when $p \to \infty$.

In the discrete setting, when the shape X is sampled at N points $\{x_1,\ldots,x_N\}$, the L_∞ criterion becomes

$$\sigma_\infty = \max_{i,j=1,\ldots,N} |d_{\mathbb{R}^m}(f(x_i),f(x_j)) - d_X(x_i,x_j)|,$$

and the discrete version of the L_p criterion can be expressed as

$$\sigma_p = \sum_{i>j} a_i a_j |d_{\mathbb{R}^m}(f(x_i),f(x_j)) - d_X(x_i,x_j)|^p, \tag{7.2}$$

where a_i and a_j are discrete area elements corresponding with the points x_i, x_j. If the shape is sampled uniformly, we can simplify σ_p by setting $a_i = 1/N$.

The canonical form obtained by means of a minimum-distortion embedding is only an approximate representation of the shape's intrinsic geometry. Nevertheless, we can still measure the similarity of shapes and the distance between their canonical forms, of course, having in mind that the distortion introduced by the embedding would influence the accuracy of such a similarity.

7.2 Multidimensional scaling

An important question is how to find the minimum-distortion embedding in practice. Assume that the shape X is uniformly sampled at points $\{x_1,\ldots,x_N\}$ and σ_2 is used as the distortion criterion. We are looking for the minimum-distortion embedding into $\mathbb{R}^m$,

$$f = \operatorname*{argmin}_{f:X\to\mathbb{R}^m} \sum_{i>j} |d_{\mathbb{R}^m}(f(x_i),f(x_j)) - d_X(x_i,x_j)|^2.$$

Denoting by $z_i = f(x_i)$ the m-dimensional Euclidean coordinates of the image of the shape sample x_i under f and arranging them into an $N \times m$ matrix $Z = (z_i^j)$, we can rewrite our distortion criterion as

$$\sigma_2(Z; D_X) = \sum_{i>j} |d_{ij}(Z) - d_X(x_i,x_j)|^2.$$

Here $D_X = (d_X(x_i,x_j))$ is an $N \times N$ matrix of geodesic distances and $d_{ij}(Z)$ is a shorthand notation for the Euclidean distance between the i-th and the j-th points on the canonical form, $d_{\mathbb{R}^m}(z_i, z_j) = \|z_i - z_j\|_2$. In this formulation, we find the coordinates of the discrete canonical form directly as the solution of a nonlinear least-squares problem,

$$Z^* = \operatorname*{argmin}_{Z \in \mathbb{R}^{N\times m}} \sigma_2(Z). \tag{7.3}$$

Problem (7.3) is a non-convex optimization problem [380] in Nm variables, in which the objective function $\sigma_2 : \mathbb{R}^{N\times m} \to \mathbb{R}$ is defined over the space of matrices $\mathbb{R}^{N\times m}$. The data in the problem are the geodesic distances, represented as an $N \times N$ matrix $D_X = (d_X(x_i, x_j))$. The solution is not unique, because, as already mentioned, applying any Euclidean isometry to Z^*, we do not change the value of σ_2.

Historically, problem (7.3) is called *multidimensional scaling* (MDS), a term coined by Torgerson [379] in 1952. Such problems have been researched in depth in the psychology and statistics communities in the relation to multidimensional data analysis and visualization. The MDS problem in a formulation similar to the one we give here is attributed to Shepard [355] and Kruskal [232, 233]; the latter also proposed a numerical algorithm for its solution. Efforts to consolidate these methods were invested during the 1960s and the 1970s. To the field of computer vision and pattern recognition, MDS methods arrived relatively late. One of the first applications to analysis of surfaces was shown by Schwartz et al. in 1989, in relation to the problem of brain surface analysis [345]. The paper showed how to use MDS in order to create a planar map of the convoluted brain cortex surface, in a manner similar to that cartographers employ to map the Earth. Later, Schweitzer applied MDS for classification of databases of images [347]. These studies have been an inspiration for the embedding-based approaches to non-rigid shape comparison we discuss here.

7.3 SMACOF algorithm

In the MDS literature, the function $\sigma_2(Z)$ we use as the distortion criterion is commonly referred to as the *(Kruskal) stress* [44]. We call $\sigma_2(Z)$ the L_2-*stress* in order to distinguish it from other distortion criteria. We can write the L_2-stress in a more convenient matrix form,

$$\sigma_2(Z; D_X) = \text{trace}(Z^{\text{T}} V Z) - 2\text{trace}(Z^{\text{T}} B(Z; D_X) Z) \\ + \sum_{i>j} d_X^2(x_i, x_j), \qquad (7.4)$$

where V is a constant $N \times N$ matrix with elements

$$v_{ij} = \begin{cases} -1 & i \neq j \\ N-1 & i = j, \end{cases}$$

and $B(Z; D_X)$ is an $N \times N$ matrix depending on Z and D_X with elements,

$$b_{ij}(Z; D_X) = \begin{cases} -d_X(x_i, x_j) d_{ij}^{-1}(Z) & i \neq j \text{ and } d_{ij}(Z) \neq 0 \\ 0 & i \neq j \text{ and } d_{ij}(Z) = 0 \\ -\sum_{k \neq i} b_{ik} & i = j. \end{cases}$$

(In the following, we will sometimes omit the dependence on D_X for brevity). A particular property of matrices B and V is that they are zero-mean, i.e., the sum of the rows and the columns is zero. We leave the derivation of the matrix form of the stress to the reader as Problem 7.2.

Solution of the MDS problem requires the minimization of $\sigma_2(Z; D_X)$ with respect to Z. The most straightforward way to do it is by using a first-order gradient descent algorithm. Recall that gradient descent consists of making a step in the negative gradient direction, which requires computing the gradient of the stress at each iteration. Using matrix function differentiation similar to that shown in Example 5.1 in Chapter 5, the gradient of $\sigma_2(Z)$ can be expressed as

$$\nabla_Z \sigma_2(Z) = 2VZ - 2B(Z; D_X)Z, \qquad (7.5)$$

see Problem 7.3. The gradient descent iteration has the following form,

$$\begin{aligned} Z^{(k+1)} &= Z^{(k)} - \alpha^{(k)} \nabla_Z \sigma_2(Z^{(k)}) \\ &= Z^{(k)} - 2\alpha^{(k)} \left(VZ^{(k)} - B(Z^{(k)}; D_X)Z^{(k)} \right). \end{aligned} \qquad (7.6)$$

The step size $\alpha^{(k)}$ can be selected either as a constant, or computed at each iteration using line search. The use of line search requires evaluating the stress and its gradient a number of times. In our problem, the complexity of computing $\sigma_2(Z)$ and $\nabla_Z \sigma_2(Z)$ is $\mathcal{O}(N^2)$, which implies that for a large value of N line search is usually disadvantageous.

Jan de Leeuw [125, 129, 126] noticed that the term $\operatorname{trace}(Z^\mathrm{T} B(Z)Z)$ of the stress can be bounded below by $\operatorname{trace}(Z^\mathrm{T} B(Q)Q)$ for all $Q \in \mathbb{R}^{N \times m}$, which leads to the inequality

$$\begin{aligned} h(Z, Q) &= \operatorname{trace}(Z^\mathrm{T} VZ) - 2\operatorname{trace}(Z^\mathrm{T} B(Q)Q) + \sum_{i>j} d_X^2(x_i, x_j) \\ &\geq \operatorname{trace}(Z^\mathrm{T} VZ) - 2\operatorname{trace}(Z^\mathrm{T} B(Z)Z) + \sum_{i>j} d_X^2(x_i, x_j) \end{aligned} \qquad (7.7)$$

(the reader is invited to prove the inequality in Problem 7.5). The function $h(Z, Q)$ is convex and quadratic with respect to Z and touches $\sigma_2(Z)$ at the point $Q = Z$, i.e., $h(Z, Z) = \sigma_2(Z)$.

By virtue of inequality (7.7), $h(Z, Q)$ serves as a majorizing function for the stress. We can resort to the iterative majorization algorithm for the solution of the MDS problem. At the $(k+1)$st iteration of the majorization algorithm, the solution $Z^{(k+1)}$ is found as the minimizer of $h(Z, Z^{(k)})$ with respect to Z. Because the majorizing function is quadratic, the minimizer can be easily expressed analytically by imposing

$$\nabla_Z h(Z, Z^{(k)}) = 2VZ - 2B(Z^{(k)}; D_X)Z^{(k)} = 0.$$

This leads to the following multiplicative update formula

$$Z^{(k+1)} = V^\dagger B(Z^{(k)}; D_X)Z^{(k)}, \qquad (7.8)$$

```
input         : N × N matrix of geodesic distances D_X.
output        : canonical form Z*.
initialization: some initial Z^(0) and k = 0.
1 repeat
2     Multiplicative update: Z^(k+1) = (1/N) B(Z^(k); D_X) Z^(k).
3     k ← k + 1.
4 until convergence
5 Z* = Z^(k).
```

Algorithm 7.2. SMACOF algorithm.

where $V^\dagger = (V^T V)^{-1} V^T$ denotes the pseudoinverse of V (as the matrix V has rank $N - 1$, it is not invertible). Further noticing that the pseudoinverse can be written as $V^\dagger = \frac{1}{N}(I - \frac{1}{N} 1_{N \times N})$, where $1_{N \times N}$ denotes an $N \times N$ matrix of ones, and keeping in mind that $B(Z^{(k)}; D_X)$ is zero mean, such that $1_{N \times N} B(Z^{(k)}; D_X) = 0$, we can rewrite the update formula (7.8) as

$$Z^{(k+1)} = \frac{1}{N} B(Z^{(k)}; D_X) Z^{(k)}.$$

This multiplicative update was given the name SMACOF, standing for *scaling by minimizing a convex function*.[2] The entire algorithm can be summarized as shown in Algorithm 7.2. SMACOF has become one of the most successful and widely used MDS methods, mainly due to its simplicity and public availability of efficiently implemented code [189].

Though not straightforward to observe, a simple manipulation of the SMACOF multiplicative update (7.8) leads to

$$\begin{aligned}
Z^{(k+1)} &= V^\dagger B(Z^{(k)}) Z^{(k)} \\
&= Z^{(k)} - Z^{(k)} + V^\dagger B(Z^{(k)}; D_X) Z^{(k)} \\
&= Z^{(k)} - \frac{1}{2} V^\dagger \left(2V Z^{(k)} - 2B(Z^{(k)}; D_X) Z^{(k)} \right) \\
&= Z^{(k)} - \frac{1}{2} V^\dagger \nabla_Z \sigma_2(Z^{(k)}).
\end{aligned}$$

Using the zero-mean property once more, we can make the final retouch to the update formula,

$$Z^{(k+1)} = Z^{(k)} - \frac{1}{2N} \nabla_Z \sigma_2(Z^{(k)}; D_X).$$

This is a somewhat surprising conclusion: the SMACOF algorithm appears to be nothing but a gradient descent with a constant step size, $\alpha^{(k)} = \frac{1}{2N}$. In a sense, the whole idea of majorization has been reduced in our problem to the specific choice of the step size in the gradient descent algorithm. On the other hand, the majorization algorithm guarantees that the sequence

$Z^{(1)}, Z^{(2)}, \ldots$ produces decreasing values of the stress, and that the iterative process converges to a (possibly local) minimum of $\sigma_2(Z)$. Such a behavior is a particularly interesting property of the MDS problem, rarely observed in other optimization problems – in general, it is uncommon that constant-step gradient descent produces a monotonic decrease of the objective function.

As a historical remark, we shall note that the SMACOF iteration was derived as early as in 1968 by Louis Guttman (1916–1987), who observed [196] that the first-order optimality condition $\nabla_Z \sigma_2(Z^*; D_X) = 0$ can be written as

$$Z^* = V^\dagger B(Z^*; D_X) Z^*.$$

This condition can be thought of as the fixed point of the multiplicative update formula (7.8). This observation is a third way toward deriving the SMACOF algorithm. Recognizing Guttman's prior work, de Leeuw and Heiser [128] dubbed their multiplicative update as the *Guttman transform*.

7.4* Second-order methods

As an alternative to the first-order SMACOF algorithm, we can use second-order methods for the solution of the MDS problem. Such methods have been studied by Kearsley et al. [220]. The basic Newton iteration in our problem takes the form

$$Z^{(k+1)} = Z^{(k)} + \alpha^{(k)} \mathcal{H}^{-1}(Z^{(k)}; D_X) \nabla_Z \sigma_2(Z^{(k)}; D_X).$$

In this notation, the Hessian $\mathcal{H}$ is a tensor, which can be thought of as a four-dimensional matrix with elements

$$h_{ij}^{kl}(Z; D_X) = \begin{cases} \tilde{h}_{ij}^{kl}(Z; D_X) & \text{if } k \neq l, \\ \tilde{h}_{ij}^{kl}(Z; D_X) + 2(v_{ij} - b_{ij}(Z; D_X)) & \text{if } k = l, \end{cases}$$

for $1 \leq k, l \leq m$ and $1 \leq i, j \leq N$, where $\tilde{h}_{ij}^{kl}(Z; D_X)$ are given by

$$\tilde{h}_{ij}^{kl}(Z; D_X) = -\begin{cases} 2(z_i^k - z_j^k)(z_i^l - z_j^l) d_X(x_i, x_j) d_{ij}^{-3}(Z) & \text{if } i \neq j, \ d_{ij}(Z) \neq 0, \\ 0 & \text{if } i \neq j, \ d_{ij}(Z) = 0, \\ \sum_{n \neq i} \tilde{h}_{in}^{kl} & \text{if } i = j. \end{cases}$$

$\mathcal{H}$ is symmetric with respect to the indices i, j and k, l. The derivation of the Hessian is tedious and is left to the reader (Problem 7.6).

Because working with such a structure can sometimes be cumbersome, it is common to convert the matrix variable into a vector one, by parsing the $N \times m$ matrix Z into an $Nm \times 1$ column vector in column-stack order. We denote this transformation by $z = \text{vec}(Z)$. The Hessian in this representation is an $Nm \times Nm$ symmetric block matrix, consisting of m^2 blocks $H^{kl}(z; D_X) = (h_{ij}^{kl}(Z; D_X))$ of size $N \times N$,

$$H(z; D_X) = \begin{pmatrix} H^{11}(Z; D_X) & H^{12}(Z; D_X) & \cdots & H^{1m}(Z; D_X) \\ \vdots & & & \vdots \\ H^{m1}(Z; D_X) & H^{m2}(Z; D_X) & \cdots & H^{mm}(Z; D_X) \end{pmatrix}. \quad (7.9)$$

The Newton iteration, in turn, can be rewritten as

$$z^{(k+1)} = z^{(k)} + \alpha^{(k)} H^{-1}(z^{(k)}; D_X) \nabla_z \sigma_2(z^{(k)}; D_X),$$

where $\nabla_z \sigma_2(z; D_X) = \text{vec}(\nabla_Z \sigma_2(Z; D_X))$. Note that the Hessian is not invertible: because each of the blocks is zero-mean, it has zero eigenvalues. The Hessian may also contain negative eigenvalues because the problem is non-convex. In order to invert such a matrix, we can modify its eigenvalues to make all the negative and zero ones to be positive, e.g., by adding a small value to its diagonal, $H + \epsilon I_{Nm \times Nm}$.[3] The step size $\alpha^{(k)}$ is selected using line search. Because of high computational complexity of the stress and its gradient, inexact line search (for example, the Armijo rule) is preferred over exact one. The main complexity of the Newton iteration is due to the Hessian construction ($\mathcal{O}(N^2 m^2)$ operations) and inversion. Because the Hessian is full, solution of the Newton system requires $\mathcal{O}(N^3 m^3)$ operations.

Example 7.2 (SMACOF vs. Newton). We exemplify the difference between the SMACOF and the Newton MDS algorithms on the problem of embedding the *Swiss roll surface* (shown in Figure 7.4) into $\mathbb{R}^3$. The Swiss roll can be thought of as a rolled piece of paper, therefore, it is isometric to the plane. In our example, the surface is given in a triangular mesh representation and the geodesic distances computed using fast marching, therefore, the isometry is only approximate. Both algorithms are implemented in MATLAB. For the Newton algorithm, Armijo rule with $\alpha, \beta = 0.3$ is used. Hessian inversion is performed using eigendecomposition of the Hessian matrix with a modification of the negative eigenvalues forcing them to be positive. Both algorithms are initialized with the same random configuration of points. The stopping condition uses the normalized gradient norm, $\|\nabla_Z \sigma_2(Z^{(k)})/N^2\| \leq 10^{-3}$ (the scaling is necessary for the stopping condition to be independent of N).

Figure 7.5 (first row) depicts the normalized gradient norm on the logarithmic scale versus time for the SMACOF and Newton algorithms in embedding of the Swiss roll sampled at $N = 50$ points. The Newton algorithm shows a classic case of quadratic convergence, whereas SMACOF has linear convergence, typical for a first-order algorithm. Overall, the convergence of the Newton algorithm is much faster: 24 iterations (1.45 sec) compared with 1094 iterations (3.6 sec) for SMACOF.

Yet, the situation changes dramatically when the number of variables becomes larger ($N = 200$, see Figure 7.5 second row). Because the complexity of the Newton system solution is proportional to N^3, the iteration complexity increases by about 64 times. As the result, the overall convergence time increases significantly to 119.4 sec (at the same time, the number of iterations of the Newton algorithm is now 43, i.e., does not grow so significantly). Unlikely,

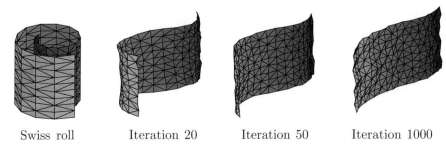

Figure 7.4. The Swiss roll surface sampled at 200 points and represented as a triangular mesh (left) and the result of its embedding into $\mathbb{R}^3$ using a few iterations of the SMACOF algorithm, starting from a random initialization.

the complexity of the SMACOF iteration is $\mathcal{O}(N^2)$, a 16 times increase. The SMACOF algorithm converges in 29.6 sec (1216 iterations).

Our first conclusion from Example 7.2 is that in large-scale MDS problems, SMACOF is advantageous over the Newton algorithm. In particular, in our problems, where typically $N \sim 1000$, performing the Newton iteration straightforwardly becomes almost impractical. Secondly, the MDS problem usually does not require reaching a very high accuracy, therefore, the condition on the gradient norm is rather inappropriate. In the MDS literature, a condition on the relative change of the stress is often used instead. If we stop the optimization when $\sigma_2(Z^{(k+1)})/\sigma_2(Z^{(k)}) \leq 0.001$, we will see that SMACOF terminates much faster than does the Newton algorithm. This phenomenon is also observed in both of our experiments. In the case of $N = 50$ points, SMACOF reaches this stopping criterion in 0.4 sec compared with 1.4 sec required for the Newton algorithm. In the second case ($N = 200$), the corresponding numbers are 4.5 sec for SMACOF and 119.4 sec for Newton.

7.5 Variations on the stress theme

Thus far, we have considered embedding obtained by minimizing the L_2-stress criterion. Let us say a few words about other possibilities. Besides $p = 2$ yielding the L_2-stress we have already seen, the cases $p = 1$ and $p = \infty$ are the most interesting among the possible choices of p in our definition of (7.2). The L_1-stress,

$$\sigma_1(Z; D_X) = \sum_{i>j} |d_{ij}(Z) - d_X(x_i, x_j)|,$$

is similar to the L_2-stress, with the exception that the sum of squared differences is replaced with the sum of absolute differences. Yet, optimization of

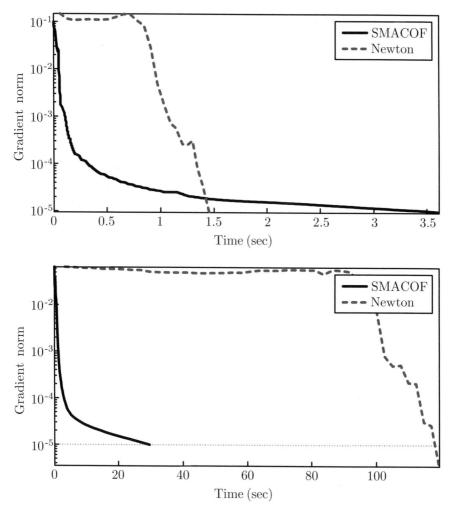

Figure 7.5. Convergence plots of the SMACOF (solid) and Newton (dashed) MDS algorithms in the problem of Swiss roll embedding with $N = 50$ (top) and $N = 200$ (bottom) points. Shown is the normalized gradient norm $\|\nabla_Z \sigma_2(Z^{(k)})\|_2 / N^2$ (bottom) as a function of CPU time. The horizontal line denotes the stopping criterion, $\|\nabla_Z \sigma_2(Z^{(k)})\|_2 / N^2 \leq 10^{-5}$.

the L_1-stress is more difficult, as the absolute value function $|t|$ is not differentiable at $t = 0$. It is possible to *smooth* the absolute value by replacing it with a differentiable function, e.g., $\varphi(t) = \sqrt{t^2 + \epsilon}$, such that $\varphi(t) \approx |t|$ for a small positive ϵ. An alternative is resorting to more complicated optimization algorithms referred to as *subgradient methods*, based on a generalization of the notion of gradient for non-differentiable functions [315].

In the case of $p = \infty$, minimization of σ_∞ is a *min-max problem*,

$$Z_\infty^* = \underset{Z}{\operatorname{argmin}} \max_{i,j=1,\ldots,N} |d_{ij}(Z) - d_X(x_i, x_j)|. \qquad (7.10)$$

A common trick in optimization is to reformulate this problem as a constrained one, introducing an *artificial variable* τ,

$$Z_\infty^* = \underset{Z,\tau}{\operatorname{argmin}} \tau \quad \text{s.t.} \quad |d_{ij}(Z) - d_X(x_i, x_j)| \leq \tau$$

$$= \underset{Z,\tau}{\operatorname{argmin}} \tau \quad \text{s.t.} \quad \begin{cases} d_{ij}(Z) - d_X(x_i, x_j) - \tau \leq 0, \\ -d_{ij}(Z) + d_X(x_i, x_j) - \tau \leq 0, \end{cases} \qquad (7.11)$$

where $i > j$. It can be shown that the two problems (7.10) and (7.11) are equivalent. In the new problem, we have $Nm + 1$ variables, a linear objective function and $\frac{1}{2}N(N-1)$ constraints. Parsing the matrix variable into an $Nm \times 1$ vector, the gradients of the constraints can be written as an $(Nm + 1) \times \frac{1}{2}N(N-1)$ sparse matrix, in which each column is a gradient of $d_{ij}(Z)$ with respect to (τ, Z).

Example 7.3 (L_2- vs L_∞-stress). We exemplify the difference between distortion criteria by showing an embedding of the horse shape into $\mathbb{R}^3$. The horse is represented as a triangular mesh with $N = 153$ points (see Figure 7.6, top). The distances are measured using fast marching on the triangular mesh. Figure 7.6 (bottom left) shows the canonical form Z_2^* obtained using the L_2-stress. The average distortion of the distances at the minimum is the average stress, $(2\sigma_2(Z_2^*)/N(N-1))^{1/2} = 0.037$. Figure 7.6 (bottom right) shows the canonical form Z_∞^* obtained using the L_∞-stress. The value of the stress obtained in this case is $\sigma_\infty(Z_\infty^*) = 0.105$.

Carefully observing the canonical forms obtained in Example 7.3, we notice that the L_2 canonical form is much smoother than its L_∞ counterpart. Such behavior is typical and highlights the difference between the two distortion criteria. Note that in our constrained problem, the value of τ is essentially the L_∞-stress, as τ gives a tight upper bound on the distance distortion. In our example, only a small number of constraints are active in the L_∞ problem. This means that there are few "problematic" pairs of points, the distances between which determine the value of the L_∞-stress. Other distances result in a smaller stress, such that the corresponding constraints are inactive.

It may often happen that a few "bad" distances (arising, for example, from numerical inaccuracies) will result in a large L_∞ measure, whereas the other distances have significantly smaller distortion. Such a situation is improbable in the L_2 problem, as a single distance does not contribute much to the L_2-stress, and the effects of numerical inaccuracies are "distributed" among all the points. In other words, the L_∞ problem is more sensitive to *outliers*, which makes the L_∞-stress disadvantageous in practical applications. On the other hand, theoretical analysis of the L_∞ problem is simpler than that of the L_2

7.5 Variations on the stress theme   151

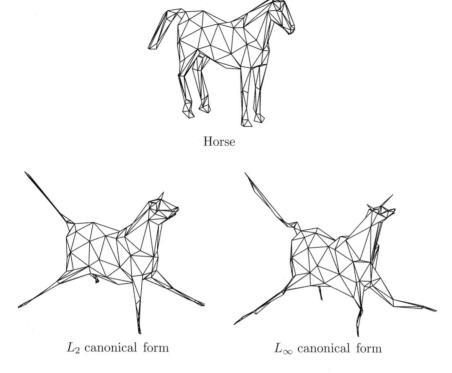

Figure 7.6. Embedding of the horse shape into $\mathbb{R}^3$ using the L_2-stress and the L_∞-stress as the distortion criterion.

one. As we will see in the next chapters, the L_∞-stress allows the use of the powerful tools of metric geometry.

Besides the L_p-stress, there exist other distortion criteria, often encountered in the MDS literature. For example, the *relative stress*,

$$\sigma_{\text{REL}}(Z; D_X) = \sum_{i>j} \frac{|d_{ij}(Z) - d_X(x_i, x_j)|^2}{d_{ij}^2(Z)},$$

can be used to measure the relative distortion of the distances. In a sense, it can be thought of as the discrete L_2 version of the dilation dil f, in the same way we think of σ_2 as of an L_2 approximation of dis f. If we wish to attribute more importance to some distances (e.g., if it is known that the accuracy of some distances may be lower), we can use the *weighted stress*,

$$\sigma_{\text{W}}(Z; D_X, W) = \sum_{i>j} w_{ij} |d_{ij}(Z) - d_X(x_i, x_j)|^2,$$

where w_{ij} are some non-negative weights. For example, if the shape is sampled non-uniformly, we can define $w_{ij} = a_i a_j$, where a_i, a_j are discrete area elements. The weights w_{ij} can be represented as a symmetric $N \times N$ matrix W. The SMACOF algorithm can still be used in this case, with the only difference that in (7.8) we use the matrix V defined as

$$v_{ij} = \begin{cases} -w_{ij} & i \neq j \\ \sum_{k \neq i} w_{ik} & i = j, \end{cases}$$

(observe that our previous definition of V is a particular case corresponding with the choice $w_{ij} = 1$).

A notable use of the weighted stress is the so-called *iteratively reweighted least squares* technique (IRLS for short), often used in statistics to approximate the solution of problems with robust norms [199, 164, 212]. To demonstrate the use of IRLS for the solution of MDS problems, we consider the following general stress function,

$$\sigma_\rho(Z; D_X) = \sum_{i>j} \rho(d_{ij}(Z) - d_X(x_i, x_j)), \qquad (7.12)$$

with $\rho(t)$ being some norm. For example, setting $\rho(t) = |t|^p$ gives the L_p norm; other notable choices are the German-McLure function

$$\rho_{\text{GM}}(t) = \frac{t^2}{t^2 + \epsilon^2}, \qquad (7.13)$$

and the quadratic-linear Huber function

$$\rho_{\text{H}}(t) = \begin{cases} \frac{t^2}{2\epsilon} & : |t| \leq \epsilon \\ |t| - 0.5\epsilon & : |t| > \epsilon, \end{cases} \qquad (7.14)$$

where ϵ is a positive constant, The necessary condition for Z to be a local minimizer of σ_ρ is

$$\nabla_Z \sigma_\rho(Z^*; D_X) = \sum_{i>j} \rho'(d_{ij}(Z^*) - d_X(x_i, x_j)) \nabla_Z d_{ij}(Z^*) = 0.$$

Instead of minimizing σ_ρ, we can minimize the weighted L_2 stress (7.12), whose minimizer has to satisfy

$$\nabla_Z \sigma(Z^*; D_X, W) = \sum_{i>j} 2 w_{ij}(d_{ij}(Z^*) - d_X(x_i, x_j)) \nabla_Z d_{ij}(Z^*) = 0.$$

If we could selecting the weights in $\sigma(Z; D_X, W)$ according to

$$w_{ij} = \frac{\rho'(d_{ij}(Z^*) - d_X(x_i, x_j))}{2(d_{ij}(Z^*) - d_X(x_i, x_j))}, \qquad (7.15)$$

the two minimizers would coincide and we could solve the weighted L_2 MDS problem instead of the general one. However, such a selection of the weights

---
input   : $N \times N$ matrix of geodesic distances D_X, the derivative of
                    ρ
output   : canonical form Z^*.
initialization: some initial $Z^{(0)}$, $w_{ij}^{(0)} = 1$, and $k = 0$.

1 **repeat**
2   Find $Z^{(k+1)} = \arg\min \sigma(Z; D_X, W^{(k)})$ using $Z^{(k)}$ as the initialization.
3   Update weights according to

$$w_{ij}^{(k+1)} = \frac{\rho'(d_{ij}(Z^{(k+1)}) - d_X(x_i, x_j))}{2(d_{ij}(Z^{(k+1)}) - d_X(x_i, x_j))}$$

4     $k \longleftarrow k + 1$.
5 **until** *convergence*
6 $Z^* = Z^{(k)}$.

Algorithm 7.3. Iteratively reweighted least squares MDS.

requires the knowledge of the minimizer Z^*, which is, of course, unknown. A possible remedy is to start by solving the un-weighted L_2 problem (all $w_{ij} = 1$), use the solution to update the weights, and iterate the process until convergence. This iteratively reweighting procedure can be summarized as shown in Algorithm 7.3.

Step 2 can be performed using the SMACOF algorithm. The reweighting in Step 3 depends on the specific selection of the function ρ. For example, if $\rho(t) = |t|^p$ is used, the update formula becomes

$$w_{ij}^{(k+1)} = \frac{1}{2} p \, |d_{ij}(Z^{(k+1)}) - d_X(x_i, x_j)|^{p-2}.$$

7.6 Multiresolution methods

High computational complexity and the risk of local convergence are problems common to all the MDS algorithms we have discussed. Trying to reduce the complexity of MDS has recently become an important research trend in the MDS community, where the growth of the data sets dealt with (sometimes exceeding $N \sim 10^6$) has led to the development of approximate large-scale MDS methods. The idea of most of these methods is reducing the complexity by considering a part of the data. For example, Chalmers [95], Morrison *et al.* [283], and Williams and Munzner [398] showed an approximation to the MDS problem, in which a distinction is made between near and far neighbors for each point. In our formulation, this approach is equivalent to removing a major part of the distances from the stress computation, leaving only the distances to the near neighbors of each point and only a few to the far ones. Faloutsos and Lin [155], Wang *et al.* [391], and de Silva and Tennenbaum [132] proposed performing MDS on a small subset of points (the so-called *landmarks*) and then finding the rest of the points by means of interpolation.[4]

The common denominator of approximate MDS approaches is posing a problem whose solution is close to the solution of the original MDS problem, but which is much simpler to solve. Such methods are applicable in problems of data visualization, where usually only qualitative results are required. In our problem of non-rigid shape comparison, the use of approximations of this kind may compromise the accuracy of shape representation [149]. Hence, in most cases, we have no choice but to solve the full MDS problem. At the same time, an approximate MDS solution can be used as a good initialization.

As an illustration, consider the landmarks method. Assume that we have $N = 1000$ points (a typical order of magnitude of the number of samples in our problems) and use 250 landmarks. We first solve the small MDS problem to embed the 250 landmarks, and then add the other 750 points by interpolation. Though the canonical form obtained this way may be insufficiently accurate for comparison of shapes, it is still close to the one obtained by solving the full MDS problem with 1000 points and can therefore be used for its initialization. This way, we spend more effort at the coarse level, where each iteration costs 16 times less, in order to save computations at the fine resolution level. Besides the complexity advantage, we also reduce the risk of local convergence: when solving the MDS problem at a coarse resolution, small local minima of the stress are usually avoided.

This idea, which we call *multiresolution optimization*, can be extended to more than two resolution levels. It is not limited to the specific choice of the distortion criterion but rather acts as an external framework, into which any iterative MDS algorithm can be incorporated. Formally, we assume to be given a hierarchy of *grids*, indexed by $l = 0, \ldots, L$, which successively approximate the full MDS problem. The Lth level is the coarsest one, while the zeroth level is the finest one, corresponding with the full problem. At each level, we work with a grid consisting of points with indices $\Omega_L \subset \Omega_{L-1} \subset \cdots \subset \Omega_0 = \{1, \ldots, N\}$. We denote the number of points at each level by $N_L < N_{L-1} < \cdots < N_0 = N$. At the lth level, the data is represented as an $N_l \times N_l$ matrix D_l, obtained by extracting the rows and columns of $D_0 = D_X$, corresponding with the indices Ω_l. The MDS problem at the lth-level is $Z_l^* = \mathrm{argmin}_{Z_l}\, \sigma(Z_l; D_l)$; we write $\sigma(Z; D)$ as a generic distortion criterion, emphasizing its dependence on the distance data D. The solution Z_l^* is transferred to the next level $l - 1$ using an *interpolation operator* P_l^{l-1}, which can be represented as an $N_{l-1} \times N_l$ matrix. The whole multiresolution MDS algorithm can be summarized as Algorithm 7.4.

Initialization of the multiresolution optimization algorithm actually conceals two problems: construction of the hierarchy of resolution levels and interpolation operators. The hierarchy of resolution levels can be constructed using the farthest point sampling strategy we have described in Chapter 3. This approach naturally allows creating a set of "nested" subsamplings, guaranteeing that every level is an r_l-covering of the shape. Typically, the number

7.6 Multiresolution methods

---

input   : $(L+1)$-level hierarchy of data $D_0, \ldots, D_L$, interpolation operators $P_1^0, \ldots, P_L^{L-1}$.
output   : canonical form Z^*.
initialization: some initial $Z_L^{(0)}$ at the coarsest grid.

1 **for** $l = L, \ldots, 0$ **do**
2   Solve the lth level MDS problem,

$$Z_l^* = \operatorname*{argmin}_{Z_l \in \mathbb{R}^{N_l \times m}} \sigma(Z_l; D_l),$$

   using $Z_l^{(0)}$ as the initialization for the optimization algorithm.
3   Interpolate the solution to the next level, $Z_{l-1}^{(0)} = P_l^{l-1} Z_l^*$.
4 **end**
5 Solve the finest level MDS problem,

$$Z^* = \operatorname*{argmin}_{Z \in \mathbb{R}^{N \times m}} \sigma(Z; D_X),$$

using $Z_0^{(0)}$ as the initialization for the optimization algorithm.

Algorithm 7.4. Multiresolution MDS algorithm.

of points at each level is chosen such that $2 \leq N_{l-1}/N_l \leq 4$. The fact that FPS is a purely intrinsic geometric algorithm allows us to employ it in the most general MDS setting, where the matrix of geodesic distances D_X is the only information available, and no extrinsic geometry can be assumed. The cost of the hierarchy construction is of $\mathcal{O}(N^2)$ complexity, usually significantly smaller compared with the MDS algorithm itself.[5]

The interpolation operators are constructed based on the hierarchy of grids. In order to pass from the lth level to the $(l-1)$st level, we need to interpolate the points of the fine grid Ω_{l-1} from the points of the coarse grid Ω_l. For each fine grid point $z_{i \in \Omega_{l-1}}$, we define the neighborhood $\mathcal{N}_l(i) \subset \Omega_l$ of coarse grid points, from which the point z_i is interpolated. Points common to the fine and the coarse grid have a trivial neighborhood (equal to the point itself, $\mathcal{N}_l(i \in \Omega_l) = \{i\}$), and are transferred unchanged. The coordinates of the rest of the points, $i \in \Omega_{l-1} \setminus \Omega_l$, are interpolated according to

$$z_i = \sum_{j \in \mathcal{N}_l(i)} p_{ij} z_j.$$

The scalars p_{ij} are called *interpolation coefficients* or *weights* and can be identified with the elements of the interpolation matrix P_l^{l-1}. Because usually the number of neighbors used for interpolation is small ($|\mathcal{N}_l(i)| \ll N_l$), the matrix P_l^{l-1} is sparse. Consequently, the interpolation operation can be carried out efficiently, with $\mathcal{O}(N_{l-1})$ complexity.

The specific choice of $\mathcal{N}_l(i)$ and p_{ij} depends on the interpolation method. The neighborhood $\mathcal{N}_l(i)$ can be defined as the K nearest neighbors of x_i on the coarse grid, or alternatively, as the metric ball of radius proportional to the sampling radius. If the shape is represented as a triangular mesh, the neighborhood can be inferred from the triangulation. In the case of parametric surfaces, the interpolation coefficients can be computed in the parameterization domain. There are numerous ways to define the interpolation coefficients, on which we will not spend our attention. The simplest choice is $p_{ij} = |\mathcal{N}_l(i)|^{-1}$, giving z_i as the center of mass of the points $\{z_j\}_{j \in \mathcal{N}_l(i)}$.

7.7* Multigrid MDS

Assume that we have a good initialization $Z_0^{(0)} \approx \mathrm{argmin}_{Z_0}\, \sigma(Z_0; D_0)$ for the MDS problem on the finest grid and wish to improve it. $Z_0^{(0)}$ can be improved by performing a few optimization steps minimizing the stress $\sigma(Z_0; D_0)$ on the fine grid, but such a minimization can be computationally expensive, especially if N is large. Instead, we would like to solve a smaller and computationally cheaper problem on the coarse grid and use the obtained solution to improve $Z_0^{(0)}$ (here, we assume two levels of grids, $L = 1$). For this purpose, we first need to transfer the solution from the fine grid to the coarse grid using a *decimation operator* P_0^1, which, much like the interpolation operator P_1^0, can be thought of as an $N_1 \times N_0$ sparse matrix. In a sense, decimation is dual to interpolation, and in many cases, the decimation operator is obtained as the transpose of the interpolation operator, $P_0^1 = (P_1^0)^\mathrm{T}$. For the sake of simplicity, we assume this relation in the following discussion.

Having transferred $Z_0^{(0)}$ to the coarse grid and obtained $Z_1^{(0)} = P_0^1 Z_0^{(0)}$, we solve the coarse grid problem, which yields the solution Z_1^*. Using Z_1^*, we can correct the fine grid solution by transferring the difference in the coarse grid solution $Z_1^* - Z_1^{(0)}$ to the fine grid,

$$\begin{aligned} Z_0^{(1)} &= Z_0^{(0)} + P_1^0 (Z_1^* - Z_1^{(0)}) \\ &= Z_0^{(0)} + P_1^0 (Z_1^* - P_0^1 Z_0^{(0)}). \end{aligned}$$

Obviously, we would like the new fine grid solution $Z_0^{(1)}$ to be at least as good as $Z_0^{(0)}$, i.e., $\sigma(Z_0^{(1)}) \leq \sigma(Z_0^{(0)})$. Yet, this would not necessarily hold, as the stress functions on the fine and coarse grids may be inconsistent: their minima do not necessarily coincide. More precisely, given the fine grid minimizer Z_0^*, for which $\nabla_{Z_0} \sigma(Z_0^*) = 0$, transferring it to the coarse grid, we will generally have $\nabla_{Z_1} \sigma(P_0^1 Z_0^*) = T_1 \neq 0$. The term T_1 is the *residual*, reflecting the inconsistency of the coarse and fine grid problems. To cancel the residual, we have to make an adjustment of the coarse grid problem by introducing a *correction term* to the stress

$$\sigma(Z_1; D_1) - \langle T_1, Z_1 \rangle = \sigma(Z_1; D_1) - \mathrm{trace}(Z_1^\mathrm{T} T_1).$$

After this correction, the coarse grid problem becomes

$$\min_{Z_1} \sigma(Z_1; D_1) - \langle T_1, Z_1 \rangle.$$

In order to guarantee consistency, T_1 must satisfy

$$\nabla_{Z_1} \sigma(Z_1; D_1) - T_1 = \nabla_{Z_1} \sigma\left(Z_0^{(0)} + P_1^0(Z_1 - P_0^1 Z_0^{(0)}); D_0\right),$$

for $Z_1 = P_0^1 Z_0^{(0)}$. Using the chain rule, we obtain

$$\begin{aligned}
\nabla_{Z_1} \sigma(Z_1; D_1) - T_1 &= \nabla_{Z_1} \sigma\left(Z_0^{(0)} + P_1^0(Z_1 - P_0^1 Z_0^{(0)}); D_0\right)\Big|_{Z_1 = P_0^1 Z_0^{(0)}} \\
&= (P_1^0)^{\mathrm{T}} \nabla_{Z_0} \sigma(Z_0^{(0)}; D_0) \\
&= P_0^1 \nabla_{Z_0} \sigma(Z_0^{(0)}; D_0),
\end{aligned} \tag{7.16}$$

which leads to the expression for the correction term,

$$T_1 = \nabla_{Z_1} \sigma(P_0^1 Z_0^{(0)}; D_1) - P_0^1 \nabla_{Z_0} \sigma(Z_0^{(0)}; D_0).$$

In particular, when we take $Z_0^{(0)}$ to be the fine grid minimizer Z_0^*, it is easy to verify that $\nabla_{Z_1} \left(\sigma(P_0^1 Z_0^*; D_1) - \mathrm{trace}((P_0^1 Z_0^*)^{\mathrm{T}} T_1)\right) = 0$, i.e., that $P_0^1 Z_0^*$ is the coarse grid minimizer.

Yet, resolving one problem, we have created a new one: the coarse grid problem $Z^* = \mathrm{argmin}_Z \sigma(Z) - \mathrm{trace}(Z^{\mathrm{T}} T)$, after the introduction of the correction term becomes unbounded. Indeed, we can add an arbitrarily large scalar to the coordinates Z without changing $\sigma(Z)$, yet, if $T \neq 0$, the term $-\mathrm{trace}(Z^{\mathrm{T}} T)$ can decrease arbitrarily. In order to overcome this difficulty, in [79, 80] Irad Yavneh and the authors introduced the *modified stress* by adding a quadratic penalty to $\sigma(Z)$,

$$\begin{aligned}
\hat{\sigma}(Z) &= \sigma(Z) + \lambda \sum_{k=1}^{m} \left(\sum_{i=1}^{N} z_i^k\right)^2 \\
&= \sigma(Z) + \lambda \, \mathrm{trace}(Z^{\mathrm{T}} 1_{N \times N} Z).
\end{aligned} \tag{7.17}$$

Here, λ is some positive constant. Recall that the solution to the MDS problem is defined up to an isometry in the embedding space. Our modification merely resolves the translation ambiguity by restricting the center of mass of the resulting canonical form to the origin. The penalty would not change the solution, and we can therefore use $\hat{\sigma}(Z)$ instead of $\sigma(Z)$ in our problem. Because the penalty term is quadratic, it grows faster than does the linear term $\mathrm{trace}(Z^{\mathrm{T}} T)$. Consequently, the function $\hat{\sigma}(Z) - \mathrm{trace}(Z^{\mathrm{T}} T)$ is bounded. The gradient of the modified stress has practically the same computational complexity as that of the standard stress and is given by

$$\nabla_Z \hat{\sigma}(Z) = \nabla_Z \sigma(Z) + 2\lambda 1_{N \times N} Z.$$

Now we can combine the pieces of the puzzle together into a scheme, usually referred as the *two-grid algorithm* (Algorithm 7.5). The scheme consists of

> **input** : two-level hierarchy of data D_0, D_1, interpolation operator P_1^0, decimation operator P_0^1.
> **output** : canonical form Z^*.
> **initialization**: some initial $Z_0^{(0)}$ at the fine grid.
>
> 1 **repeat**
> 2   Compute an improved fine grid solution $Z_0^{(1)}$ by performing N_R optimization iterations on $\sigma(Z_0)$, initialized with $Z_0^{(0)}$.
> 3   Decimate the fine grid solution: $Z_1^{(1)} = P_0^1 Z_0^{(1)}$.
> 4   Compute the correction: $T_1 = \nabla_{Z_1} \hat{\sigma}(Z_1^{(1)}; D_1) - P_0^1 \nabla_{Z_0} \hat{\sigma}(Z_0^{(1)}; D_0)$.
> 5   Solve the coarse grid problem,
>
> $$Z_1^* = \operatorname*{argmin}_{Z_1 \in \mathbb{R}^{N_1 \times m}} \hat{\sigma}(Z_1; D_1) - \operatorname{trace}(Z_1^T T_1),$$
>
> using $Z_1^{(1)}$ as the initialization.
> 6   Correct the fine grid solution, $Z_0^{(2)} = Z_0^{(1)} + \alpha_1 P_1^0 (Z_1^* - P_0^1 Z_0^{(1)})$.
> 7   Set $Z_0^{(0)} \longleftarrow Z_0^{(2)}$.
> 8 **until** *convergence*
> 9 $Z^* = Z_0^{(0)}$.

Algorithm 7.5. Two-grid MDS algorithm.

repeatedly improving the fine grid solution by performing a few optimization steps on the fine grid, decimating the result to the coarse grid and solving the coarse grid problem with an appropriate correction term, and then using the coarse grid solution to improve the fine grid result.

At the correction stage (Step 6 of the two-grid algorithm), a step size α_1 must be determined using line search (for example, Armijo rule) in order to guarantee a descent direction [291]. Step 5 is generic, and the choice of the optimization algorithm usually depends on the stress used. For example, if the L_2-stress is used as the distortion criterion, we can employ SMACOF-type iteration,

$$\begin{aligned} Z^{(k+1)} &= Z^{(k)} - \frac{1}{2N} \nabla_Z \hat{\sigma}_2(Z^{(k)}) \\ &= \frac{1}{N} B(Z^{(k)}) Z^{(k)} - T + 2\lambda 1_{N \times N} Z^{(k)}. \end{aligned}$$

Applying the two-grid algorithm repeatedly, we converge to its fixed point, the fine grid minimizer Z_0^*.

The two-grid algorithm we have described appears to be a particular case of a *multigrid* algorithm. The idea of multigrid dates back to the papers of Fedorenko [157] and Bakhvalov [18] in the 1960s, though the method as we know it today was formulated a decade later by Brandt [49, 50] and Hackbusch [197]. Originally, multigrid methods were applied to partial differential equations and introduced relatively late to the field of numerical optimization

7.7* Multigrid MDS

---

input : $(L+1)$-level hierarchy of data $D_0, \ldots, D_L$, interpolation operators $P_1^0, \ldots, P_L^{L-1}$, decimation operators $P_0^1, \ldots, P_{L-1}^L$, constants N_R, N_R'.
output : canonical form Z^*.
initialization: some initial $Z^{(0)}$ and $k = 0$.

1 **repeat**
2 $\quad Z^{(k+1)} \leftarrow$ V-cycle$(Z^{(k)}, 0, D_0, N_R, N_R')$.
3 $\quad k \leftarrow k+1$.
4 **until** *convergence*
5 $Z^* = Z^{(k)}$.

Algorithm 7.6. MG-MDS algorithm with V-cycle outer iterations.

[135, 291]. In the multigrid jargon, Step 5 is called *relaxation*; interpolation and decimation are often referred to respectively as *prolongation* and *restriction*.

The two-grid algorithm can be easily extended by applying it on the coarse grid repeatedly in a recursive manner. In a general multigrid MDS algorithm (referred to as MG-MDS for brevity), we have a hierarchy of grids $\Omega_0, \ldots, \Omega_L$ and data $D_0, \ldots, D_L$, interpolation operators $P_1^0, \ldots, P_L^{L-1}$, and decimation

---

1 **if** $l = L$ *(coarsest resolution)* **then**
2 $\quad$ Solve the Lth level problem,

$$Z_L^* = \underset{Z_L \in \mathbb{R}^{N_L \times m}}{\operatorname{argmin}} \hat{\sigma}(Z_L; D_L) - \operatorname{trace}(Z_L^\mathrm{T} T_L),$$

$\quad$ using $Z_L^{(0)}$ as the initialization.
3 $\quad$ **return** Z_L^*.
4 **else**
5 $\quad$ Relaxation: apply N_R optimization iterations on
$\quad$ $\hat{\sigma}(Z_l; D_l) - \operatorname{trace}(Z_l^\mathrm{T} T_l)$ initialized with $Z_l^{(0)}$, obtaining $Z_l^{(1)}$.
6 $\quad$ Restriction: $Z_{l+1}^{(1)} = P_l^{l+1} Z_l^{(1)}$.
7 $\quad$ Recursively apply the V-cycle on the coarser level,

$$Z_{l+1}^{(2)} \leftarrow \text{V-cycle}(Z_{l+1}^{(1)}, T_{l+1}, D_{l+1}, N_R, N_R').$$

8 $\quad$ Correction: $Z_l^{(2)} = Z_l^{(1)} + \alpha_l P_{l+1}^l (Z_{l+1}^{(2)} - Z_{l+1}^{(1)})$.
9 $\quad$ Relaxation: apply N_R' optimization iterations on
$\quad$ $\hat{\sigma}(Z_l; D_l) - \operatorname{trace}(Z_l^\mathrm{T} T_l)$ initialized with $Z_l^{(2)}$, obtaining $Z_l^{(3)}$.
10 $\quad$ **return** $Z_l^{(3)}$.
11 **end**

Algorithm 7.7. Function V-cycle$(Z_l^{(0)}, T_l, D_l, N_R, N_R')$

160   7 Multidimensional Scaling

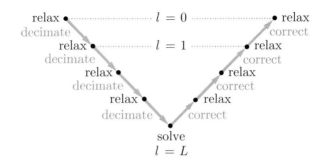

Figure 7.7. Illustration of the V-cycle iterative process.

operators $P_0^1, \ldots, P_{L-1}^L$. The algorithm first goes down from the finest grid to the coarsest grid, performing N_R relaxation iterations at each level, and then back to the finest grid, with N_R' relaxation iterations (typically, N_R and N_R' range from 1 to 5). Because of the resemblance of the outer iterative process to the letter "V" (Figure 7.7), it has been named the *V-cycle*, or, being more specific, the $V(N_R, N_R')$-cycle. As in the two-grid algorithm, the V-cycles are repeated until convergence (Algorithms 7.6 and 7.7).

The cost of a V-cycle is usually comparable with a few SMACOF iterations. Yet, the MG-MDS algorithm requires much fewer iterations to converge. More importantly, the number of cycles (outer iterations) is approximately constant with N, unlike SMACOF, whose number of iterations tends to grow with N. This behavior is often observed in multigrid methods. For this reason, the use of multigrid offers a significant speedup of MDS algorithms, especially pronounced in large-scale problems.

Example 7.4 (multigrid acceleration). We demonstrate the MG-MDS method on the problem we have already seen in Example 7.2, Swiss roll embedding into $\mathbb{R}^3$. The Swiss roll surface is sampled at $N = 2145$ points, which results in a relatively large-scale MDS problem. We use a multigrid $V(3,3)$-cycle with SMACOF-type relaxation and compare it with the standard SMACOF algorithm. Both algorithms are initialized by the coordinates of the points on the original surface. To ensure a fair comparison, we first run SMACOF with the stopping criterion $\sigma_2(Z^{(k+1)})/\sigma_2(Z^{(k)}) \leq 0.01$, and then let the MG-MDS algorithm reach the same stress. Figure 7.8 depicts the L_2-stress on a logarithmic scale versus time for the SMACOF and multigrid algorithms. The MG-MDS algorithm converges in 229.46 sec (7 cycles), compared with 2.03×10^3 sec (341 iterations) of the SMACOF algorithm – almost an order of magnitude speedup.

7.8* Vector extrapolation

As an alternative way to accelerate the convergence of the SMACOF iterations, Guy Rosman proposed using *vector extrapolation* [327, 326]. Such meth-

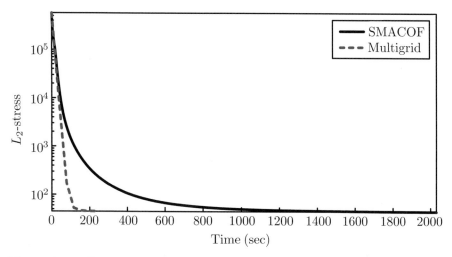

Figure 7.8. Convergence plots of the standard SMACOF algorithm (solid) and the multigrid V(3,3)-cycle with SMACOF-type relaxation (dashed) in the problem of Swiss roll embedding with $N = 2145$. Shown is the L_2-stress value $\sigma_2(Z^{(k)})$ as a function of CPU time.

ods were originally derived for speeding up slowly converging linear iterative minimization and numerical approximation schemes,[6] though in practice they also work well for nonlinear problems [51], like certain processes in computational fluid dynamics [359] and reconstruction in tomography [321, 386].

Assume that we have a sequence of iterates $Z^{(0)}, Z^{(1)}, \ldots$ produced by some optimization algorithm (e.g., SMACOF), which converges to the minimizer $Z^* = \lim_{k \to \infty} Z^{(k)}$. The key idea of vector extrapolation is the following: if $Z^{(k)}$ converges slowly, we can construct another sequence, denoted by $\hat{Z}^{(k)}$, which will converge to the same minimizer Z^*, but much faster. Writing the original sequence as $Z^{(k)} = Z^* + E^{(k)}$, where $E^{(k)}$ is the *remainder*, we are looking for a new sequence $\hat{Z}^{(k)} = Z^* + \hat{E}^{(k)}$ such that

$$\lim_{k \to \infty} \frac{\|\hat{E}^{(k)}\|}{\|E^{(k)}\|} = 0. \tag{7.18}$$

The new sequence is given as some transformation of the form $\hat{Z}^{(k)} = T(Z^{(k)}, \ldots, Z^{(k+K)})$, applied to $K+1$ iterates $Z^{(k)}, \ldots, Z^{(k+K)}$ produced by a standard optimization algorithm. The extrapolation is used as a new initialization to restart the optimization. The entire process is repeated for a few cycles (Algorithm 7.8). Because in general there is no guarantee for the extrapolation to succeed, a safeguard (Steps 4–8) must be used to ensure that the stress value does not increase.

It is common to construct $\hat{Z}^{(k)}$ as a linear combination of $K+1$ iterates,

$$\begin{aligned} \hat{Z}^{(k)} &= \gamma_0 Z^{(k)} + \ldots + \gamma_K Z^{(k+K)} \\ &= (\gamma_0 + \cdots + \gamma_K) Z^* + \gamma_0 E^{(k)} + \ldots + \gamma_K E^{(k+K)} \end{aligned} \tag{7.19}$$

```
input        : N × N matrix of geodesic distances $D_X$.
output       : canonical form $Z^*$.
initialization: some initial $Z^{(0)}$ and $k = 0$.
1  repeat
2      Generate the sequence of iterates $Z^{(k+1)}, \ldots, Z^{(k+K)}$ by some
       optimization algorithm on $\sigma(Z; D_X)$, using $Z^{(k)}$ as initialization.
3      Extrapolate $\hat{Z}^{(k)}$ from the iterates $Z^{(k)}, Z^{(k+1)}, \ldots, Z^{(k+K)}$.
4      if $\sigma(\hat{Z}^{(k)}; D_X) > \sigma(Z^{(k+K)}; D_X)$ then
5          $Z^{(k+K+1)} \longleftarrow Z^{(k+K)}$
6      else
7          $Z^{(k+K+1)} \longleftarrow \hat{Z}^{(k)}$
8      end
9      $k \longleftarrow k + K + 1$.
10 until convergence of $Z^{(k)}$
11 $Z^* = Z^{(k)}$.
```

Algorithm 7.8. MDS algorithm with vector extrapolation acceleration.

where γ_i are some coefficients. Various vector extrapolation algorithms differ in the definition of these coefficients. Ideally, the new sequence should be $\hat{Z}^{(k)} = Z^*$, such that one can try to select the coefficients γ_i in order to satisfy this requirement. Usually this is impossible to achieve, thus we will try to reduce the residual terms $\gamma_0 E^{(k)} + \ldots + \gamma_K E^{(k+K)}$ as much as possible, at the same time requiring $\gamma_0 + \cdots + \gamma_K = 1$ in equation (7.19).

Note that $E^{(k)} = Z^{(k)} - Z^*$ depends on the unknown Z^*. In order to eliminate this dependence, we can consider the difference $\Delta\hat{Z}^{(k)} = \hat{Z}^{(k+1)} - \hat{Z}^{(k)}$, which ideally should vanish. This leads to a constrained linear system,

$$\gamma_0 \Delta Z^{(k)} + \ldots + \gamma_K \Delta Z^{(k+K)} = 0 \quad \text{s.t.} \quad \gamma_0 + \ldots + \gamma_K = 1, \quad (7.20)$$

with Nm equations and $K+1$ variables. Because typically $K \ll Nm$, the system is over-determined and usually cannot be satisfied. We therefore compute the coefficients γ_i by finding a least-square solution to (7.20). This method of vector extrapolation is known as *reduced rank extrapolation* (RRE) [270, 144].[7]

If we parse the $N \times m$ matrices $\Delta Z^{(k)}, \ldots, \Delta Z^{(k+K)}$ into $Nm \times 1$ vectors $\Delta z^{(i)} = \text{vec}(\Delta Z^{(i)})$ and denote $A_{K+1} = (\Delta z^{(k)} \ldots \Delta z^{(k+K)})$, we can formulate the RRE method as finding a least-squares solution to the following constrained over-determined linear system,

$$A_{K+1}\gamma = 0 \quad \text{s.t.} \quad \gamma_0 + \ldots + \gamma_K = 1. \quad (7.21)$$

which can be solved by first solving

$$A_{(K+1)}^\mathrm{T} A_{(K+1)} \tilde{\gamma} = 1_{(K+1) \times (K+1)}, \quad (7.22)$$

and then setting

$$\gamma = \frac{\tilde{\gamma}}{\tilde{\gamma}_0 + \ldots \tilde{\gamma}_K}$$

(prove this result in Problem 7.11).

For a numeric solution, the matrix A_{K+1} can be represented as $A_{K+1} = Q_{K+1} R_{K+1}$ using *QR factorization*, where Q_{K+1} is an $Nm \times (K+1)$ unitary matrix and R_{K+1} is a $(K+1) \times (K+1)$ upper triangular matrix [359]. This type of factorization can be carried out efficiently using the *modified Gram-Schmidt* algorithm [177]. Because $Q_{K+1}^T Q_{K+1} = I$, equation (7.22) becomes $R_{K+1}^T R_{K+1} \tilde{\gamma} = 1_{(K+1) \times (K+1)}$. Because of the triangular form of the matrix R_{K+1}, we can employ forward and backward substitutions, similarly to the solution of the Newton system with Cholesky factorization we have encountered in Chapter 5. The entire RRE method can be summarized as shown in Algorithm 7.9.

Example 7.5 (RRE acceleration). We demonstrate the vector extrapolation method on the problem of Swiss roll embedding into $\mathbb{R}^3$, with the same data and the same settings as in the previous Example 7.4. We compare the SMACOF algorithm with and without vector extrapolation acceleration. In the accelerated version, we use the RRE method with $K = 10$. Figure 7.9 (top) depicts the L_2-stress on a logarithmic scale versus time for both algorithms. The SMACOF algorithm without acceleration converges after 2.03×10^3 sec (341 iterations) compared with 256.59 sec (4 cycles) for the RRE-accelerated version. Figure 7.9 (bottom) shows the inner iterations of the RRE algorithm. The "jumps" in the stress values correspond with extrapolated values in each cycle.

The RRE approach gives nearly the same acceleration to the SMACOF algorithm as the multigrid scheme but is advantageous being significantly simpler and easier to implement. In addition to RRE, many other vector extrapolation methods exists, the most popular choices being *minimal polynomial extrapolation* (MPE) [91] and the *topological ϵ-algorithm* (TEA) [51]. We refer the reader to the above references for additional details.

---

input : sequence of iterates $\Delta Z^{(k)}, \ldots, \Delta Z^{(k+K)}$.
output : extrapolation $\hat{Z}^*$.

1 Compute the matrix $A_{K+1} = (\Delta z^{(k)} \ldots \Delta z^{(k+K)})$ and find its QR factorization $A_{K+1} = Q_{K+1} R_{K+1}$.
2 Forward substitution: solve $R_{K+1}^T y = 1_{(K+1) \times (K+1)}$ for y.
3 Backward substitution: solve $R_{K+1} \tilde{\gamma} = y$ for $\tilde{\gamma}$.
4 Compute $\gamma = \tilde{\gamma}/(\tilde{\gamma}_0 + \ldots + \tilde{\gamma}_K)$.
5 Compute the extrapolation $\hat{Z}^{(k)} = \gamma_0 Z^{(k)} + \ldots + \gamma_K Z^{(k+K)}$.

Algorithm 7.9. Reduced Rank Extrapolation (RRE).

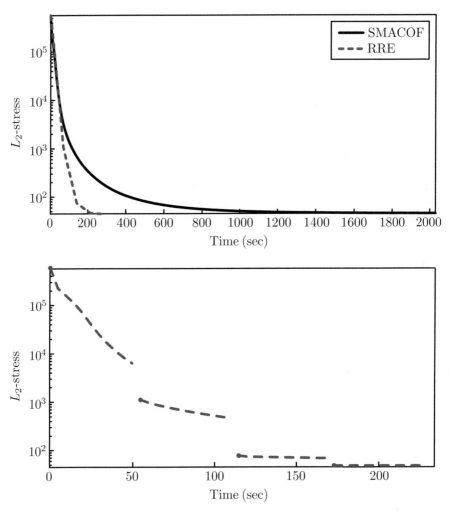

Figure 7.9. Convergence plots of the SMACOF algorithm with (top, dashed) and without (top, solid) RRE acceleration in the problem of Swiss roll embedding with $N = 2145$. Shown is the L_2-stress value in outer cycles as a function of CPU time. Values of the stress in inner iterations (bottom, dotted) and outer cycles (bottom, bold dots) of the RRE algorithm are shown in the bottom plot.

7.9 A trouble with topology

Using intrinsic similarity criterion for the comparison of non-rigid shapes, we have been tacitly assuming that the shapes have similar *topology*. Said differently, we did not allow our deformations to make holes or "glue" parts of the shapes. In certain situations, however, this assumption does not necessarily hold.

For example, when acquiring three-dimensional objects using range cameras, a phenomenon called *topological noise* is a well-known problem. A range camera estimates the three-dimensional coordinates of the object at a discrete set of points, producing a point cloud. The point cloud is then triangulated in order to obtain a triangular mesh representation of the object. Yet, if no additional information is assumed, the problem of constructing a mesh out of a point cloud is not well-defined. The same object can be triangulated differently, in some cases producing meshes with different topology. Typically, topological noise is manifested in the form of connections between vertices that should not be connected or vice versa.

Topological noise affects the intrinsic geometry of the shape. Connectivity changes can significantly alter the geodesics, and consequently, if we try to compute the canonical forms of two topologically different shapes, they may differ substantially (see example in Figure 7.10). We therefore conclude that intrinsic similarity can be used only for shapes undergoing topology-preserving deformations.

At the other end, extrinsic similarity is insensitive to topological noise, yet, as we have already seen, is sensitive to non-rigid deformations. In [74], we showed that it is possible to define a new criterion of similarity insensitive to some types of topological noise by combining extrinsic and intrinsic similarity criteria. We must note, however, that the general problem of shape similarity invariant under deformation not preserving topology appears as a very challenging problem, to which, to the best of our knowledge, no complete solution has been proposed thus far.

Suggested reading

For a comprehensive overview of MDS problems and algorithms, the reader is referred to the books by Cox and Cox [119] and Borg and Groenen [44]. A good treatment of MDS with a theoretical emphasis is the book by Young and Hamer [405]. In the fields of machine learning and neural networks, the concept of *self-organizing maps* introduced by Teuvo Kohonen [231] is closely related to MDS. A classic reading on multigrid methods is Briggs [54], McCormick [264], and Wesseling [393]. A more recent overview is the book by Trottenberg et al. [381]. A detailed description and experimental validation of the multigrid MDS method can be found in [79, 80]. For an introduction and overview of vector extrapolation methods, we refer to review papers by Sidi and coauthors [366, 362, 360, 363, 361] A different approach to vector extrapolation through the Shanks-Schmidt transformation is presented in [339, 351, 52]. The use of vector extrapolation for MDS problems is described in [327]. A different acceleration of the SMACOF algorithm was proposed by de Leeuw [127]. An attempt to address the problem of comparison of topologically different shapes using a combination of extrinsic and intrinsic similarity criteria is presented by the authors in [74].

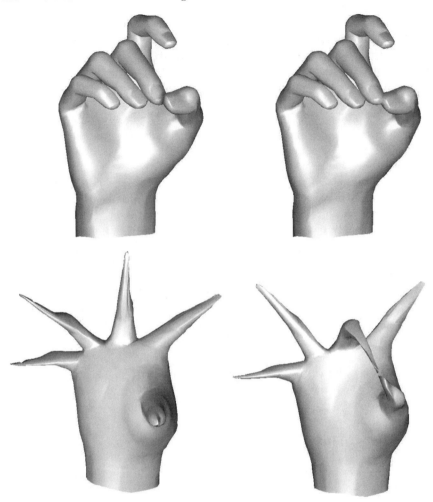

Figure 7.10. Illustration of the effect of the topological noise on the computation of canonical forms. Gluing the fingertips of the hand (right) changes the intrinsic geometry of this shape and as a result, the canonical form is substantially different from that of the original shape (left).

Software

A free MATLAB implementation of the SMACOF algorithm with vector extrapolation acceleration is distributed as part of the TOSCA toolbox. Mark Steyvers' Nonmetric MDS for MATLAB allows minimization of different variants of stress. The function `mdscale` supports different types of MDS problems and is distributed as part of the commercial MATLAB Statistics toolbox.

Problems

7.1. Verify that the L_2-stress function is non-convex.

7.2. Derive the matrix form of the L_2-stress in equation (7.4).

7.3. Derive the matrix form of the gradient of the L_2-stress in equation (7.5).

7.4. How does the fact that the matrix $B(Z)$ is zero-mean relate to the degrees of freedom in the MDS problem?

7.5. Prove the bound on the L_2-stress in inequality (7.7).

7.6. Derive the Hessian of the L_2-stress.

7.7. For the case $m = 2$, show a way to invert the Hessian matrix (7.9) using the inversion formula for block matrices.

7.8. Show that the constrained optimization problem (7.11) is equivalent to the L_∞-stress minimization optimization problem (7.10).

7.9. Derive the matrix form of the modified stress (7.17) and show that the addition of the quadratic penalty does not change the solution of the MDS problem.

7.10. Derive the relation (7.16).

7.11. Derive the equation (7.22).

Notes

[1] One peculiar example of the distortion introduced by cartographic projections is the fact that river Lena appears to be the longest river in the world when viewed on a map using a Mercator projection. In reality, Amazon and Nile hold this title.

[2] Borg and Groenen [44] give a different meaning to this acronym, *scaling by majorizing a complex function*. We find this a slight misnomer, as the term "complex function" usually refers to a complex-valued function.

[3] The presence of zero eigenvalues is directly related to the degrees of freedom in the MDS problem. An alternative way to resolve the degrees of freedom is to fix the positions of three non-collinear points [220].

[4] It was shown by Platt [314] that these three methods are equivalent.

[5] As we have already mentioned in Chapter 3, this cost can be reduced to $\mathcal{O}(N \log N)$ using efficient data structures. The complexity of MDS is often given as $\mathcal{O}(N^3)$, though, more rigorously, it should be proportional to N^2 multiplied by the number of iterations.

[6] There are evidences that prototypes of extrapolation methods are very old. Brezinsky [53] traces the history of speeding up convergence of sequences by means of extrapolation back to Christian Huygens (1629–1695).

[7] The derivation of RRE is done assuming that the sequence $Z^{(k+1)}, \ldots, Z^{(k+K)}$ is produced by a linear iteration formula [366]. In the nonlinear case, the algorithm can still be used, but some of its properties are lost.

8
Spectral Embedding

> Can you hear the shape of the drum?
>
> M. KAC

In the previous chapter, we showed how to represent the intrinsic geometry of a non-rigid shape in a Euclidean space using canonical forms. Given a shape (X, d_X) sampled at N points $\{x_1, \ldots, x_N\}$, we first computed the $N \times N$ matrix D_X of geodesic distances between the samples. Then, we employed an iterative MDS algorithm in order to find the minimum-distortion embedding of the shape into the m-dimensional Euclidean space. As a result, we obtained an $N \times m$ matrix Z^* representing the canonical form of X – the best possible approximation of the intrinsic geometry of X in $\mathbb{R}^m$ with the restricted Euclidean metric.

We start this chapter with a mathematical exercise that will bring us to a new class of MDS algorithms and set the scene for a somewhat different, algebraic viewpoint on the problem of non-rigid shape similarity. For simplicity, we first assume that X is isometrically embeddable into $\mathbb{R}^m$, that is, we can find a canonical form Z^* such that $d_X(x_i, x_j) = \|z_i^* - z_j^*\|_{\mathbb{R}^m}$ for all $i, j = 1, \ldots, N$. Consequently, we can also write

$$d_X^2(x_i, x_j) = d_{ij}^2(Z^*) = \langle z_i^* - z_j^*, z_i^* - z_j^* \rangle_{\mathbb{R}^m}$$
$$= \langle z_i^*, z_i^* \rangle_{\mathbb{R}^m} - 2\langle z_i^*, z_j^* \rangle_{\mathbb{R}^m} + \langle z_j^*, z_j^* \rangle_{\mathbb{R}^m}.$$

This equality can be rearranged as

$$2\langle z_i^*, z_j^* \rangle_{\mathbb{R}^m} = \langle z_i^*, z_i^* \rangle_{\mathbb{R}^m} + \langle z_j^*, z_j^* \rangle_{\mathbb{R}^m} - d_X^2(x_i, x_j). \tag{8.1}$$

Recall that isometric embedding into $\mathbb{R}^m$ is defined up to Euclidean isometry. Therefore, we can translate the canonical form by an arbitrary vector, for example, by setting the origin at $z_1^* = 0$. We can subtract the vector z_1^* from all the coordinates of the canonical form Z^* without changing the distances between them. Such a transformation allows us to rewrite the right-hand side of equation (8.1): because the embedding is isometric, the terms $\langle z_i^* - z_1^*, z_i^* - z_1^* \rangle_{\mathbb{R}^m}$ and $\langle z_j^* - z_1^*, z_j^* - z_1^* \rangle_{\mathbb{R}^m}$ can be identified with the squared geodesic distances $d_X^2(x_1, x_i)$ and $d_X^2(x_1, x_j)$, respectively. Therefore, we have

$$2\langle z_i^*, z_j^*\rangle_{\mathbb{R}^m} = \langle z_i^* - z_1^*, z_i^* - z_1^*\rangle_{\mathbb{R}^m} + \langle z_j^* - z_1^*, z_j^* - z_1^*\rangle_{\mathbb{R}^m} - d_X^2(x_i, x_j)$$
$$= d_X^2(x_1, x_i) + d_X^2(x_1, x_j) - d_X^2(x_i, x_j). \tag{8.2}$$

We notice that on the left-hand side, up to the factor of two, we have the elements of the matrix $Z^*(Z^*)^{\mathrm{T}}$, referred to as the *Gram matrix*. Defining an $N \times N$ matrix K_X with the elements

$$k_{ij} = \frac{1}{2}\left(d_X^2(x_1, x_i) + d_X^2(x_1, x_j) - d_X^2(x_i, x_j)\right), \tag{8.3}$$

we can rewrite condition (8.2) as $K_X = Z^*(Z^*)^{\mathrm{T}}$.

Using the properties of Gram matrices (see Problem 8.1), from condition (8.2) it follows that K_X is positive semidefinite and of rank m, which implies that it has m positive and $N - m$ zero eigenvalues. Denoting by Λ_m an $m \times m$ diagonal matrix of the positive eigenvalues and by U_m the $N \times m$ matrix of the corresponding eigenvectors, we can write K_X in the following way,

$$K_X = U_m \Lambda_m U_m^{\mathrm{T}} = U_m \Lambda_m^{\frac{1}{2}} \Lambda_m^{\frac{1}{2}} U_m^{\mathrm{T}} = (U_m \Lambda_m^{\frac{1}{2}})(U_m \Lambda_m^{\frac{1}{2}})^{\mathrm{T}}.$$

8.1 Classic MDS

Let us draw some conclusions from this mathematical exercise. The assumption that the shape X can be isometrically embedded into $\mathbb{R}^m$ leads to the fact that K_X can be realized as a Gram matrix of rank m, which is positive semidefinite by definition. Conversely, given a positive semidefinite matrix K_X or rank m, we can write it as $K_X = (U_m \Lambda_m^{\frac{1}{2}})(U_m \Lambda_m^{\frac{1}{2}})^{\mathrm{T}}$. This gives us a simple recipe for computing the canonical form by setting $Z^* = U_m \Lambda_m^{\frac{1}{2}}$, and implies that X is isometrically embeddable into $\mathbb{R}^m$. Combining both directions, we can say that X is isometrically embeddable into $\mathbb{R}^m$ if and only if $K_X \succeq 0$. This result was proved by Isaac Schoenberg in 1935 [340], of which the statistician John Clifford Gower commented "a surprisingly late date for such a fundamental property of Euclidean geometry" [184].[1]

Unfortunately, the geometry of the objects we are dealing with is non-Euclidean: in Chapter 7, we saw that usually it is impossible to isometrically embed a shape into $\mathbb{R}^m$, in contradiction to what we assumed throughout our exercise. For a general shape that is not isometrically embeddable into a Euclidean space, some of the eigenvalues of K_X will be negative. As a result, we can no more write K_X as a Gram matrix. If the reader still insists on writing K_X as a Gram matrix of rank m, we can resort to the following approximation: define a new $N \times N$ matrix

$$\tilde{K}_X = U \tilde{\Lambda} U^{\mathrm{T}},$$

where

8.1 Classic MDS

---

input : $N \times N$ matrix of geodesic distances D_X, dimensionality m.
output: canonical form Z^*.

1. Compute the matrix K_X according to (8.3).
2. Perform an eigendecomposition of K_X to compute the $m \times m$ matrix Λ_m of the m largest nonnegative eigenvalues and the $N \times m$ matrix of corresponding eigenvectors U_m.
3. Compute the canonical form as $Z^* = U_m \Lambda_m^{\frac{1}{2}}$.

Algorithm 8.1. Classic MDS.

$$\tilde{\Lambda} = \begin{pmatrix} \Lambda_m & \\ & 0_{(N-m)\times(N-m)} \end{pmatrix}.$$

is a matrix produced by taking the m largest positive eigenvalues of K_X (assuming that K_X has at least m positive eigenvalues) and removing all the rest. $\tilde{K}_X$ is the best rank-m approximation of K_X; it allows us to define the canonical form as $Z^* = U\tilde{\Lambda}^{\frac{1}{2}} = U_m \Lambda_m^{\frac{1}{2}}$, which satisfies condition (8.2) approximately: $\tilde{K}_X \approx Z^*(Z^*)^{\mathrm{T}}$. This brings us to the way to compute the canonical form shown in Algorithm 8.1.

This approach, usually referred to as *classic MDS*,[2] is due to Torgerson [379] and Gower [183] and is based on the theoretical results of Schoenberg [341] and Eckart, Young and Householder [143, 406]. It is one of the earliest practical MDS methods, derived from purely algebraic considerations and belonging to a family of methods known as *spectral embeddings* or *eigenmaps* – methods in which the embedding coordinates are computed as eigenvectors of some matrix. Classic MDS has some very appealing properties: it is based on eigendecomposition, a procedure extensively researched in numerical linear algebra, for which many efficient numerical algorithms are available. Therefore, classic MDS is typically significantly faster compared, for example, with the SMACOF algorithm. More importantly, unlike the iterative MDS methods discussed in Chapter 7, classic MDS does not suffer from local convergence.

At the same time, the Achilles' heel of classic MDS is the lack of geometric meaning. To understand why, let us quantify the error introduced by taking $\tilde{K}_X$ instead of K_X,

$$\|\tilde{K}_X - K_X\|_{\mathrm{F}}^2 = \|U(\Lambda - \tilde{\Lambda})U^{\mathrm{T}}\|_{\mathrm{F}}^2 = \lambda_{m+1}^2 + \cdots + \lambda_N^2, \qquad (8.4)$$

measured by the *Frobenius norm*, which is defined as $\|A\|_{\mathrm{F}}^2 = \mathrm{trace}(A^{\mathrm{T}} A)$. It is easy to see (a formal proof is left as Problem 8.2) that defining $\tilde{\Lambda}$ by taking the largest m eigenvalues of Λ produces the smallest error in the sense of the Frobenius norm in (8.4). Therefore, we can think of classic MDS as of a minimization problem, in which the objective is the *strain* function,

$$\sigma_{\mathrm{F}}(Z) = \|ZZ^{\mathrm{T}} - K_X\|_{\mathrm{F}}^2.$$

The strain measures how "close" a Gram matrix of rank m can be to K_X.[3] Such a criterion does not have much geometric sense: we cannot interpret it as the average or the maximum distortion of the geodesic distances, a meaning we could attribute to σ_2 and σ_∞. In practice, it also appears that shape similarity based on canonical forms computed using classic MDS usually produces inferior results compared with that based on least-squares MDS [149].

As previously, the canonical form is defined ambiguously: the liberty we have in the definition of K_X reflects the degrees of freedom related to translation ambiguity in the MDS problem. In equation (8.3), we defined K_X based on the assumption that the point z_1^* is located at the origin. This definition is not unique, and we can define K_X in different ways by choosing a different origin for the embedding space (see Problem 8.3). The rotation and reflection ambiguity is concealed in the fact that eigenvectors are defined up to a unitary transformation: if U are the eigenvectors of the matrix K_X, then RU are eigenvectors as well (here R is a rotation and reflection matrix). This makes our definition of the canonical form ambiguous up to a rotation and reflection transformation, that is, we can write $Z^* = R U_m \Lambda_m^{\frac{1}{2}}$.

Another observation is that if we have a canonical form in $\mathbb{R}^m$ and wish to compute a canonical form in $\mathbb{R}^{m'}$, $m' > m$, there is no need to re-compute it (which we actually need to do in the case of iterative MDS methods). Because each dimension is considered separately, it is enough to compute the $(m+1)$st to m'th eigenvalues and eigenvectors. This interesting property is called *nested dimensions*; as we will see, it appears to be common to spectral embedding algorithms.

Our final remark is about numerical algorithms for classic MDS. The core of this approach is the eigendecomposition of a large-scale symmetric matrix K_X with non-negative elements. Because in most cases $N \gg m$, computing all the eigenpairs of K_X may be computationally expensive or even prohibitive. Instead, we can compute a few largest eigenvalues and the corresponding eigenvectors. Algorithms especially suitable for eigendecomposition problems include the Arnoldi [9] and Lanczos [240] methods and their variants [246, 92, 16].[4] These algorithms find the largest-magnitude eigenvalues, therefore, it may happen that the first computed eigenvalues are negative. This problem can usually be avoided by computing more than m eigenvalues and taking only the m largest positive ones.

Example 8.1. We exemplify the classic MDS on the problem of embedding of the horse shape from Example 7.3 into $\mathbb{R}^3$. The eigenvalues of K_X, computed using the Arnoldi method (MATLAB function `eigs`) are shown in Figure 8.1. The first three eigenvalues ($\lambda_1 = 42.82$, $\lambda_2 = 10.34$ and $\lambda_3 = 3.87$) capture about 99% of the information about the shape. About half of the eigenvalues are negative. The canonical form obtained is shown in Figure 8.2 (left).

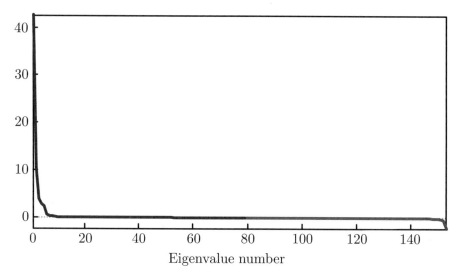

Figure 8.1. Eigenvalues of the matrix K_X in Example 8.1. Note the negative eigenvalues, which are evidence of the inability to embed the shape without distortion.

8.2 Local methods

Thus far, we have derived approaches for constructing the canonical forms while trying to preserve the geodesic distances measured between all pairs of shape samples. This is a *global* approach. Instead of thinking globally, we could formulate a *local* criterion and apply it to overlapping small regions of the shape. Combined together, these local criteria form a global one, or, as Saul and Roweis [336] formulated it paraphrasing the famous phrase of the environmentalist David Brower,[5] "think globally, fit locally."

The local criterion must guarantee the preservation of local metric properties of the shape. Roughly speaking, we would like our embedding $f : (X, d_X) \to (\mathbb{R}^m, d_{\mathbb{R}^m})$ to map neighboring points to neighboring points. For example, *locally linear embedding* (a map that is piecewise linear on small neighborhoods in X) preserves such local structures. Speaking more formally, assume that the shape is represented as a graph, with the connectivity matrix E_X, defining the neighborhood $\mathcal{N}(x_i)$ for each shape sample (as we saw in Chapter 3, there are many ways to define the neighborhood, for example, taking a fixed number of nearest neighbors or points within a fixed radius). If $x_j \in \mathcal{N}(x_i)$, that is, $d_X(x_i, x_j)$ is small, we want the corresponding distance in the embedding space to be small as well. We can therefore think of $d_{ij}^2(Z)$ as a local criterion, which incurs large penalty if neighboring points are mapped apart by the embedding. Summing up the local criteria at every point, we have a global criterion,

$$\sigma_{\text{LOC}}(Z) = \frac{1}{2} \sum_{i=1}^{N} \sum_{j \in \mathcal{N}(x_i)} d_{ij}^2(Z) = \sum_{i>j} w_{ij} d_{ij}^2(Z), \quad (8.5)$$

where

$$w_{ij} = \begin{cases} 1 & x_j \in \mathcal{N}(x_i) \\ 0 & \text{else.} \end{cases} \quad (8.6)$$

Yet, at the same time, we want to avoid the trivial solution of all the neighbor points collapsed to a single point (which would give $d_{ij}^2(Z) = 0$). This can be done, for example,[6] by introducing the constraint $Z^T Z = I$. Following steps similar to those we have seen in the case of the L_2-stress (see Problem 8.5), we rewrite the embedding error criterion as

$$\sigma_{\text{LOC}}(Z) = \text{trace}(Z^T L_X Z), \quad (8.7)$$

where L_X is an $N \times N$ matrix with elements,

$$l_{ij} = \begin{cases} -w_{ij} & i \neq j \\ \sum_{k \neq i} w_{ik} & i = j. \end{cases}$$

L_X is often referred to as the *Laplacian matrix* (or simply *Laplacian*) in graph theory [105, 280] and is positive semidefinite. Notice further that it is zero-mean, similarly to the matrices we had in the least-squares MDS.

Combining everything together, we formulate a constrained optimization problem

$$Z^* = \underset{Z \in \mathbb{R}^{N \times m}}{\text{argmin}}\; \sigma_{\text{LOC}}(Z) \quad \text{s.t.} \quad Z^T Z = I, \quad (8.8)$$

which, in turn, is equivalent to finding eigenvectors corresponding with the smallest eigenvalues of L_X (the proof is left to the reader as Problem 8.7). Therefore, this approach also belongs to the family of spectral embedding algorithms. The resulting canonical form Z^* is called *Laplacian eigenmap*, first introduced by Belkin and Niyogi [21]. Laplacian eigenmap computation is summarized in Algorithm 8.2.

Let us make a few remarks about the computational aspects of the Laplacian eigenmap algorithm. First, we note that the matrix L_X is sparse. Usually, this offers a significant computational advantage in most eigendecomposition algorithms. Second, we need to compute the lower part of the spectrum of L_X. Standard eigendecomposition techniques (which compute the largest eigenvalues) can be applied in this case using a spectral shift transformation [177]. Finally, because L_X is zero-mean, it has a zero eigenvalue corresponding with the constant eigenvector, which we would definitely like to exclude.

The reader may notice that in our description of the Laplacian eigenmap algorithm, the definition of w_{ij} in (8.6) is quite arbitrary. It is possible to choose different weights w_{ij}, e.g., inversely related to the distances $d_X(x_i, x_j)$, in order to increase the influence of small distances. Belkin and Niyogi suggest using the Gaussian weights,

$$w_{ij} = \begin{cases} \exp\left(-\frac{d_X^2(x_i,x_j)}{t^2}\right) & x_j \in \mathcal{N}(x_i) \\ 0 & \text{else,} \end{cases} \quad (8.9)$$

which include the previous case defined in (8.6) as $t = \infty$. Another important example is a family of Laplacians constructed by limiting the weights to the 1-*ring support* (i.e., $j \in \mathcal{N}(i)$ if i and j share an edge) and requiring them to be non-negative. In discrete geometry, such Laplacians are referred to as *umbrella operators*.

Let us make a further step toward the generalization of the Laplacian eigenmap approach. For this purpose, we write formula (8.5) in a slightly different way,

$$\sigma_{\text{LOC}}(Z) = \text{trace}(Z^T L_X Z)$$
$$= \frac{1}{2} \sum_{k=1}^{m} \sum_{i=1}^{N} \sum_{j=1}^{N} l_{ij} z_i^k z_j^k = \frac{1}{2} \sum_{k=1}^{m} \sum_{i=1}^{N} a_i(z^k),$$

where z^k is the kth column of Z. We can think of $a_i(z)$ as of an operator, applied to the coordinates of the canonical form, and measuring its local variation in the neighborhood of the point z_i, independently for each dimension (recall the nested dimensions property). It appears that such a view unites many spectral embedding algorithms. In general, we can define $a_i(z)$ to be an arbitrary quadratic positive semidefinite form on the columns of Z, and use the respective definition of $\sigma_{\text{LOC}}(Z)$ in our optimization problem (8.8). For example, choosing $a_i(z^k)$ to measure the Frobenius norm of the Hessian of the kth coordinate of the embedding at z_i, we have the Donoho and Grimes *Hessian Locally Linear Embedding* (HLLE) algorithm [187]. Defining it as a measure of deviation from a local linear representation of the points, we have the *Locally Linear Embedding* (LLE) [328] and the *Local Tangent Space Alignment* (LTSA) [410] algorithms. Using the diffusion operator instead of L_X leads to the *diffusion map* approach, introduced by Lafon et al. [109].

---

input : $N \times N$ matrix of geodesic distances D_X, dimension m.
output: canonical form Z^*.
1 Compute the Laplacian L_X.
2 Perform eigendecomposition of L_X to compute the $m \times m$ matrix Λ_m of the m smallest positive eigenvalues and the $N \times m$ matrix of corresponding eigenvectors U_m.
3 Compute the canonical form as $Z^* = U_m$.

Algorithm 8.2. Laplacian eigenmap.

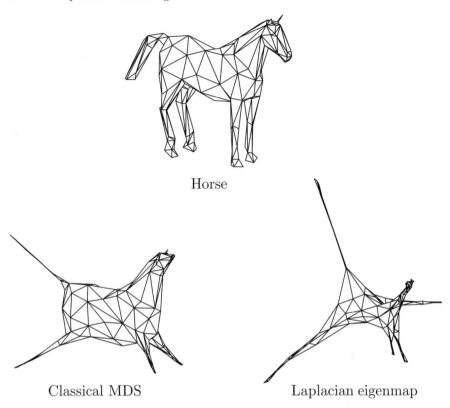

Figure 8.2. Embedding of the horse shape into $\mathbb{R}^3$ using classic MDS (left) and Laplacian eigenmaps (right).

In practice, all these algorithms are reduced to constructing a sparse matrix representing some local operator and performing its eigendecomposition.[7]

Example 8.2. To exemplify the difference between local and global spectral embedding methods, we compare the Laplacian eigenmap and the classic MDS on the horse shape from Example 7.3. The connectivity is defined using metric ball of fixed radius. Gaussian weights with $t = 1$ are used for the computation of L_X, and the eigendecomposition is performed using the Arnoldi method. We ignore the zero eigenvalue of L_X. The canonical form obtained by Laplacian eigenmap is shown in Figure 8.2 (right). It appears to be "twisted," which demonstrates the disadvantage of the local methods: while keeping the local properties, there is nothing that preserves the global ones.[8]

8.3 The Laplace-Beltrami operator

Discrete spectral embedding methods have an important relation to the continuous geometry of surfaces. In order to understand this relation, we need

8.3 The Laplace-Beltrami operator

to recall the basics of differential geometry from Chapter 2 and prepare some additional machinery. Let X be our shape, equipped with the Riemannian metric g, and $f : X \to \mathbb{R}$ be a C^2 scalar field on the shape. In our problem, f represents a coordinate of the embedding of X into $\mathbb{R}^m$ (due to the nested dimensions property, we can consider each coordinate separately). We define the *differential* $df_x(dv)$ as a linear form $df : T_xX \to \mathbb{R}$, approximating the change in the function f at the point x, if we make an infinitesimal step dv in the tangent space T_xX. We can express the differential as an inner product,

$$df_x(dv) = \langle \nabla f(x), dv \rangle_x = g_x(\nabla f(x), dv),$$

where $\nabla f(x) : X \to T_xX$ is the gradient of f at the point x. Note that our definition is a natural extension of the gradient we have encountered in Chapter 5.[9] In a similar manner, we can define the Hessian as $\nabla^2 f = \nabla df$.

Let us assume that we are given a point x on the shape and another point x', obtained by making an infinitesimal displacement dv from x. The absolute value of the differential measures how our embedding maps away the points x and x',

$$|f(x') - f(x)| \approx |df_x(dv)| = |\langle \nabla f(x), dv \rangle_x|.$$

At the same time, by the Cauchy-Schwartz inequality, we have

$$|\langle \nabla f(x), dv \rangle_x| \leq \|\nabla f(x)\|_x \cdot \|dv\|_x.$$

We therefore see that if $\|\nabla f(x)\|_x$ is small, then points close to x are mapped to points close to $f(x)$. This is the continuous analog of our discrete local embedding error criterion. Averaging the criterion on the whole shape, we arrive at the following optimization problem,

$$\min_{f \in L_2(X)} \int_X \|\nabla f(x)\|_x^2 da \quad \text{s.t.} \quad \|f\|_{L_2(X)}^2 = 1, \qquad (8.10)$$

which replaces problem (8.8). Here $\|f\|_{L_2(X)}^2 = \langle f, f \rangle_{L_2(X)} = \int_X |f(x)|^2 da$ is the norm on the space $L_2(X)$ of square-integrable functions on X and da denotes the area element. This condition corresponds with the normalization constraint $z^T z = 1$ in the discrete case.

In order to unveil the relation to eigendecomposition of the Laplacian matrix, we use the fact that

$$\int_X \langle \nabla f(x), \nabla f(x) \rangle_x da = - \int_X f(x) \operatorname{div}(\nabla f(x)) da.$$

This is a generalization of the *Stokes theorem*, known from basic calculus, to Riemannian manifolds. In Riemannian geometry, the operator

$$\Delta_X f = -\operatorname{div}(\nabla f(x)),$$

is called the *Laplace-Beltrami operator*, or sometimes the *Laplacian* for short.[10] The Laplace-Beltrami operator is expressible solely in terms of the Riemannian metric and is therefore an intrinsic property of the shape. It can be alternatively defined as $\Delta_X f = \text{trace}(\nabla^2 f)$, from which it follows that the Laplace-Beltrami operator is the non-Euclidean analog of the Laplacian operator: in the Euclidean case, we have $\Delta f = -(f_{xx} + f_{yy})$.

Using Δ_X, we can rewrite problem (8.10) as

$$\min_{f \in L_2(X)} \int_X f(x) \Delta_X f(x) da \quad \text{s.t.} \quad \|f\|_{L_2(X)}^2 = 1. \tag{8.11}$$

Note that now the integral $\int_X f(x) \Delta_X f(x) da$ in (8.11) replaces the term $z^T L_X z$ we had in the discrete case, and the Laplace-Beltrami operator Δ_X takes the role of the Laplacian matrix L_X.

8.4 To hear the shape of the drum

In 1787, the German physicist Ernst Chladni, considered by many the father of acoustics, published the book *Entdeckungen über die Theorie des Klanges* ("Discoveries concerning the theory of sound") [102]. In his book, Chladni described a famous experiment for the visualization of vibrations produced by acoustic waves: covering a thin plate with sand and making it vibrate by running a violin bow. The sand was observed to accumulate in certain regions, producing patterns of beautiful complexity. A modern physicist would call these shapes *stationary waves*. Mathematically, the behavior of such waves is governed by the *stationary Helmholtz equation* (representing the spatial part of the wave equation solutions),[11] which in our notation reads

$$\Delta_X f = \lambda f.$$

It follows that the beautiful patterns observed by Chladni are *eigenfunctions* (the continuous analog of eigenvectors) of the Laplace-Beltrami operator Δ_X.[12] An example of such eigenfunctions is shown in Figure 8.3.

For compact shapes, the spectrum of Δ_X is discrete, that is, there exists a countable set of the solutions to the equation $\Delta_X f = \lambda f$. Because the Laplace-Beltrami operator is an intrinsic property of the shape, its spectrum (the set of all eigenvalues) is isometry-invariant (this property is exemplified in Figure 8.4). Based on this property, Reuter *et al.* proposed an isometry-invariant description of shapes by numerically approximating the Laplace-Beltrami spectrum [323] (a related approach was also described in [330]).

However, does the Laplace-Beltrami spectrum completely characterize the intrinsic geometry of a shape? More formally, we want to know whether two shapes with the same spectrum (such shapes are called *isospectral*) are necessarily isometric. The question whether a Riemannian surface can be determined by its spectrum was asked for the first time in the early 1960s by

Figure 8.3. The first non-trivial eigenfunctions of the Laplace-Beltrami operator of a human shape. Colors and contours visualize the values of the eigenfunctions at each point of the shape.

Leon Green. Because of the acoustic interpretation of the Laplace-Beltrami spectrum, Mark Kac has posed this question metaphorically in a 1966 paper entitled "Can one hear the shape of the drum?" [215]. Kac addressed the particular case of the Laplace-Beltrami spectrum on planar domains and is

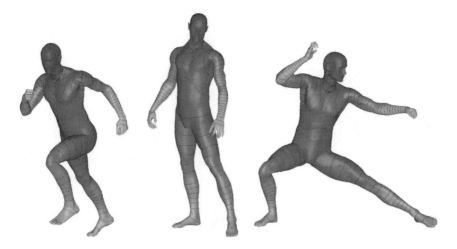

Figure 8.4. Empirical evidence of the fact that the Laplace-Beltrami operator is an intrinsic property of a shape. Shown is one of the eigenfunctions of the Laplace-Beltrami operator, which remains approximately without changes despite the near-isometric deformations of the shape.

quoted to have said, "Personally, I believe that one cannot hear the shape, but I may well be wrong and I am not prepared to bet large sums either way."

In Riemannian geometry, it is known that the area of two isospectral surfaces is equal, which implies that the area can be "heard" from the Laplace-Beltrami spectrum [394, 27]. Other "audible" properties include the total Gaussian curvature $\int_X K\,da$ and Euler characteristic χ [265]. Yet, generally, a Riemannian manifold cannot be determined by its spectrum. In other words, one cannot hear the shape of the drum. This fact is supported by numerous counterexamples of manifolds that are isospectral but not isometric [389, 90, 82, 83, 180].

On the other hand, in order to apply methods based on the Laplace-Beltrami spectra to the problem of non-rigid shape representation, it is crucial to know how different the classes of isospectral and isometric surfaces are. That is, whether there exist large classes of non-isometric but isospectral surfaces, and how different (non-isometric) can isospectral surfaces be. For the time being, this is an open research question.

8.5* Discrete Laplace-Beltrami operator

Before concluding this chapter, there are a few words to say about the discrete approximation of the Laplace-Beltrami operator. The resemblance of the Laplace-Beltrami operator to the graph Laplacian may create a wrong impression that the two are equivalent. First, note the Laplace-Beltrami operator is uniquely defined, whereas the discrete Laplacian can be defined in many ways, some of which were mentioned in this chapter. Secondly, excepting some special cases (see, for example, Belkin and Niyogi [22], Singer [364], and Hein et al. [201]), the discrete Laplacian would not converge to the Laplace-Beltrami operator. In discrete geometry, a distinction is usually made between *discrete* (or *combinatorial*) and *discretized* Laplacian. Discrete Laplacian arises from a discrete object, a triangular mesh representing the shape. Discretized Laplacian, on the other hand, is a consistent numerical approximation of the Laplace-Beltrami operator of the shape, consistency in this context implying that the discrete approximation preserves certain geometric properties of its continuous counterpart. Because we are interested in intrinsic geometric properties of the underlying continuous surface, we want the Laplacian to be independent or at least minimally dependent on the triangular mesh and thus need the discretized rather than the combinatorial Laplacian.

Surprisingly, it appears that consistently discretizing the Laplace-Beltrami operator is not an easy task. To be more specific, we need a list of properties of the continuous Laplace-Beltrami operator that a consistent discretization must satisfy. Our analysis closely follows [392]; the reader is referred to this paper for more details. Let X be a Riemannian manifold (possibly with boundary) and $f, h : X \to \mathbb{R}$ smooth functions on the manifold. We further assume that f and h are smooth and vanishing around the boundary. The continuous Laplace-Beltrami operator Δ_X has the following properties:

(L1) *Constant eigenfunction*: $\Delta_X f = 0$ for any $f = const$;
(L2) *Symmetry*: $\langle \Delta_X f, h \rangle_{L_2(X)} = \langle h, \Delta_X f \rangle_{L_2(X)}$;
(L3) *Locality*: $\Delta_X f(x)$ is independent of $f(x')$ for any points $x \neq x'$ on X;
(L4) *Euclidean case*: if X is a part of $\mathbb{R}^2$, for any linear function of the form $f(x, y) = ax + by + c$, $\Delta_X f = 0$.
(L5) *Positive semidefiniteness*: $\langle \Delta_X f, f \rangle_{L_2(X)} \geq 0$.

Property (L3) implies that the action of the Laplace-Beltrami is local, and a change in the value of the function f at a point will not influence the value of $\Delta_X f$ at another point. Property (L4) stems from the fact that in the Euclidean case, Δ_X is the Laplace operator and thus for any linear function of the form $f(x, y) = ax + by + c$, $\Delta_X f = -(f_{xx} + f_{yy}) = 0$.

For the construction of a discrete version of the Laplace-Beltrami operator (which we denote here by L_X), we assume that the shape X is sampled at N points $\{x_1, \ldots, x_N\}$ and represented as a triangular mesh T_X. A function on the shape is discretized and given as a vector with elements $f_i = f(x_i)$ for $i = 1, \ldots, N$. The discrete Laplacian is defined as a linear operator of the form

$$(L_X f)_i = \sum_{j=1}^{N} w_{ij}(f_i - f_j)$$

on the vertex-based function f and can be represented in a matrix-vector notation as $L_X f$.

Note that for $f = const$, we have $L_X f = 0$, and thus the discrete analog of (L1) is satisfied automatically. The discrete equivalent of other properties (L2)–(L5) can be stated as follows:

(DL2) *Symmetry*: $L_X = L_X^T$;
(DL3) *Locality*: $w_{ij} = 0$ if i and j do not share an edge[13];
(DL4) *Euclidean case*: if X is a part of $\mathbb{R}^2$, then

$$(L_X x)_i = \sum_{j=1}^{N} w_{ij}(x_i - x_j) = 0$$

for all interior vertices (here $x_j \in \mathbb{R}^2$ denotes the planar coordinates of the jth sample).
(DL5) *Positive semidefiniteness*: $L_X \succeq 0$.

Properties (DL2) and (DL5) guarantee that the eigenvalues of the matrix L_X are real and non-negative, and its eigenvectors orthogonal. It is also common to add the following constraint,

(DL6) *Positive weights*: $w_{ij} \geq 0$ for all $i \neq j$, and for each i, there exists at least one j such that $w_{ij} > 0$.

The convenience of this requirement is that combined with Property (DL2), it gives (DL5); however, the converse is not necessarily true. Finally, it is important that the discrete Laplacian converges to its continuous counterpart, in the following sense:

(DL7) *Convergence*: solution to the discrete PDE involving L_X converges to the solution of the smooth PDE involving Δ_X as $N \to \infty$, assuming appropriate boundary conditions and refinement scheme.

For a rigorous formulation of Property (DL7), we refer the reader to [205]. For us, it implies that the eigenvectors of L_X are a good numerical approximation of the eigenfunctions of Δ_X.

The "quality" of a discretization of the Laplace-Beltrami operator can be judged by how many of the above properties it satisfies, with an ideal wish to satisfy all of them. For example, umbrella operators and their variants violate property (DL4) and thus are not geometric. The Laplacian with Gaussian weights (8.9) used by Belkin and Niyogi violates the locality property (DL3). An even worse piece of news is that these discrete Laplacians do not converge to their continuous counterpart [403, 402], violating property (DL7). In our application, this is especially disadvantageous, as we are interested in a good approximation of the eigenfunctions of the Laplace-Beltrami operator.

In [134], Desbrun *et al.* proposed a discretization that was later shown to be convergent, under certain conditions, to the continuous Laplace-Beltrami

8.5* Discrete Laplace-Beltrami operator

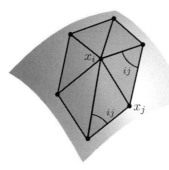

Figure 8.5. Discretization of the Laplace-Betrami operator using cotangent weights. Shaded is the area a_i.

operator [403, 402]. This fact makes such a discretization a good choice for shape description [250]. According to Desbrun et al., the value of $\Delta_X f$ at point i is approximated as

$$(L_X f)_i = \frac{3}{a_i} \sum_{j \in \mathcal{N}(i)} \frac{\cot\alpha_{ij} + \cot\beta_{ij}}{2}(f_i - f_j),$$

where a_i is the area of the triangles sharing the vertex x_i, and α_{ij} and β_{ij} are angles shown in Figure 8.5. In discrete geometry, this approximation is commonly known as *cotangent weights*.

In matrix notation, the discretized Laplacian is obtained by choosing the following weights,

$$w_{ij} = \begin{cases} 1.5(\cot\alpha_{ij} + \cot\beta_{ij})a_i^{-1} & i \text{ is adjacent to } j \\ 0 & \text{else.} \end{cases}$$

Note that as in general a_i varies from vertex to vertex, the matrix of weights is not symmetric, and consequently, the Laplacian matrix L_X violates the symmetry property (DL2) [330]. This gives rise to an unpleasant drawback: because the matrix is not symmetric, there is no guarantee that the eignvectors and eigenvalues of L_X are real. Lévy [250] suggested treating this problem by symmetrizing the matrix, i.e., using $0.5(L_X + L_X^T)$ instead of L_X.

Before ending this chapter, we should satisfy the reader's curiosity regarding a discrete Laplacian meeting all the aforementioned requirements. Most recently, Wardetzky et al. have put an end to the search for such an "ideal" Laplacian. In the paper entitled "Discrete Laplace operators: no free lunch" [392], they showed that for a general mesh, it is theoretically impossible to satisfy properties (DL2)–(DL7) at the same time, and thus the ideal discretization does not exist. Retrospectively, this result also explains why there exists such a large diversity of discrete Laplacians, each having a subset of the above properties that make it suitable for certain applications and unsuitable for others.

Suggested reading

For a traditional presentation of classic MDS, we refer the reader to Borg and Groenen [44]. An interesting view from the perspective of kernel PCA is presented by Williams [397]. Dattorro presents a comprehensive overview of distance matrix properties in [123]. The basics of spectral graph theory are presented in the book of Chung [105]; alternatively, see the review article by Mohar [280]. Berger [26] devotes Chapter 9 entirely to the spectral properties of the Laplace-Beltrami operator and gives various examples of isospectral non-isometric manifolds. For details on computational eigendecomposition algorithms, the reader is referred to the classic numerical linear algebra book by Golub and van Loan [177]. An insightful review of Laplace-Beltrami operator discretization techniques is presented in [392]; convergence of different discretizations is analyzed in [403, 402] . For additional applications of the Laplace-Beltrami operator for shape editing and compression, see [250] and [218].

Software

An implementation of the classic scaling algorithm is distributed as part of the TOSCA toolbox. The function cmdscale invokes classic scaling in the commercial MATLAB Statistics toolbox. Classic scaling is also used in the Isomap algorithm [377]. Laurens van der Maaten's MATLAB Toolbox for Dimensionality Reduction includes LLE, HLLE, LTSA, diffusion maps, and many other algorithms discussed in this chapter.

Problems

8.1. Let XX^T be a Gram matrix, where X is an $N \times m$ matrix. Prove the following properties:

1. $XX^\mathrm{T} \succeq 0$,
2. $\mathrm{rank}(XX^\mathrm{T}) = \mathrm{rank}(X)$.

8.2. Show that the classic MDS solution minimizes the strain $\sigma_\mathrm{F}(Z)$.

8.3. Show that one way to obtain the matrix K_X is

$$K_X = -\frac{1}{2} J(D_X \odot D_X) J,$$

where $\odot$ denotes the Hadamard (element-wise) matrix product, such that $(D_X \odot D_X)_{ij} = d_X^2(x_i, x_j)$ and $J = \left(I_N - N^{-1} 1_{N \times N}\right)$ is the *centering matrix*. Extend this result by replacing J by a general projection matrix $P = I_N - 1_{N \times N} w^\mathrm{T}$, where w is a vector satisfying $w^\mathrm{T} 1_{N \times 1} = 1$.

8.4. Establish the duality between classic MDS and PCA.

8.5. Derive the expression (8.7).

8.6. Show that the Laplacian matrix L_X is positive semidefinite.

8.7. Show that the minimum eigenvalue problem $Ax = \lambda_{\min}x$, where $A \succeq 0$, is equivalent to the constrained optimization problem

$$\min_{x \in \mathbb{R}^N} x^T A x \quad \text{s.t.} \quad x^T x = 1.$$

Extend this result by showing that

$$\min_{X \in \mathbb{R}^{N \times m}} \text{trace}(X^T A X) \quad \text{s.t.} \quad X^T X = I,$$

is equivalent to finding the m smallest eigenvalues of A.

8.8. Show that ∇f projected onto $T_x X$ coincides with $\nabla_X f$.

Notes

[1] More precisely, Schoenberg showed in [340] that in order for D_X to represent lengths of the edges of an N-simplex lying in $\mathbb{R}^m$ but not in $\mathbb{R}^{m-1}$, the matrix K_X must be positive-semidefinite and of rank m.

[2] In some references, classic MDS is referred to as *classic scaling*, *CMDS* [405], *Young-Householder algorithm*, *Torgerson algorithm*, or *Torgerson-Gower scaling* [44]. In manifold learning and dimensionality reduction applications, the name *Isomap*, due to Tenenbaum et al. [377], is sometimes used referring to an algorithm that is essentially classic MDS applied on the distances measured using the Dijkstra algorithm. In computer vision, this approach was introduced earlier by Schwartz et al. [345].

[3] Speaking in signal processing terminology, we can say that the strain measures the *energy* of the error introduced by our approximation to K_X. Dividing it by the energy of all eigenvalues, $\sum_{k=1}^{m} \lambda_k^2 / \sum_{k=1}^{N} \lambda_k^2$, we have a criterion of how much "information" of the shape the m-dimensional Euclidean embedding captures.

[4] Prof. Michael Saunders of Stanford University suggests that one of the most efficient public domain MATLAB codes for our problem is the *implicitly restarted block-Lanczos* (IRBL) method by Baglama et al. [16].

[5] David Brower is credited with coining the phrase "think globally, act locally" as the founding motto of the environmentalist organization *Friends of Earth*.

[6] Belkin and Niyogi [21] use the constraint $Z^T D Z = I$, where D is an $N \times N$ diagonal matrix with elements of the diagonal of L_X. This results in the generalized eigenvalue problem $L_X Z = D Z \Lambda$.

[7] Ham et al. [198] showed that all these algorithms can be considered as instances of *kernel PCA*.

[8] A way to overcome this problem of local methods by introducing stiffness constraints was shown by Brand [48].

[9] A 1-form defined as $v^*(u) = g_x(v, u)$ is called *dual* to v. In our case, we can think (pointwise) of the the differential df as the dual to the gradient ∇f.

[10] Here, we use sign convention to define the Laplace-Beltrami operator as positive definite. A more common definition is $\Delta_X f = \text{div}(\nabla f(x))$. We do not go into details on the exact definition of the div operator, appealing to the analogy with classic results known from analysis.

[11] This is, of course, an over-simplified mathematical description of Chladni figure formation. We tacitly ignore the material properties and assume a Dirichlet boundary condition $f(\partial X) = 0$.

[12] Chladni's experiments were performed on planar domains, on which the Laplace-Beltrami operator equals the Laplacian. It is more difficult to repeat these experiments for curved shapes, as sand will not remain on their surface. Yet, the intuitive interpretation of the results is similar.

[13] In other words, the support of the discrete operator is a 1-ring.

9

Non-Euclidean Embedding

> Out of nothing I have created a strange new universe.
>
> J. BOLYAI, *on the creation of non-Euclidean geometry.*

In the previous two chapters, we explored a variety of numerical techniques used for embedding of surfaces. A common property that all these techniques share is that the embedding space was always chosen to be Euclidean. In other words, a complicated metric structure of a surface seeks for the best possible housing (in the sense of some distortion criterion) in a flat space. We have also seen that such a Euclidean embedding is rarely distortionless. The "suffering" of the sphere from Example 7.1 in Chapter 7 struggling to find a comfortable accommodation in $\mathbb{R}^m$ clearly demonstrates the fact that the intrinsic geometry of many surfaces may significantly differ from the Euclidean one. As a consequence, an attempt to embed a surface that looks more like a sphere than a plane into a Euclidean space is likely to result in high distortions of the metric. On the other hand, a surface resembling a sphere would feel more comfortable in the curved rooms of $\mathbb{S}^m$. Said in a broader manner, embedding a surface into a non-Euclidean space may decrease the distortion. In this chapter, we consider generalizations of multidimensional scaling that allow for non-Euclidean embedding.

9.1 Spherical embedding

A non-Euclidean embedding problem can be formulated in the MDS flavor as

$$\min_{z_1,\ldots,z_N \in \mathbb{Z}} \sum_{i>j} |d_X(x_i, x_j) - d_{\mathbb{Z}}(z_i, z_j)|^p,$$

where $(\mathbb{Z}, d_{\mathbb{Z}})$ is some metric space. As in the Euclidean case, given a discrete metric space $X_N = \{x_1, \ldots, x_N\}$ equipped with $d_X|_{X_N}$, we replace this metric with $d_{\mathbb{Z}}$ and aim at finding a configuration of points $\{z_1, \ldots, z_N\}$ in $\mathbb{Z}$ that represents the original geometry as accurately as possible. Several practical issues arise at this point. First, in order to allow for a formulation of the problem in terms of continuous minimization, $\mathbb{Z}$ should be a manifold and have a convenient representation of points using some, preferably global, system of continuous coordinates. Second, the metric $d_\mathbb{Z}$ participating in the stress function should have a simple analytic form or, at least, there should be a way to compute it efficiently. Third, because our final goal is non-rigid surface matching, the primary use of the embedding is to replace the complicated isometry group of the embedded surface with the isometry group of $\mathbb{Z}$. It is therefore advantageous that such a group be as simple as possible and easy to handle. Practically, this implies the existence of an efficient algorithm for rigid surface matching in $\mathbb{Z}$.

The simplest choice of a non-Euclidean geometry for the embedding space $\mathbb{Z}$ is the spherical geometry. Formally, an m-dimensional[1] *unit sphere* is the m-dimensional manifold embedded in $\mathbb{R}^{m+1}$, described by the set of unit vectors $\mathbb{S}^m = \{z \in \mathbb{R}^{m+1} : \|z\|_2 = 1\}$. A sphere of radius r is obtained by scaling $\mathbb{S}^m$ by r. A one-dimensional sphere is simply a circle and a two-dimensional sphere is the surface of revolution created by rotating a circle. $\mathbb{S}^m$ is a very special manifold, as it has constant positive curvature. Moreover, it can be shown that every space of constant positive curvature is isometric to a sphere.

Because we live on the surface of an approximately spherical planet, spherical geometry should be familiar to us. As we have already seen in previous chapters, a point z on a two-dimensional sphere can be parameterized by a pair of coordinates $(u^1, u^2) \in [0, 2\pi) \times [-\frac{\pi}{2}, \frac{\pi}{2}]$, usually referred to as longitude and latitude. The corresponding vector in $\mathbb{R}^3$ is given by

$$z^1(u^1, u^2) = r\cos u^1 \cos u^2;$$
$$z^2(u^1, u^2) = r\sin u^1 \cos u^2;$$
$$z^3(u^1, u^2) = r\sin u^2,$$

as shown in Figure 9.1.

Centuries ago, cartographers learned that the shortest path between two points on the face of the Earth is a segment of a planar section of the sphere, which is called a *great circle* (we leave the proof of this fact as an exercise in Problem 9.1). The geodesics on the sphere are therefore arcs centered at the origin. For example, the great circle connecting Moscow and San Francisco passes approximately over the North Pole, which explains the route of aircraft flying between these two distant cities.[2] The geodesic distance between two points u and u' on a two-dimensional sphere is simply given by the length of the arc $\Gamma(u, u')$ connecting them,

$$d(u, u') = r \cdot L(\Gamma(u, u')) = r \cdot \mathrm{acos}\left(\frac{z(u)^\mathrm{T} z(u')}{r^2}\right),$$

as visualized in Figure 9.2. The same rules apply in higher dimensions: an m-dimensional sphere can be parameterized by an m-tuple of coordinates; if $(u^1, \ldots, u^{m-1})$ represent a vector x on an $m-1$-dimensional sphere of radius r, then $(z^1 \cos u^m, \ldots, z^{m-1} \cos u^m, r \sin u^m)^\mathrm{T}$ represents a vector on

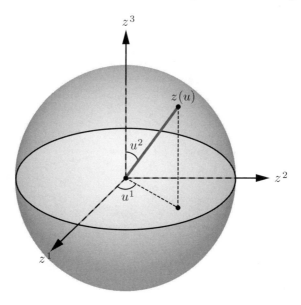

Figure 9.1. Parameterization of a point on the two-dimensional sphere $\mathbb{S}^2$.

an m-dimensional sphere. For example, the three-dimensional sphere $\mathbb{S}^3$ is parameterized as

$$z^1(u^1, u^2, u^3) = r \cos u^1 \cos u^2 \cos u^3;$$
$$z^2(u^1, u^2, u^3) = r \sin u^1 \cos u^2 \cos u^3;$$
$$z^3(u^1, u^2, u^3) = r \sin u^2 \cos u^3;$$
$$z^4(u^1, u^2, u^3) = r \sin u^3.$$

As in the two-dimensional case, the geodesic distances are measured as arc lengths.

The spherical MDS problem can be written as

$$\min_{u_1, \ldots, u_N} \sum_{i>j} |d_X(x_i, x_j) - d_{\mathbb{S}^m}(u_i, u_j)|^p. \tag{9.1}$$

Observe that the embedding space $\mathbb{Z} = \mathbb{S}^m$ admits a global parameterization (in our case, on $[0, 2\pi) \times [-\frac{\pi}{2}, \frac{\pi}{2}]^{m-1}$) and therefore allows a convenient representation of points as vectors of coordinates $(u^1, \ldots, u^m)$. Moreover, the geodesic distances in $\mathbb{S}^m$ are given analytically by

$$d_{\mathbb{S}^m}(u, u') = \operatorname{acos}\left(z(u)^{\mathrm{T}} z(u')\right). \tag{9.2}$$

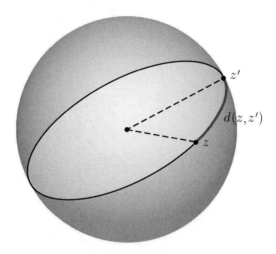

Figure 9.2. The geodesic distance on the sphere is measured as the length of a segment of a great circle.

Here we assume $r = 1$; without loss of generality, we can limit our attention to unit spheres, as the effect of changing the sphere radius is equivalent to scaling the distance terms $d_X(x_i, x_j)$ by $1/r$. Note that by taking r to infinity, we end up with $\mathbb{R}^m$. Consequently, if we are given the control over the radius, $\mathbb{S}^m$ constitutes a richer embedding space, which includes the Euclidean space as a limit case. Typically, for every class of surfaces, there exists some optimal value of r that minimizes the embedding error, and in most cases this value is smaller than infinity (see Example 9.1 below).

Example 9.1. In order to visualize the dependence of the distortion on the sphere radius, we embedded four objects (cat, dog, horse, and man) into a two-dimensional sphere with the radius ranging from 1 to 10^4 by minimizing the stress (9.1) with $p = 2$. The obtained results are depicted in Figure 9.3. For most surfaces, there exists some "sweet spot" for the radii ranging from 500 to 1000 (depending on the surface geometry and scale), where the distortion reaches its minimum. Increasing or decreasing the radius usually increases the distortion, sometimes by an order of magnitude. An extreme case of such a behavior is exhibited by a sphere, for which the distortion would drop drastically (to zero, in theory) for embedding into a sphere with a matching radius. An exception from this rule would be a convex planar patch, for which the embedding distortion decreases monotonically with the radius and achieves smaller values as r increases.

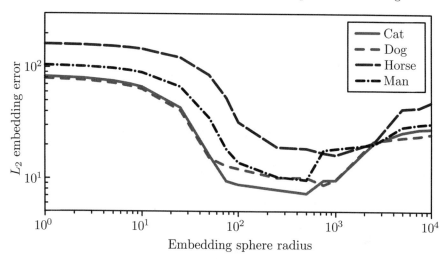

Figure 9.3. Distortion (in terms of the L_2 stress) produced by embedding different surfaces into $\mathbb{S}^2$ with varying radius. Each surface has its optimal radius. Uniformly scaling a surface scales the optimal radius by the same amount.

The spherical MDS problem (9.1) can be solved using one of the continuous minimization techniques discussed in Chapter 5. However, unlike the Euclidean case where the optimization variables are coordinates in $\mathbb{R}^m$ and can therefore assume any value, here the optimization variables must be restricted to the parameterization domain. One possibility is to impose the inequality constraints $0 \leq u_i^1 < 2\pi$ and $-\frac{\pi}{2} \leq u_i^k \leq \frac{\pi}{2}$ for $k = 2, 3, \ldots, m$. Yet, because the parameterization is periodic and defined up to an integer product of 2π, no catastrophes happen even if some optimization variables go outside the domain $[0, 2\pi) \times [-\frac{\pi}{2}, \frac{\pi}{2}]^{m-1}$, as they would still represent valid points on $\mathbb{S}^m$. Consequently, the preferred numerical trick often used in this case is a *projected descent*: after each iteration of an unconstrained descent algorithm, the current solution is projected onto the parameterization domain by applying a *projection operator*, $u^{(k)} = P\left(u^{(k)}\right)$. In the case of a two-dimensional case, the projection operator is defined as

$$P(u) = \begin{cases} (u^1 \bmod 2\pi, (u^2 + \frac{\pi}{2}) \bmod 2\pi - \frac{\pi}{2})^{\mathrm{T}} & : \quad (u^2 + \frac{\pi}{2}) \bmod 2\pi \leq \pi \\ (u^1 \bmod 2\pi, \frac{3\pi}{2} - (u^2 + \frac{\pi}{2}) \bmod 2\pi)^{\mathrm{T}} & : \quad (u^2 + \frac{\pi}{2}) \bmod 2\pi > \pi \end{cases}$$

(the general case is left to the reader as Problem 9.4). Applying P to the optimization variables $\{u_i\}$ forces them to reside in $[0, 2\pi) \times [-\frac{\pi}{2}, \frac{\pi}{2}]^{m-1}$ while clearly not changing the position of the points on the sphere they represent. Multi-resolution and multi-grid methods can also be exploited for solving the spherical MDS problem.

Spherical embedding replaces the rich isometry group of the embedded surface with the isometry group of $\mathbb{S}^m$, which includes all Euclidean isometries

192   9 Non-Euclidean Embedding

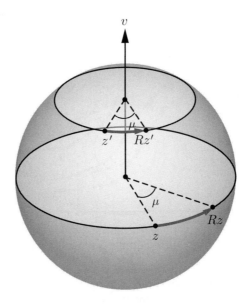

Figure 9.4. A rigid isometry $R \in O(3)$ on the two-dimensional sphere can be represented as rotation by an angle θ about an axis $\hat{v}$.

in $\mathbb{R}^{m+1}$ that leave the origin fixed (Figure 9.4). Such a group, generated by $(m+1) \times (m+1)$ unitary matrices, is called the *orthogonal group* and is denoted by $O(m+1)$. From the first sight, comparison of the spherical canonical forms appears straightforward using an analog of ICP, which minimizes the distance between two surfaces in $\mathbb{S}^m$ over all rotations and inversions about the origin defined by $m+1$ degrees of freedom. However, in order to measure the distance between two surfaces in $\mathbb{S}^m$, $d_{\mathbb{S}^m}$ has to be used, which makes the transition from $\mathbb{R}^m$ to $\mathbb{S}^m$ far from trivial. Unfortunately, unlike the exceptionally well-developed algorithms available for rigid surface matching in $\mathbb{R}^3$, there are no such tools readily available for spherical spaces. Comparison of surfaces in $\mathbb{S}^m$ invariant to rigid isometries is an interesting open research question.

9.2 Generalized multidimensional scaling

Replacing the Euclidean geometry of the embedding space by the spherical one usually leads to smaller metric distortions (and, consequently, to better isometry-invariant representation of surfaces) while maintaining practically the same computational complexity compared with the Euclidean MDS algorithms. However, this apparent improvement pales facing the difficulty of rigid surface matching in spherical spaces, despite the simplicity of the isometry group. On one hand, non-Euclidean embeddings seem very attractive because they allow reducing the metric distortion. On the other hand, there is little hope to achieve anything useful out of such an embedding, as even geometries as simple as the spherical one appear hardly tractable when rigid surface matching is considered.

A way out of this seemingly hopeless situation is to use an embedding space $\mathbb{Z}$ with a trivial isometry group. In this case, the canonical form of X in $\mathbb{Z}$ will be uniquely defined. We still have to measure the distance between two canonical forms, but now this distance does not involve any minimization at all, as there is no more rigid isometry ambiguity. However, we would like to go further and avoid completely the need to compute the distance between two canonical forms in $\mathbb{Z}$. The ultimate way to achieve this goal is by choosing one of the surfaces, say Y, as the embedding space. In other words, we would like to embed X into Y by solving the following problem

$$\min_{y'_1,\ldots,y'_N \in Y} \sum_{i>j} \left| d_X(x_i, x_j) - d_Y(y'_i, y'_j) \right|^p, \qquad (9.3)$$

which we refer to as the *generalized MDS* or *GMDS* for short [70, 67]. We intentionally denote the images of x_i in Y as y'_i in order to avoid confusion with the samples of the surfaces Y, which are denoted by y_i. The minimum stress value quantifies how much the metric of X has to be distorted in order to fit into Y.

At this point, several conceptual shifts are required in order to fully understand this apparently small change. First, note that we no more need any "intermediate" metric space. Instead of embedding X and Y into some common embedding space $\mathbb{Z}$ that introduces inevitable distortions, we embed X directly into Y. If the two surfaces are isometric, such an embedding will be distortionless (up to numerical errors); otherwise, the distortion will measure the dissimilarity between X and Y. Thus far, the embedding distortion has been an enemy that was likely to lower the sensitivity of the canonical form method; now it has become a friend that tells us how different the surfaces are.

Second, there is no more need for measuring distances between canonical forms in $\mathbb{Z}$ invariant to the isometries of $\mathbb{Z}$ (recall that this was an obstacle in using spherical and other non-Euclidean embeddings). On one hand, GMDS constitutes the best non-Euclidean embedding, in the sense that it allows us to completely avoid unnecessary representation errors stemming from embedding into an intermediate space $\mathbb{Z}$. On the other hand, GMDS renders superfluous the need to measure distances between sets in spaces with complicated geometries. Strictly speaking, we do not use canonical forms anymore; the distance between two surfaces is obtained from the solution of the embedding problem itself.

Third, recall that the canonical form method requires the embedding space to be at least three-dimensional (or, in fancier terminology, the *co-dimension* of the embedding has to be at least one). In GMDS, a two-dimensional space is embedded into another two-dimensional space. Last but not least, observe that given a subset $X' \subset X$ with $d_X|_{X'}$, the embedding error of X' in Y is not higher than that of X. This feature enables *partial matching* between non-rigid surfaces: if X is similar to Y, then a subset of it is also similar to Y. We defer the discussion of this important property to Chapter 11.

Generalized multidimensional scaling can be viewed as minimization of the *generalized L_p-stress*,

$$\sigma_p^p(y_1',\ldots,y_N') = \sum_{i>j} |d_X(x_i,x_j) - d_Y(y_i',y_j')|^p, \qquad (9.4)$$

or, for $p = \infty$,

$$\sigma_\infty(y_1',\ldots,y_N') = \max_{i>j} |d_X(x_i,x_j) - d_Y(y_i',y_j')|, \qquad (9.5)$$

that looks pretty much like the standard MDS problem.

Although GMDS seems a *panacea* for isometry-invariant surface matching, it introduces several challenges. First, in order to be able to keep on using efficient minimization algorithms, $\{y_i'\}$ has to be a set of continuous variables. In other words, the issue of the representation of points in Y and its implications on the minimization algorithm have to be addressed. Second, unlike the Euclidean or the spherical cases, we have no more the luxury of computing the distance terms $d_Y(y_i', y_j')$ analytically. Because Y is an arbitrarily complicated surface, the geodesic distances in the embedding space are unlikely to have any closed form expression at all, and we have to devise a method to efficiently approximate d_Y.

9.3 Representation issues

The requirement of the points on Y to be represented in continuous coordinates implies that Y must be a continuous surface. On the other hand, in order to be computationally tractable, Y has to be represented by a finite discrete sampling $Y_M = \{y_1,\ldots,y_M\}$. The simplest representation of the embedding space that satisfies both conditions is a triangular mesh $T(Y_M)$. Let us start with the simple case where the mesh obeys a global parameterization $y : [0,1]^2 \to Y$. In this case, every point on the triangular faces of Y can be represented as a pair of continuous coordinates $(u^1, u^2) \in [0,1]^2$, and the triangulation of Y induces a triangulation of the parameterization domain.

Recall that because the images y_i' of x_i in Y may fall "between the samples" on the mesh, we have to approximate the geodesic distances between any pair of arbitrary points u, u' in the parameterization domain. Clearly, these distances cannot be pre-computed, as y'_i are the optimization variables. However, it is possible to pre-compute all pairwise geodesic distances $d_Y(u_i, u_j)$ between the points of Y_M and use them to approximate $d_Y(u, u')$. As we will see shortly, for this purpose we need to determine the vertices on Y_M of the two triangles enclosing u and u'. Thus, given a point in the parameterization domain, we have to find to which triangle it belongs.

For a surface represented as a geometry image, the parameterization domain is sampled on a regular Cartesian grid, producing the trivial regular triangulation depicted in Figure 9.5 (top). In this case, the enclosing triangle can be found in $\mathcal{O}(1)$. The regularity of geometry images makes them a very appealing choice for representing the embedding space. Surfaces reconstructed by many range acquisition devices such as structured and coded light scanners are often readily representable as geometry images. Other, more general, objects can be re-parameterized and re-sampled on a regular grid [192, 335, 258]. However, the process introduces inevitable inaccuracies. Much worse is the fact that it is hard to estimate the effect of such inaccuracies on the intrinsic geometry of $T(Y_M)$.

Relaxing the regularity of the grid (Figure 9.5, bottom), we can still work in the continuous parameterization domain, yet now the search for the enclosing triangles becomes a non-trivial task. Using binary space partition trees or similar structures, it can be done with logarithmic complexity. This adds $\mathcal{O}(N \log N)$ operations to the evaluation of all distance terms $d_Y(y'_i, y'_j)$, which is still negligible compared with the $\mathcal{O}(N^2)$ complexity of the generalized stress computation.

However, objects are often given as general triangular meshes rather than parametric surfaces and therefore have to be parameterized. Finding an accurate continuous global parameterization is a challenging task, especially for surfaces with complicated topology. A preferable solution is to use local parameterization, that is, represent the points y'_i in some local coordinate systems. A good candidate for such a representation is barycentric coordinates, which we have already encountered in Chapter 3. In barycentric coordinates, each point on the mesh $T(Y_M)$ is represented as a convex combination of the vertices of the triangle enclosing it. This way, each y'_i is represented by a triangle index t_i and a triplet $u_i = (u_i^1, u_i^2, u_i^3)$. In practice, one of the coordinates is redundant and can be inferred from the relationship $u_i^1 + u_i^2 + u_i^3 = 1$; this allows us to actually represent the point as a pair of coordinates $u_i = (u_i^1, u_i^2)$.

At first sight, the fact that t_i is a discrete index may appear problematic, as we desire to deal with continuous optimization variables only. However, the problem can be resolved by devising a way to smoothly switch between different local charts of Y, or using a more relaxed terminology, to travel along a path on the mesh. In the following sections, we will explore the implications that such a solution has on the minimization algorithm.

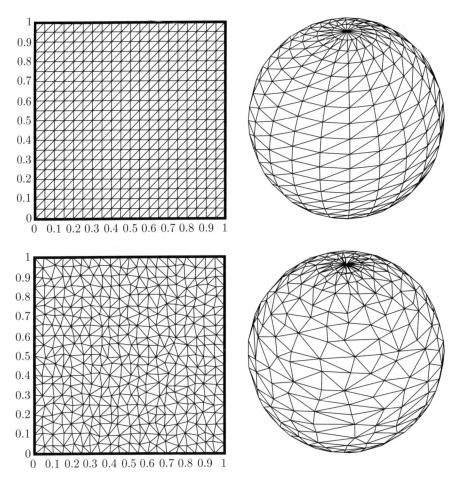

Figure 9.5. Global parameterization of a surface on a unit square. Top: the parameterization domain is sampled on a regular Cartesian grid inducing a trivial regular triangulation. Bottom: arbitrary triangulation.

To summarize, when Y is given as a geometry image or a parametric surface, we prefer the global parametric approach, whereas for general triangular meshes, local parameterization appears to be more suitable. At the same time, the representation issue is less relevant for the surface X we are embedding into Y, as the only information needed from X is the matrix of all pair-wise geodesic distances $d_X(x_i, x_j)$. In fact, we can think of the GMDS problem as of embedding any metric structure represented by the matrix of geodesic distances D_X into an arbitrary two-dimensional surface represented as a mesh $T(Y_M)$.

9.4 Geodesic distance computation

The generalized stress (9.4) involves the geodesic distance terms $d_Y(y_i', y_j')$. However, because Y is an arbitrarily complex surface, generally there is no analytic expression for the geodesic distances. Recall that this is where GMDS differs the most from standard MDS algorithms. For this reason, the computation of the geodesic distances between arbitrary points in Y is the numerical core of GMDS. Let us now address this important issue. Formally, we are given a triangular mesh $T(Y_M)$ approximating the surface Y. We assume that all pair-wise geodesic distances $d_{ij} = d_Y(y_i, y_j)$ between the points on $T(Y_M)$ are available (for example, computed using fast marching). Given two arbitrary points on the mesh, $y = (t, u)$ and $y' = (t', u')$ in barycentric coordinates (points in global representation can always be translated into barycentric coordinates), our goal is to find $\hat{d}_Y(y, y') \approx d_Y(y, y')$ that approximates the true geodesic distance between y and y'.

There is a set of properties $\hat{d}_Y$ should satisfy. First, because $d_Y(y_i, y_j)$ are first-order approximations to the true geodesic distances, we require that $\hat{d}_Y$ is also first-order accurate. Second, $\hat{d}_Y$ has to be consistent with the data, meaning that $\hat{d}_Y(y_i, y_j) = d_Y(y_i, y_j)$ for any pair of samples of Y. Third, $\hat{d}_Y(y, y')$ has to be symmetric, that is, $\hat{d}_Y(y, y') = \hat{d}_Y(y', y)$. Fourth, because $\hat{d}_Y(y, y')$ is used in an optimization algorithm, it has to be at least $\mathcal{C}^1$ with respect to y and y', and we have to be able to compute its derivatives analytically. We may, though, allow $\hat{d}_Y(y, y')$ not to be $\mathcal{C}^1$ at some points or along some lines. Last, as each computation of the generalized stress involves $\mathcal{O}(N^2)$ distance terms, $\hat{d}_Y$ and its derivatives have to be computed efficiently with constant complexity, independent of M and N.

We will now explore an interpolation method we first proposed in [67] under the name of *three point geodesic distance interpolation*. Let us start with the particular case where y' is one of the mesh vertices. Assume without loss of generality that triangle t to which y belongs is formed by the vertices y_1, y_2, and y_3 of $T(Y_M)$, and the coordinates of y can be expressed as

$$y = u^1 y_1 + u^2 y_2 + u^3 y_3, \tag{9.6}$$

where $u^1 + u^2 + u^3 = 1$. We will switch freely between y and u. Because $y' \in Y_M$, the distances $d_1 = d_Y(y_1, y')$, $d_2 = d_Y(y_2, y')$, and $d_3 = d_Y(y_3, y')$ are known. Using this information as well as the geometry of the triangle t, we have to approximate $d_Y(y, y')$. Recall that this was exactly the problem we encountered in Chapter 3 while discussing the fast marching algorithms. We borrow the fast marching wavefront propagation model to compute $\hat{d}_Y(y, y')$.

Using the planar wavefront model, the distance map from y' on the mesh is piecewise linear, meaning that for any point u in the triangle t, it can be expressed as a linear function of u. Such a function has three degrees of freedom, which can be resolved by substituting the data $\hat{d}_Y(y_1, y') = d_1$, $\hat{d}_Y(y_2, y') = d_2$, and $\hat{d}_Y(y_3, y') = d_3$. This yields the following simple interpolation formula

$$\hat{d}_Y(y, y') = u^1 d_1 + u^2 d_2 + u^3 d_3 = d^\mathrm{T} u, \tag{9.7}$$

where $d = (d_1, d_2, d_3)^\mathrm{T}$.

Let us now remove the restriction that y' coincides with one of the vertices of Y_M and consider the general case where $y' = (t', u')$ is an arbitrary point on the mesh. The major difference is that now the distances $d_1 = d_Y(y_1, y')$, $d_2 = d_Y(y_2, y')$, and $d_3 = d_Y(y_3, y')$ are unknown. Yet, assuming that the triangle t' to which y' belongs is formed by the vertices y_4, y_5, and y_6, we may apply the previously described interpolation method in the triangle t with $d = (d_{14}, d_{24}, d_{34})^\mathrm{T}$, $d_{ij} = d_Y(y_i, y_j)$, to obtain $\hat{d}_4 \approx d(y, y_4)$. In a similar manner, $\hat{d}_5 \approx d(y, y_5)$ and $\hat{d}_6 \approx d(y, y_6)$ are obtained. This can be expressed as

$$\begin{pmatrix} \hat{d}_4 \\ \hat{d}_5 \\ \hat{d}_6 \end{pmatrix} = \begin{pmatrix} d_{14} & d_{24} & d_{34} \\ d_{15} & d_{25} & d_{35} \\ d_{16} & d_{26} & d_{36} \end{pmatrix} u = D_Y(t, t') u, \tag{9.8}$$

where the matrix $D_Y(t, t')$ depends on the triangle indices t and t' only. Now, we may apply the interpolation once again, this time in the triangle t' with $d = (\hat{d}_4, \hat{d}_5, \hat{d}_6)^\mathrm{T}$, obtaining the sought interpolant

$$\hat{d}_Y(u, u') = u'^\mathrm{T} D_Y(t, t') u. \tag{9.9}$$

This final step completes the picture: now we have a computational tool for the interpolation of the geodesic distances on $T(Y_M)$. Figure 9.6 visualizes the four steps. We leave as an exercise to the reader (Problem 9.9) the proof of the fact that $\hat{d}_Y$ satisfies the five properties stated in the beginning of the section.

9.5 Minimization of the generalized stress

Substituting the interpolated distance terms $\hat{d}_Y(u_i, u_j) = u_i^\mathrm{T} D_Y(t_i, t_j) u_j$ to the generalized L_p stress function, we obtain

$$\sigma_p^p(t_1, u_1, \ldots, t_N, u_N) = \sum_{j > i} \left(d_X(x_i, x_j) - u_i^\mathrm{T} D_Y(t_i, t_j) u_j \right)^p. \tag{9.10}$$

We now have all the ingredients ready for constructing a numerical scheme for the solution of the GMDS problem. In what follows, we will focus our attention on the L_2 case only; other cases can be addressed, for example, by using the iterative reweighting scheme described in Chapter 7. To simplify notation, we will write σ instead of σ_2^2.

The choice of the L_2 stress has a significant advantage: Because the summation excludes the cases where $i = j$, the interpolated distance terms

9.5 Minimization of the generalized stress

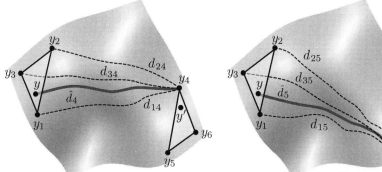

Step 1: $\hat{d}_4 \approx d_Y(y, y_4)$ is computed from d_{14}, d_{24}, d_{34} in the triangle y_1, y_2, y_3.

Step 2: $\hat{d}_5 \approx d_Y(y, y_5)$ is computed from d_{15}, d_{25}, d_{35} in the triangle y_1, y_2, y_3.

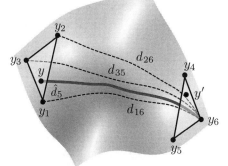

Step 3: $\hat{d}_6 \approx d_Y(y, y_6)$ is computed from d_{16}, d_{26}, d_{36} in the triangle y_1, y_2, y_3.

Step 4: $\hat{d} \approx d_Y(y, y')$ is computed from $\hat{d}_4, \hat{d}_5, \hat{d}_6$ in the triangle y_4, y_5, y_6.

Figure 9.6. Four steps performed to interpolate the geodesic distance $\hat{d}_Y(y, y')$ between two arbitrary points on the mesh Y.

$u_i^T D_Y(t_i, t_j) u_j$ are linear in u_i. This implies that fixing all the t_j's and all the u_j's except for some u_i, we obtain a *quadratic* stress function

$$\sigma(u_i) = \sum_{j \neq i} \left(d_X(x_i, x_j) - u_i^T D_Y(t_i, t_j) u_j \right)^2, \quad (9.11)$$

which we write for brevity as $\sigma(u_i) = u_i^T A_i u_i + 2 b_i^T u_i + c_i$, denoting by

$$A_i = \sum_{j \neq i} D_Y(t_i, t_j) u_j u_j^T D_Y(t_i, t_j)^T, \quad (9.12)$$

$$b_i = -\sum_{j \neq i} d_X(x_i, x_j) D_Y(t_i, t_j) u_j, \quad (9.13)$$

and
$$c_i = \sum_{j \neq i} d_X^2(x_i, x_j) \tag{9.14}$$

the quadratic form coefficients. We leave to the reader to show that the matrix A_i is positive semi-definite, that is, $\sigma(u_i)$ is convex.[3]

The quadratic form of $\sigma(u_i)$ allows us to express its minimizer in a closed form,
$$u_i^* = \arg\min_{u_i} \sigma(u_i) = -A_i^{-1} b_i. \tag{9.15}$$

However, such a solution may happen to be outside the triangle, rendering the barycentric representation invalid. In order to obtain a valid point, we constrain u_i^* to lie within the triangle:
$$u_i^* = \arg\min_{u_i} \sigma(u_i) \quad \text{s.t.} \quad \begin{cases} u_i \geq 0, \\ u_i^1 + u_i^2 + u_i^3 = 1. \end{cases} \tag{9.16}$$

This linearly constrained quadratic problem still has a closed-form solution, the derivation of which is left as an exercise to the reader.

Observe that replacing u_i with u_i^* results in a decrease of the stress value. We can use this fact to construct an algorithm that fixes all the variables except for some u_i, updates u_i with u_i^*, then picks a different u_i while fixing the rest of the variables, and continues in the same way, producing a sequence of updates. A reasonable choice of the point to update can be based on the gradient of the stress function with respect to each u_i,
$$g_i = \nabla \sigma(u_i) = 2 A_i u_i + 2 b_i. \tag{9.17}$$

It is desirable to select the u_i with the largest gradient in order to obtain the steepest decrease in the stress function value when the point is updated. The resulting algorithm is similar in its spirit to the coordinate descent (L_1 steepest descent) method from Chapter 5, where each time only one coordinate corresponding with the largest derivative was updated. Here, we operate on a single point u_i, resulting in an update of a block of two coordinates, u_i^1 and u_i^2, in the vector of optimization variables. For this reason, our minimization algorithm belongs to the family of *block coordinate descent* algorithms.

It is important to note that the updated point u_i^* may happen to lie on an edge or a vertex of the triangle t_i (that is, at least one constraint in (9.16) is active). In this case, the need to update the triangle index t_i may arise. If u_i^* lies on a triangle edge shared with some other triangle t_i', we translate the vector of barycentric coordinates u_i^* in the coordinate system of t_i to another vector $u_i'^*$ in the coordinate system of t_i'. This translation does not change the value of $\sigma(u_i)$, yet, as the stress function is not $\mathcal{C}^1$ on the triangle boundaries, the gradient direction may change. We evaluate the new gradient direction in the triangle t_i' and update t_i to be t_i' only if the negative gradient

```
1  for k = 0, 1, 2, ... do
2      Compute the parameters A_i, b_i and the gradients g_i of the stress
       function σ(t_1^(k), u_1^(k), ..., t_N^(k), u_N^(k)).
3      Select i corresponding to max ||g_i||.
4      if ||g_i|| is sufficiently small then stop
5      Solve the constrained quadratic problem
```

$$u_i^* = \arg\min_{u_i \geq 0} \sigma(u_i) \quad \text{s.t.} \quad u_i^1 + u_i^2 + u_i^3 = 1$$

```
       with the rest of the u_j's fixed to u_j^(k).
6      if u_i* is on an edge of the triangle t_i then
7          let N(u_i*, t_i) be the triangle sharing the edge with t_i, or ∅ if the
           edge is on the shape boundary.
8      else if u_i* is on a vertex of t_i then
9          let N(u_i*, t_i) be the set of triangles sharing the vertex with t_i.
10     else
11         set N(u_i*, t_i) = ∅.
12     if N(u_i*, t_i) = ∅ then update u^(k+1) = u_i*, and go to Step 2.
13     forall t' ∈ N(u_i*, t_i) do
14         Translate u_i* in the triangle t_i to u_i'* in the triangle t'.
15         Evaluate the gradient g_i = ∇_{u_i} σ(t', u_i'*).
16         if −g_i is directed inside the triangle t' then
17             update t_i^(k+1) = t', u^(k+1) = u_i'*, and go to Step 2.
18         end
19     end
20 end
```

Algorithm 9.1. Least squares GMDS.

direction points inside t_i'. In this case, subsequent minimization of σ with respect to u_i will guarantee a further decrease of the stress function value. If the triangle edge is not shared with another triangle (i.e., the edge is part of the mesh boundary), no index update is performed. The case where u_i^* lies on a triangle vertex is treated in a similar way. The entire minimization procedure is summarized in Algorithm 9.1.

It is worthwhile noting that the described algorithm is also suitable when the surface Y admits a global parameterization. In this case, the point coordinates u_i are expressed in the global system of coordinates, and the triangle indices t_i are inferred from the point locations. As a concluding remark before we proceed to the next topic, we should mention that Algorithm 9.1 is by no means the only way to solve the GMDS problem. For example, in [67] we proposed a different numerical scheme based on a *path search* – a generalization of line search, where instead of a line in a high-dimensional Euclidean space the minimum is searched over poly-linear paths on the mesh $T(Y_M)$.

9.6 Multiresolution encore

Like in the Euclidean MDS case, we can use the multiresolution strategy to solve the GMDS problem. We define $\Omega_L \subset \Omega_{L-1} \subset \cdots \subset \Omega_0 = \{1,\ldots,N\}$ to be a hierarchy of $L+1$ grids on the surface X, such that at the lth level we have N_l points. The data at each level is the $N_l \times N_l$ matrix D_l. Under the assumption that Y is given as a parametric surface, the coordinates of the lth level solution $y'_1,\ldots,y'_{N_l}$ are represented by $u_1,\ldots,u_{N_l}$ and arranged into the $N_l \times 2$ matrix U_l. The lth level GMDS problem is formulated as

$$U_l^* = \operatorname*{argmin}_{U_l \in \mathbb{R}^{N_l \times 2}} \sigma(U_l; D_l, D_Y);$$

This notation emphasizes the dependence on the data D_l and on the embedding space (expressed in terms of its geodesic distances D_Y). Going up to the finer $(l-1)$st level is performed using the data interpolation operator P_l^{l-1}, applied to U_l. The whole process is essentially equivalent to the multiresolution MDS algorithm we described in Chapter 7, with the exception of the fact that now our variables are parametric coordinates of the points on Y and not their Euclidean coordinates.

In case of local parameterization, that is, using barycentric coordinates, the interpolation operation P_l^{l-1} is not straightforward and, particularly, cannot be represented as a matrix. When going up in the resolution level, the points $\{y'_i : i \in \Omega_l\}$ are transferred "as are" to the finer grid Ω_{l-1}. The rest of the points, $\{y'_i : i \in \Omega_{l-1} \setminus \Omega_l\}$, can be chosen in such a way that the geodesic distances from y'_i to nearby points are as close as possible to the geodesic distances from x_i to its nearby points. Formally, we define

$$y'_i = \arg\min_y \sum_{j \in \mathcal{N}(x_i)} \left(d_Y(y, y'_j) - d_X(x_i, x_j)\right)^2,$$

for all $i \in \Omega_{l-1} \setminus \Omega_l$. Here $\mathcal{N}(i)$ is a neighborhood of x_i on X. Note that practically, the minimum can be found by exhaustively searching over a subset of the samples of Y, with the complexity of $\mathcal{O}(M)$. This interpolation does not have to be very accurate, as it is subsequently refined by the minimization of the stress.

In the GMDS problem, the idea of multiresolution optimization can be taken another step further by noting that the embedding space itself (that is, the surface Y) can also be represented at different resolution levels. At coarse levels, the embedding is performed into a coarse embedding space (i.e., Y approximated using a smaller number of triangles), which is more efficient computationally. When going upwards in the hierarchy of resolutions, the embedding space is gradually refined. For this purpose, we construct a hierarchy of $L+1$ grids $\Omega'_L \subset \Omega'_{L-1} \subset \cdots \subset \Omega'_0 = \{1,\ldots,M\}$ on Y, in a manner equivalent to the construction of the data hierarchy. The number of points at each level is denoted by $M_L < M_{L-1} < \cdots < M_0 = M$ and the matrices

---

1. Construct the hierarchy of data $D_0, \ldots, D_L$, data interpolation operators $P_1^0, \ldots, P_L^{L-1}$ and the hierarchy of embedding spaces with distance matrices $D'_0, \ldots, D'_L$.
2. Start with some initial parametric coordinates $U_L^{(0)}$ at the coarsest grid, and $l = L$.
3. **while** $l \geq 0$ **do**
4.   Solve the lth level GMDS problem,

$$U_l^* = \operatorname*{argmin}_{U_l \in \mathbb{R}^{N_l \times 2}} \sigma(U; D_l, D'_l),$$

    using $U_l^{(0)}$ as the initialization for the optimization algorithm.
5.   Interpolate the solution to the next resolution level, $U_{l-1}^{(0)} = P_l^{l-1} U_l^*$
6. $l \longleftarrow l - 1$
7. **end**

Algorithm 9.2. Multiresolution GMDS.

of geodesic distances by $D'_L, \ldots, D'_0 = D_Y$. At the lth resolution level, we minimize the generalized stress $\sigma(U_l; D_l, D'_l)$, that is, embedding N_l points into an M_l-point approximation of Y. The whole scheme can be summarized as shown in Algorithm 9.2.

Though the multiresolution framework improves the convergence of the GMDS algorithm, the importance of its initialization is hard to overestimate. Like in the case of ICP algorithms, there probably exists no universally good answer to this issue. However, some globally optimal initialization ideas in the spirit of [170] can be adopted for GMDS. The reader is referred to [59, 322] for further details.

Suggested reading

A brief explanation of spherical multidimensional scaling techniques can be found in Cox and Cox [119]. For a more detailed presentation, the reader is referred to [118]. For the uses of spherical embedding for representation of cortical surfaces and human faces, the reader is referred to [148] and [69], respectively. A different numerical scheme for GMDS based on the spherical wavefront model is discussed in [67]. In [59, 322], a globally optimal *branch and bound* technique is proposed as a means for the initialization of GMDS.

Software

An implementation of the multiresolution GMDS algorithm is available in the TOSCA toolbox.

Problems

9.1. Write the differential arclength in $\mathbb{S}^m$. Given a path, formulate a functional measuring its length and using the Euler-Lagrange equations, show that the geodesics on a sphere are portions of great circles.

9.2. Derive the gradient for the stress function in problem (9.1).

9.3. Show that the spherical embedding problem can be formulated as an algebraic problem similar to classic MDS.

9.4. Derive the projection operator for the general m-dimensional sphere.

9.5. Show that the isometry group in $\mathbb{S}^m$ is the orthogonal group $O(m+1)$.

9.6. Express a rigid isometry in $\mathbb{S}^2$ as a transformation in the parameterization domain.

9.7 (Research question). Derive an ICP-like algorithm for rigid alignment and matching of surfaces in $\mathbb{S}^3$.

9.8 (Research question). Derive a method for rigid surface matching in $\mathbb{S}^m$ based on moment signatures.

9.9. Prove that our interpolation $\hat{d}_Y(u, u') = u'^{\mathrm{T}} D_Y(t, t') u$ satisfies the five properties stated in the beginning of Section 9.4. Is $\hat{d}_Y(u, u')$ continuously differentiable everywhere?

9.10. Derive a different geodesic distance interpolation scheme based on a spherical wavefront model.

9.11. Prove that the L_2 stress is convex with respect to each u_i.

9.12. Derive a closed-form expression for the minimizer of the linearly constrained quadratic stress function $\sigma(u_i)$ in (9.16).

9.13 (Research question). Show a way to introduce point or curve constraints into the GMDS problem, forcing all points lying on some $\Gamma_X \subset X$ to be mapped onto some $\Gamma_Y \subset Y$.

Notes

[1] The term *m-dimensional sphere* refers to the dimension of the manifold (e.g., a two-dimensional sphere is a two-dimensional manifold). However, in the literature $\mathbb{S}^m$ is sometimes referred to as an $m+1$-*dimensional sphere*, referring to the dimension of the ambient space, $\mathbb{R}^{m+1}$. We find the latter notation slightly misleading, and stick to the former one.

[2] In reality, aircraft routes are also influenced by the so-called *jet streams* – narrow air currents found just under the boundary of the stratosphere, at altitudes around 11 kilometers. Wind speeds inside a jet stream may reach hundreds of km/h, making them an important factor in planning the flight course.

[3] The fact that the stress is convex with respect to each u_i does not contradict the fact that it is not convex with respect to all the u_i's together.

10
Isometry-Invariant Similarity

> It is incredible what Gromov can do just with the triangle inequality!
>
> D. SULLIVAN, *quoted by* M. BERGER.

In the previous three chapters, we explored computational tools for the comparison of non-rigid shapes. The discussed methods had one property in common: they gave a quantitative measure of similarity between two shapes. It is natural to think of such a similarity as of a *distance* function on some abstract *shape space*. Because our goal is to compare objects in a way insensitive to isometric deformations, a key property of such a distance is *isometry invariance*. In this chapter, we put the non-rigid shape matching problem on a more solid theoretical ground. We start by defining a set of desired properties that a good isometry-invariant distance should satisfy and show that the previously presented approaches satisfy them only partially. We then proceed by defining the important notion of the Gromov-Hausdorff distance and show how it fits to the same computational framework.

10.1 Equivalence, similarity, and distance

Let us examine an object X and its non-rigid deformation Y; as a visualization, the reader may consider different folding of a piece of paper or postures of his or her hand. An observer, judging two such objects only by their intrinsic properties, will find them indistinguishable, though the objects may differ significantly in their extrinsic geometric properties (we emphasize that for us such a short-sightedness is rather beneficial, as we would like to "undo" the richness of non-rigid deformations). On one hand, we know that the two objects X and Y are not identical (incongruent), and at the same time we deem them "the same" (isometric). This idea can be expressed by saying that rather than being equal, the two objects are *equivalent*. Formally, an equivalence relation is a relation between pairs of objects, which is

(E1) *reflexive*, meaning that X is equivalent to itself;
(E2) *symmetric*, meaning that if X and Y are equivalent, then Y and X are also equivalent; and

(E3) *transitive*, meaning that if X and Y are equivalent, and Y and Z are equivalent, then X and Z are equivalent.

An equivalence relation is usually denoted as $X \sim Y$, pronounced as "X is equivalent to Y." The set of all shapes in $\mathbb{M}$, which are equivalent (in our case, isometric) to $X \in \mathbb{M}$, is called the *equivalence class* of X. We say that the equivalence relation *partitions* the space $\mathbb{M}$ into equivalence classes. Considering each equivalence class as a single unit, $\mathbb{M}$ can be transformed into a new space, referred to as the *quotient space* of $\mathbb{M}$ under the equivalence relation $\sim$ (denoted as $\mathbb{M}^*$ or $\mathbb{M}\backslash \sim$). A point in $\mathbb{M}^*$ is a surface together with all its isometries.

Equivalence is a binary relation, which can be described by a function $d: \mathbb{M} \times \mathbb{M} \to \{0,1\}$. Each pair of objects $X, Y \in \mathbb{M}$, is either isometric ($d(X,Y) = 0$) or not ($d(X,Y) = 1$). However, ideal equivalence is rare to find in our imperfect world, and we would therefore like to relax this notion by allowing a surface X to be "almost equivalent" to another surface Y. As we have already seen, a natural way to do so is by resorting to almost isometries. We say that two objects are *similar* if they are almost isometric. Clearly, the relation of being similar is, in a sense, a superset of being equivalent and allows some mismatch in the intrinsic geometries. Unlike the binary answer we get from the equivalence relation on whether or not X and Y are isometric, if X and Y are ϵ-isometric, we would like to associate the number ϵ with the pair (X,Y). This brings us from the informal concept of similarity to a somewhat more rigorous notion of *distance* – a non-negative function $d: \mathbb{M} \times \mathbb{M} \to \mathbb{R}$ quantifying the degree of similarity of a pair of objects in $\mathbb{M}$. Because there exists no unique recipe for defining such a function, let us now examine the list of properties that a good distance should satisfy.

First, as similarity is a reflexive relation, we require d to be *symmetric*, that is, $d(X,Y) = d(Y,X)$. Second, similarity is also transitive, which can be expressed by requiring d to obey the *triangle inequality* $d(Z,X) \leq d(Y,X) + d(Z,Y)$. This inequality guarantees that if X is similar to Y and Y is similar to Z, then X and Z cannot be dissimilar. Third, because we look for *isometry invariance*, we would like d to associate zero distance to pairs of isometric objects, which are indistinguishable in terms of intrinsic geometry. On the other hand, we would like to distinguish between non-isometric objects, and thus require d to be strictly positive for any pair of such objects. Formally, this demand is expressed as $d(X,Y) = 0$ if and only if X and Y are isometric. We can briefly summarize the former three requirements by saying that we want our distance to be a *metric* on the quotient space $\mathbb{M}^*$.

The same way the last property associated equivalence with zero distance, we would like to associate similarity with small distance by requiring that if $d(X,Y) \leq \epsilon$, then X and Y are $c\epsilon$-isometric, and vice versa, if X and Y are ϵ-isometric, then $d(X,Y) \leq c\epsilon$, where c is a positive constant independent of X, Y, and ϵ. Observe that if $d(X,Y) > \epsilon$, then X and Y are not ϵ/c-isometric, because if they were, one would have $d(X,Y) \leq \epsilon$. Similarly, if X and Y are not ϵ-isometric, then $d(X,Y) > \epsilon/c$. This gives us the converse association of dissimilarity with large distance.

Thus far, we tacitly assumed to be able to work with continuous surfaces from $\mathbb{M}$. However, in practice we can only compute an approximation $\hat{d}$ of our distance, operating on sampled surfaces $X_N^r = \{x_1, \ldots, x_N\}$ and $Y_M^r = \{y_1, \ldots, y_M\}$ that constitute r-coverings of X and Y, respectively. In order to make such a numerical approximation practical, we demand $\hat{d}$ to be *consistent to sampling*, that is

$$\lim_{r \to 0} \hat{d}(X_N^r, Y_M^r) = d(X,Y).$$

This guarantees that the approximated distance approaches the true one as the sampling density increases. Also, practical considerations require $\hat{d}(X_N^r, Y_M^r)$ to be computable *efficiently*, which we interpret as polynomial time complexity with respect to the sample sizes M and N.

10.2 Embedding distance

In Chapter 9, we have seen that finding the minimum distortion embedding of X into Y allows us to quantify the dissimilarity of the intrinsic geometries of the two surfaces – that was precisely the *raison d'être* of generalized multidimensional scaling. The lowest achievable distortion can be viewed as the distance

$$d_{\mathrm{E}}(Y, X) = \inf_{\varphi: X \to Y} \mathrm{dis}\, \varphi, \qquad (10.1)$$

which we term as the *embedding distance*. In practice, we compute an approximation $\hat{d}_{\mathrm{E}}(Y_M^r, X_N^r) \approx d_{\mathrm{E}}(Y, X)$ by solving the GMDS problem. We leave the reader to verify that $|\hat{d}_{\mathrm{E}}(Y_M^r, X_N^r) - d_{\mathrm{E}}(Y, X)| = \mathcal{O}(r)$, implying that the embedding distance is consistent to sampling.

Observe that if X and Y are ϵ-isometric, then there exists a mapping $\varphi: X \to Y$ with dis $\varphi \leq \epsilon$, meaning that $d_{\mathrm{E}}(Y, X) \leq \epsilon$ (and, consequently, if $d_{\mathrm{E}}(Y, X) > \epsilon$, X and Y are not ϵ-isometric). Unfortunately, the converse is generally not true, as $d_{\mathrm{E}}(Y, X) \leq \epsilon$ implies only that X can be embedded with low distortion into Y but does not guarantee that such an embedding is ϵ-surjective. In fact, X and Y may have arbitrarily different intrinsic geometries while having small d_{E} (an extreme example is X being a very small part of Y). Another observation suggests that $d(Y, X)$ is not symmetric. Indeed, if X is a small portion of Y, it is embeddable into the latter without distortion ($d(Y, X) = 0$); however, Y cannot be embedded into X without distortion, and thus $d(X, Y) > 0$. Though d_{E} still satisfies the triangle inequality $d(Z, X) \leq d(Y, X) + d(Z, Y)$, unlike a true metric, the other side of the inequality, $d(X, Z) \leq d(Y, X) + d(Z, Y)$ does not generally hold.

We conclude that while being reasonably efficient to compute and consistent to sampling, the embedding distance does not satisfy all of the desired properties. In particular, it is not a metric, and, which is by far worse, the connection between similarity and the distance values works only in one direction.

10.3 Gromov-Hausdorff distance

Another type of similarity we met in Chapter 7 was the canonical form distance, computed as the Hausdorff distance between the minimum-distortion embeddings of two shapes into some common metric space $(\mathbb{Z}, d_{\mathbb{Z}})$. For example, we used the Euclidean space $\mathbb{R}^m$ or the m-dimensional sphere $\mathbb{S}^m$. As we mentioned, this approach suffers from an inherent inaccuracy due to the fact that usually a zero-distortion embedding of a surface into a given metric space is impossible to obtain.

Instead of having $\mathbb{Z}$ fixed for all surfaces, we can let $\mathbb{Z}$ be the best suitable space for the comparison of two given surfaces X and Y by introducing it as a variable into our optimization problem. Formally, we can write the following distance,

$$d_{\mathrm{GH}}(Y, X) = \inf_{\substack{\mathbb{Z} \\ f:X \to \mathbb{Z} \\ g:Y \to \mathbb{Z}}} d_{\mathrm{H},\mathbb{Z}}(f(X), g(Y)), \qquad (10.2)$$

where the infimum is taken over all metric spaces $\mathbb{Z}$ and isometric embeddings f and g from X and Y, respectively, to $\mathbb{Z}$. d_{GH} is called the *Gromov-Hausdorff distance* and can be thought of as an extension of the Hausdorff distance. The Gromov-Hausdorff distance was introduced in 1981 by the Russian-born mathematician Mikhail Gromov [190] and first applied to the field of pattern recognition by Facundo Mémoli and Guillermo Sapiro in 2004 [269].

At this point, the reader might wonder whether the space $\mathbb{Z}$ in (10.2) exists at all. In fact, we demand that $\mathbb{Z}$ has two metric subspaces $f(X)$ and $g(Y)$ with the restriction of $d_{\mathbb{Z}}$ that are isometric to X and Y, respectively, which seems like a very strong property. However, it appears that such a space always exists; moreover, we can even reduce it to the disjoint union of X and Y. More precisely, we may let $\mathbb{Z} = X \sqcup Y$ and define a (semi-) metric $d_{\mathbb{Z}}$ such that its restrictions to X and Y coincide with d_X and d_Y (clearly, $d_{\mathbb{Z}}$ is not unique). Using this reduction, the Gromov-Hausdorff distance can be reformulated in terms of the infimum over all the metrics $d_{\mathbb{Z}}$ on $X \sqcup Y$,

$$d_{\mathrm{GH}}(Y, X) = \inf_{d_{\mathbb{Z}}} d_{\mathrm{H},(X \sqcup Y, d_{\mathbb{Z}})}(Y, X). \qquad (10.3)$$

The Gromov-Hausdorff distance brings us excellent news, as it satisfies all the desired theoretical properties: it is a metric on $\mathbb{M}^*$, and it satisfies the similarity property with the constant $c = 2$ (namely, $d_{\mathrm{GH}}(X,Y) < \epsilon$ implies that X and Y are 2ϵ-isometric, and X and Y are ϵ-isometric implies that $d_{\mathrm{GH}}(X,Y) < 2\epsilon$). We leave the proof as an exercise for the reader (Problem 10.1).

At the first glance, the theoretical properties of the Gromov-Hausdorff distance make it a perfect choice for comparing non-rigid shapes. Yet, from definition (10.2), d_{GH} seems alarmingly impractical, as the minimization over all metric spaces $\mathbb{Z}$ or over all metrics $d_{\mathbb{Z}}$ on $X \sqcup Y$ is intractable. Fortunately, the Gromov-Hausdorff distance can be reformulated in terms of distances in X and Y, without resorting to the embedding space $\mathbb{Z}$:

$$d_{\mathrm{GH}}(Y,X) = \frac{1}{2} \inf_{\substack{\varphi: X \to Y \\ \psi: Y \to X}} \max\{\mathrm{dis}\,\varphi, \mathrm{dis}\,\psi, \mathrm{dis}\,(\varphi, \psi)\}. \tag{10.4}$$

Let us understand the notation first. The first two terms,

$$\mathrm{dis}\,\varphi = \sup_{x,x' \in X} |d_X(x,x') - d_Y(\varphi(x), \varphi(x'))|;$$

$$\mathrm{dis}\,\psi = \sup_{y,y' \in Y} |d_Y(y,y') - d_X(\psi(y), \psi(y'))|,$$

denote the familiar distortion of the embeddings φ and ψ, respectively. On the other hand, the term $\mathrm{dis}\,(\varphi, \psi)$ is new and is used to denote

$$\mathrm{dis}\,(\varphi, \psi) = \sup_{x \in X, y \in Y} |d_X(x, \psi(y)) - d_Y(y, \varphi(x))|$$

(see Figure 10.1). This reformulation of the Gromov-Hausdorff distance can be interpreted in the following way: we try to jointly embed X into Y and Y into X such that the distortions of the embeddings φ and ψ are as low as possible. In addition, we would like φ and ψ to be as close as possible one to the inverse of the other, in the sense that the compositions $\psi \circ \varphi : X \to Y$ and $\varphi \circ \psi : Y \to X$ are as close as possible to identity mappings (Figure 10.2). We leave to the reader to prove this extraordinary transfiguration of d_{GH} (Problem 10.2).

In its alternative formulation (10.4), the Gromov-Hausdorff distance closely resembles the embedding distance. The only difference is that now we have a slightly more complicated problem involving two embeddings and three distortion terms, yet it can be solved in the same spirit. Indeed, discretizing X and Y, we obtain

$$d_{\mathrm{GH}}(Y_M, X_N) = \frac{1}{2} \min_{\substack{y'_1,\dots,y'_N \in Y \\ x'_1,\dots,x'_M \in X}} \max \left\{ \begin{array}{l} |d_X(x_i, x_j) - d_Y(y'_i, y'_j)|, \\ |d_Y(y_k, y_l) - d_X(x'_k, x'_l)|, \\ |d_X(x_i, x'_k) - d_Y(y_k, y'_i)| \end{array} \right\}, \tag{10.5}$$

where $i, j = 1, \dots, N$, and $k, l = 1, \dots, M$. This problem can be viewed as a simultaneous solution of two L_∞ GMDS problems, coupled together by the

210   10 Isometry-Invariant Similarity

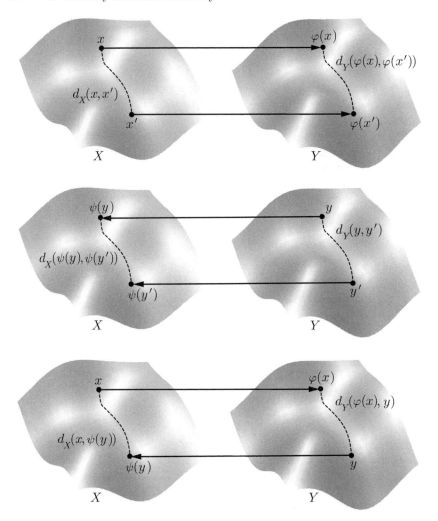

Figure 10.1. An illustration of the three distortion terms participating in the Gromov-Hausdorff distance: dis φ (top row), dis ψ (middle row), and dis (φ, ψ) (bottom row).

distortion terms $|d_X(x_i, x'_k) - d_Y(y_k, y'_i)|$. As in the GMDS, the minimization is performed over the images $y'_i = \varphi(x_i)$ and $x'_k = \psi(y_k)$, instead of the mappings φ and ψ themselves. The only difference is that now we have two sets of variables: one on the surface Y and the other on the surface X.

One of the ways to solve (10.5) is by introducing an artificial variable $\epsilon \geq 0$ and casting the min-max problem to the following constrained minimization problem,

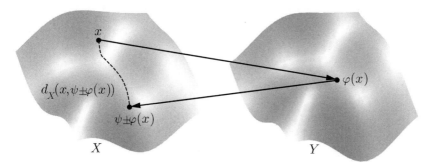

Figure 10.2. The distortion term dis (φ, ψ) tells us how far is φ from the inverse of ψ and vice versa, how far is ψ from the inverse of φ.

$$d_{\text{GH}}(Y_M, X_N) = \min_{\substack{\epsilon \geq 0 \\ y'_1,\ldots,y'_N \in Y \\ x'_1,\ldots,x'_M \in X}} \frac{\epsilon}{2} \quad \text{s.t.} \quad \begin{cases} |d_X(x_i, x_j) - d_Y(y'_i, y'_j)| \leq \epsilon \\ |d_Y(y_k, y_l) - d_X(x'_k, x'_l)| \leq \epsilon \\ |d_X(x_i, x'_k) - d_Y(y_k, y'_i)| \leq \epsilon. \end{cases}$$
(10.6)

Example 10.1. As a visualization of the performance of d_{GH}, we reproduce here a numerical experiment from [67], performed on the set of non-rigid shapes. The Gromov-Hausdorff distances were computed numerically between each pair of objects. The matching results are visualized in Figure 10.3 as dissimilarities in $\mathbb{R}^2$ (i.e., the closer are the points, the smaller is the distance between the corresponding objects).

It should be noted that our analysis was done with the L_∞ distortion. The reader may wonder whether similar results can be made for some L_p formulation of the distortion. Unfortunately, it appears that the properties of the Gromov-Hausdorff distance change dramatically when L_∞ is substituted by L_p. For example, the beautiful connection to the Hausdorff distance ceases to exist (although, the intriguing question whether an L_p version of d_{GH} is connected to some L_p version of the Hausdorff distance is still open). Having said that, an L_p formulation of the Gromov-Hausdorff distance is clearly useful for practical non-rigid surface matching and may be even more useful than its L_∞ counterpart due to its lower sensitivity to noise.

10.4 Intrinsic symmetry

This chapter dedicated to shape similarity would be incomplete without mentioning an interesting and beautiful particular case of shape *self-similarity*

Figure 10.3. Comparison of non-rigid surfaces using the Gromov-Hausdorff distance. Each point in the plot represents a surface; Euclidean distances between the points approximate the computed d_{GH}. Surfaces were represented as triangular meshes with 2000 vertices, and 50 points were embedded.

or *symmetry*. Symmetry, self-similarity, and invariance are synonyms of the same cornerstone of nature, exhibiting itself through the shapes of natural creations and ubiquitous laws of physics. As it was noted by Hermann Weyl "symmetry, as wide or as narrow as you may define its meaning, is one idea by which man through the ages has tried to comprehend the created order, beauty, and perfection" [395]. These words of one of the greatest twentieth century mathematicians reflect the importance symmetry has in all aspects of our life.

The interest in symmetry of shapes dates back to the dawn of human civilization. Early evidences that our predecessors attributed importance to

10.4 Intrinsic symmetry

symmetries can be found in many cultural heritages, ranging from monumental architecture of the Egyptian pyramids to traditional ancient Greek decorations. Johannes Kepler was among the first who attempted to give a geometric formulation to symmetries in his treatise *On the six-cornered snowflake* [223] in as early as 1611. A few centuries later, the study of symmetric shapes became a foundation of crystallography. Finally, symmetries of more complicated higher-dimensional objects underlie modern physics theories about the nature of matter, space and time.

Because many natural objects are symmetric, symmetry breaking can often be an indication of some anomaly or abnormal behavior. Therefore, detection of asymmetries arises in numerous practical problems, among which medical applications are probably the first to come in mind. For example, detection of tumors in medical images can be based on deviations from otherwise symmetric body organs and tissues. Facial symmetry is important in craniofacial plastic surgery, as symmetric facial features are often associated with beauty and aesthetics [266]. Furthermore, facial asymmetry can also be an indication of various syndromes and disorders. Conversely, the assumption of symmetry can be used as a prior knowledge in many problems. It may facilitate, for example, the reconstruction of surfaces, face detection, recognition, and feature extraction.

In pattern recognition and computer vision, there exists a wealth of literature dedicated to finding symmetries in images, two-dimensional and three-dimensional shapes (see, e.g., [401, 13, 5, 278]). Traditionally, symmetries are considered as *extrinsic* geometric properties of shapes that related to the way the shape is represented in the Euclidean space. Though adequate for rigid shapes, such a point of view is inappropriate for non-rigid or deformable ones. Because of the deformations such shapes can undergo, the extrinsic symmetries may be lost while *intrinsically* the shape still remains symmetric. Consider as an example the human body. Extrinsic bilateral symmetry of the body is broken when the body assumes different postures. Yet, from the point of view of intrinsic geometry, the new shape remains almost identical; as such, a deformation does not significantly change its metric structure (Figure 10.4). In this sense, intrinsic symmetries are a superset of the extrinsic ones. Considering intrinsic rather than extrinsic symmetries allows us to characterize the object self-similarity that is invariant to nearly isometric deformations. Using our terminology, intrinsic symmetries are the group of self-isometries on the metric space formed by the non-rigid shape with its intrinsic geometry. We will speak more about such self-isometries in Chapter 12.

Whereas extrinsic symmetry computation is a well-established subject, intrinsic symmetries remain largely *terra incognita*. In [322], Dan Raviv and the authors used the generalized multidimensional scaling to compute approximate self-isometries of non-rigid shapes. For the time being, this seems to be the only study dedicated to intrinsic symmetry.

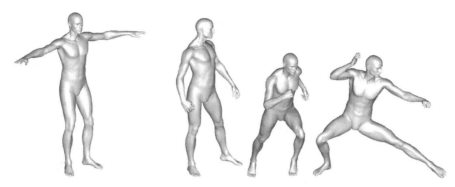

Figure 10.4. Symmetric or not? Extrinsically symmetric shape is also intrinsically symmetric (left), however, isometries of the shape are intrinsically symmetric but extrinsically asymmetric (three rightmost shapes).

Suggested reading

For a comprehensive overview of the theoretical properties of the Gromov-Hausdorff distance, the reader is referred to the textbook *Course on Metric Geometry* by Burago et al. [88] and to the original work of Gromov [190]. For practical aspects of using the Gromov-Hausdorff distance in applications of shape analysis, the reader is referred to [269, 67]. For the presentation of theoretical and computational frameworks for the treatment of intrinsic symmetries of shapes, the reader is referred to [322].

Problems

10.1. Prove that the Gromov-Hausdorff distance obeys all the desired properties listed in the beginning of the chapter. Hint: use the equivalent definition (10.4) of the Gromov-Hausdorff distance.

10.2. Prove the equivalence of the two definitions (10.2) and (10.4) of the Gromov-Hausdorff distance. Hint: use the correspondences φ and ψ to define a valid metric on $X \sqcup Y$.

10.3. Show that the Gromov-Hausdorff distance is consistent to sampling.

10.4. Consider the following distance

$$d(Y, X) = \frac{1}{2} \inf_{\substack{\varphi: X \to Y \\ \psi: Y \to X}} \max\{\operatorname{dis}\varphi, \operatorname{dis}\psi\},$$

which resembles d_{GH} without the term $\operatorname{dis}(\varphi, \psi)$. What is the relation between the new distance and the Gromov-Hausdorff distance? Does the new distance obey the similarity property?

10.5. Consider the following distance

$$d(Y, X) = \frac{1}{2} \inf_{\substack{\varphi: X \to Y \\ \psi: Y \to X}} \text{dis}\,(\varphi, \psi),$$

What is the relation between the new distance and the Gromov-Hausdorff distance? Does the new distance satisfy the similarity property?

10.6 (Research question). Consider an L_p formulation of the distortion

$$\text{dis}\,\varphi = \int_{X \times X} |d_X(x, x') - d_Y(\varphi(x), \varphi(x'))|^p d\mu \times \mu(x, x'),$$

where μ is some measure on the surface X (for example, the standard *Hausdorff* or *area measure*). Using the new distortion, devise the L_p Gromov-Hausdorff distance and investigate its properties. Does it satisfy axioms (D)? Devise a new notion of surface similarity satisfied by the L_p Gromov-Hausdorff distance.

11

Partial Similarity

> The whole is more than the sum of its parts.
>
> ARISTOTLE, *Metaphysica*

In Chapter 2, we drew inspiration from Greek science as the foundation of modern geometry. Let us now resort to ancient Greece again, this time for a mythological rather than scientific illustration. In the pantheon of strange living things in Greek mythology, the most famous and strange are centaurs,[1] half-equine half-human creatures (Figure 11.1). Arguing whether a centaur is similar to a horse or to a man is as useless as asking whether a zebra is white or black. In a sense, it is similar to both, because its upper part is similar to that of a human, and the lower part makes it similar to a horse.

Note, however, that here we attribute a different meaning to the term "similar." The methods we have discussed thus far, whether iterative closest point and moment methods for rigid object comparison or MDS-based techniques and the Gromov-Hausdorff distance for the comparison of non-rigid objects, considered the shape as a whole, or in other words, measured *global* or *full* similarity. In the centaur example, on the other hand, we encounter a new type of similarity, which we call *partial*. There is sufficient evidence that such a type of similarity plays an important role in human judgment. In fact, our everyday experience is a proof that the human visual system has a very developed ability of judging partial similarity: seeing a small fragment of a picture, we are usually able to recognize the entire object in spite of it having large missing parts [206, 357, 358, 207]. Often, even a single significant part is sufficient to recognize the entire object.

The properties of full and partial similarities are substantially different. In general, partial similarity is a weaker criterion, as it does not require the whole shapes to be similar, but rather parts of them. Typically, partially similar objects have similar parts but are globally dissimilar. Furthermore, the partial similarity relation is *intransitive*. In our mythological example, whereas a centaur is partially similar to a man and a horse, a man and a horse are dissimilar (Figure 11.2). As a result, if we consider partial similarity as an abstract distance on the space of our objects, as we did in the previous chapter, it will not be a metric, as the triangle inequality does not hold.[2]

218   11 Partial Similarity

Figure 11.1. Centaurs (Greek κένταυροι) are a mythological race of half-human half-horse creatures. Marble depiction of centaur Nessus by Laurent-Honoré Marqueste, 1892, Tuileries Gardens, Paris.

11.1 Recognition by parts

Conceptually, the human visual perception can be thought of as a three-stage process: division of the shapes into meaningful parts, comparison of the parts

11.1 Recognition by parts 219

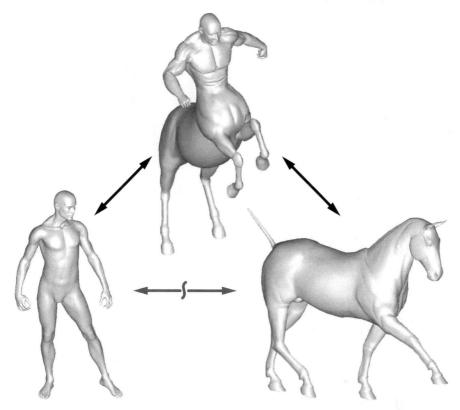

Figure 11.2. Partial similarity is an intransitive relation.

separately, and merging of the "partial" similarities (Figure 11.3). Alex Pentland referred to such a mechanism as "recognition by parts" [310]. For us, all the stages of this task seem so natural that we almost never notice it in everyday life, unless it is lost due to neurological disorders – sometimes, in quite peculiar ways. A remarkable case was described by the British psychiatrist Oliver Sacks in his book *The Man Who Mistook His Wife for a Hat* [331]. A patient, suffering from a rare disease called *visual agnosia*, was able to recognize parts of objects and describe them in detail, but at the same time, failed to recognize the entire objects.

More formally, partial similarity can be derived from a full similarity criterion in the following way. Given two objects X and Y, we first decompose them into parts $X_1, \ldots, X_K$ and $Y_1, \ldots, Y_L$. Then, each pair of parts is compared by using some full similarity criterion d_F, which is selected depending on the type of the objects being compared and the desired properties of such comparison (for instance, comparing rigid objects, we will select d_F as the Hausdorff distance, and comparing nonrigid objects, we will use the Gromov-Hausdorff distance). The partial similarity can be defined as some aggregation of part-wise full dissimilarities, for example,

11 Partial Similarity

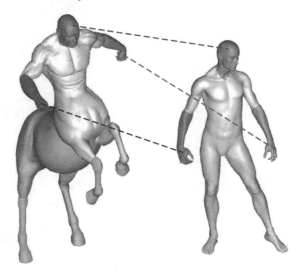

Figure 11.3. Recognition by parts.

$$d_{\mathrm{P}}(X,Y) = \min_{\substack{i=1,\ldots,K \\ j=1,\ldots,L}} d_{\mathrm{F}}(X_i, Y_j).$$

Unfortunately, we do not have a clear understanding of how our brain partitions the objects we see into meaningful parts [19]. Consequently, we cannot provide a precise recipe for finding $X_1, \ldots, X_K$ and $Y_1, \ldots, Y_L$ in the "algorithm" described above. Many attempts to imitate the human "recognition by parts" in the literature on object recognition use *ad hoc* definitions [112, 237, 325, 358, 42]. Some examples include parts described as convex or near-convex subsets [206, 230], primitive geometric objects [34, 33, 17, 310], or parametric description derived from a model of the shape class [81, 202]. The meaningfulness of parts described in such ways is not always obvious, and therefore, such approaches are generally limited.

In a recent paper, Latecki *et al.* [241] proposed a partial similarity criterion that avoids the necessity to use arbitrary shape partition.[3] A meaningful part is defined as the most similar common part of two shapes and is practically found by simplifying the shapes until they look the most similar, in the sense of some full similarity d_{F}. The easiest way to perform shape simplification is by removing parts from it. The partial similarity by Latecki *et al.* (referred to by the authors as the *optimal partial similarity*) can be formulated as follows:

$$d_{\mathrm{OPS}}(X,Y) = \inf_{X' \in \Sigma_X} d_{\mathrm{F}}(Y, X'^c).$$

Here $X'^c = X \setminus X'^c$ is the simplified shape obtained by removing the part X' from it. Σ_X is the set of all the possible parts of X, a subset of the powerset 2^X.[4]

Figure 11.4. Grotesque medieval monsters composed of human body parts are an example of how the human perception of partial similarity can be misleading (woodcuts from folio XIIr of Hartman Schedel's 1493 *Liber Chronicarum* [337], Morse Library, Beloit College. Reproduced by courtesy of Constantine T. Hadavas).

Note that the object Y is used entirely, whereas parts are cropped only from X; this makes the distance d_{OPS} non-symmetric. Latecki *et al.* give an analogy with text search according to keywords: the large object X can be thought of as text and a small query object Y as a keyword. The optimal partial similarity approach relies on a tacit assumption that the query object Y is carefully selected in order to be sufficiently representative, very much like text search is sensitive to the selection of keywords.

However, in many situations, the knowledge of parts similarity does not allow us to infer information about the similarity of the whole objects – we may be comparing small parts that happen to be similar, yet, belong to objects that are completely different. For example, a leg is supposed to be a representative part according to which we would recognize an object as a human body; this is a prior information we have acquired during our lives and use in our judgment of similarity. Yet, applied to grotesque medieval monsters depicted in Figure 11.4, this judgment could be wrong: the shapes, though including legs, are not of human beings. Normally, one would label such objects as "weird" or "strange," implying that they look differently from what we are used to.[5]

11.2 Paretian approach to partial similarity

The above example teaches us that often two dissimilar shapes have many common parts, yet, using these parts to conclude that the shapes are similar would be wrong. The existence of common similar parts appears to be insufficient *per se*: such a criterion does not describe how *significant* such parts are. It is clear that some parts carry more information that allows us to recognize the entire object, hence they are more significant than other parts.

The most straightforward way to quantitatively define significance is by making it proportional to the "size" of the parts: the larger is the part, the more it is significant. For the following formulation, it is convenient to define *partiality* $\lambda(X', Y')$, which represents how small X' and Y' are compared with the entire surfaces X and Y (the larger is the partiality, the smaller are the parts). In shape comparison problems, it is natural to use

$$\lambda(X', Y') = \mu_X(X'^c) + \mu_Y(Y'^c) \qquad (11.1)$$
$$= (\mu_X(X) + \mu_Y(Y)) - (\mu_X(X') + \mu_Y(Y'))$$

as the partiality, where

$$\mu_X(X') = \int_{X'} da \qquad (11.2)$$

is the measure of area derived from the Riemannian structure of X. The partiality can also be interpreted as a measure of "reliability" of our judgment of similarity: if there exist two parts of the objects that are similar, but these parts are small, our conclusion about the similarity of the entire objects is unreliable.

Following this logic, in order for two objects to be partially similar, they must have *significant similar parts*. The computation of partial similarity between X and Y can be therefore formulated as a problem of finding a pair of parts (X', Y') with minimum dissimilarity $d_F(X', Y')$ and minimum partiality $\lambda(X', Y')$. More formally, we define a *multicriterion optimization problem*,[6] in which a *vector objective* function $\Phi(X', Y') = (\lambda(X', Y'), d_F(X', Y'))$ is minimized over $\Omega = \Sigma_X \times \Sigma_Y$. Because we optimize over all the possible combinations of parts, the headache of finding a meaningful shape partition is avoided – we obtain it as a by-product of our solution, like in the method by Latecki et al. [241].

It is important to understand that partiality and dissimilarity are competing, such that no solution simultaneously optimal for both can be found, unless X and Y are fully similar. One consequence is that the notion of optimality used in traditional scalar optimization must be replaced by a new one, adapted to the multicriterion problem. Recall that for a scalar objective function $f : \mathbb{R}^N \to \mathbb{R}$, we define a global minimizer as a point x^*, for which the value of the objective is "the best," or, said differently, there does not exist another x such that $f(x^*) > f(x)$. In the vector case, we cannot straightforwardly apply this definition, as there does not exist a *total order* relation between vectors: we cannot say, for example, whether $(0.5, 1)$ is better than $(1, 0.5)$ or vice versa. At the same time, there is no doubt that $(0.5, 0.5)$ is better than $(1, 1)$, because it has both criteria smaller. We can therefore define *partial order* between vectors, saying that $(\lambda_1, \epsilon_1) < (\lambda_2, \epsilon_2)$ if $\lambda_1 < \lambda_2$ and $\epsilon_1 < \epsilon_2$ simultaneously.

Using this relation, we define a minimizer of our vector objective Φ as a pair (X^*, Y^*), such that there is no other pair $(X', Y') \in \Omega$ for which $\Phi(X^*, Y^*) > \Phi(X', Y')$, where the inequality is understood in the vector sense. Such a point is called *Pareto optimal*,[7] after the Italian economist Vilfredo Pareto

11.2 Paretian approach to partial similarity

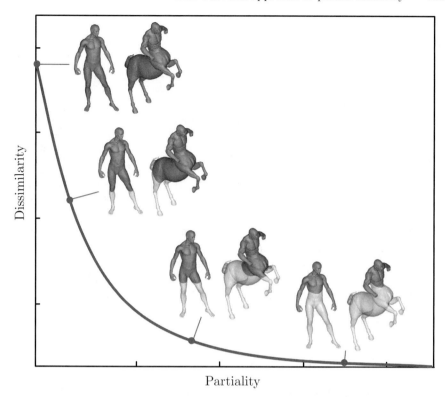

Figure 11.5. Illustration of the multicriterion optimization problem used for the computation of partial similarity. The Pareto frontier is shown as a curve.

(1842–1923), who first introduced this notion [307]. An intuitive explanation of Pareto optimality is that no criterion can be improved without compromising the other: if we try to reduce the dissimilarity, we necessarily increase the partiality, and vice versa. Note that Pareto optimum is not unique. If we denote by $\Omega^* \subseteq \Omega$ the set of all pairs (X^*, Y^*) satisfying the above condition, the set of the corresponding criteria values $\Phi(\Omega^*)$ can be represented as a planar curve referred to as the *Pareto frontier* (see Figure 11.5).

The notion of Pareto optimality brings us to a somewhat unorthodox concept of partial similarity: the *entire* Pareto frontier is used as generalized, *set-valued distance*. We denote $d_P(X,Y) = \Phi(\Omega^*)$ and call it the *Pareto distance*. The dissimilarity value at the point $\lambda = 0$ on $d_P(X,Y)$ coincides with the value of the full similarity $d_F(X,Y)$. Consequently, the information contained in the set-valued Pareto distance is a superset of that contained in the traditional full similarity. The following example accentuates this difference.

Example 11.1 (set-valued distances). Consider the shapes depicted in Figure 11.6: a man, a spear-bearer, and a centaur. We would like to find their partial intrinsic similarity. The Gromov-Hausdorff distance (acting as d_F in this example and corresponding with the point with $\lambda = 0$ on the Pareto frontier) between the man and the spear-bearer is large, due to the distortion caused when trying to embed the spear into the spear-less human shape. Similarly, the Gromov-Hausdorff distance between the man and the centaur is large due to the distortion caused by the bottom part of the horse body, which has approximately the same diameter as the spear. Hence, from the point of view of global intrinsic similarity, the man is approximately as dissimilar to the spear-bearer as it is dissimilar to the centaur. This is the only information we can infer from the scalar-valued distance.

However, if we consider the entire Pareto frontier, we see that the curve representing the set-valued distance between the man and the spear-bearer decays much faster compared with the one representing the distance between the man and the centaur. The reason is that in order to make a spear-bearer similar to a man, we have to remove only a small part (the spear), whereas in order to make a centaur similar to a man, we have to remove large parts (the horse body from the centaur and the legs from the man). Thus, from the set-valued distances, we can infer that the man is more similar to the spear-bearer than to the centaur, which corresponds with our intuition.

11.3 Scalar partial similarity

Although containing more information than scalar-valued distances, the inconvenience of Pareto distances is that they are not always mutually comparable. This problem stems from the absence of a total order relation between vectors: we can say that $d_\mathrm{P}(X,Y) < d_\mathrm{P}(X,Z)$ (meaning that X and Z are more partially dissimilar than X and Y) only if the curve $d_\mathrm{P}(X,Y)$ is entirely below $d_\mathrm{P}(X,Z)$ (in Example 11.1, the man–spear-bearer Pareto distance was below the man–centaur distance, therefore, we could say that a man is more similar to a spear-bearer than to a centaur). In order to define a total order between partial similarities, we have to convert our set-valued Pareto distance into a scalar-valued one. We refer to such a "scalarized" partial dissimilarity criterion as a *scalar partial distance* and denote it by $d_\mathrm{SP}(X,Y)$.

Straightforwardly, a scalar partial distance can be obtained by selecting a single point on the Pareto frontier. For example, with a fixed partiality λ_0, we can set a minimum threshold on the area of the parts we compare. This way, small values of λ_0 will make our criterion more reliable but at the same time, more restrictive: in order to say that two shapes are partially similar, they must have larger parts. The extreme case of $\lambda_0 = 0$ brings us back to the full similarity.

The scalar partial distance in this case is computed as the solution to a constrained optimization problem with a scalar valued objective,

$$d_\mathrm{SP}(X,Y) = \min_{(X',Y')\in\Omega} d_\mathrm{F}(X',Y') \quad \text{s.t.} \quad \lambda(X',Y') \leq \lambda_0,$$

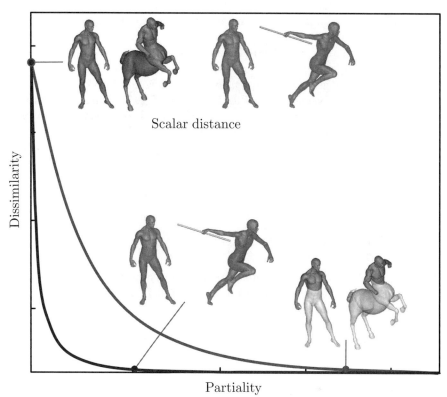

Figure 11.6. Visualization of the difference between scalar-valued and set-valued distances.

which can be reformulated as an unconstrained optimization problem,

$$d_{\mathrm{SP}}(X,Y) = \min_{(X',Y')\in\Omega} d_{\mathrm{F}}(X',Y') + \eta\lambda(X',Y'), \qquad (11.3)$$

by introducing the Lagrange multiplier η. Alternatively, we can set the dissimilarity to ϵ_0 and define the scalar partial similarity as the area of the parts we have to remove from the objects X and Y in order to make the remaining parts ϵ_0-dissimilar. Such a criterion may be useful for numerical computation if we know in advance a bound on the accuracy of sampling and geodesic distance measurement and say that below the value of ϵ_0, our distance is unable to discern between the shapes.

We should however note that in most cases, it is impossible to pre-set a value of dissimilarity or partiality suitable for partial comparison of different objects, but these values should be chosen adaptively (in Example 11.1, for instance, we could fix a small λ_0 to compare the man and the spear-bearer but would be forced to choose a large λ_0 to compare the man and the centaur). A pretty generic approach for such a choice was proposed by Mindia

Salukwadze[8] in the context of control theory [333]. Recall that in our multi-criterion optimization problem, we wish to bring to minimum the partiality λ and the dissimilarity d_F. Ideally, both of them should be zero, which corresponds with the "utopia point" $(0,0)$; however, this point is usually not achieved. Salukwadze proposed quantifying the degree of non-optimality of a solution (X', Y') by measuring the distance of $\Phi(X', Y')$ from the utopia point. A solution yielding the minimum distance is called *Salukwadze optimal*. It can be shown that Salukwadze optimum is necessarily realized on Ω^* (the proof is straightforward and we leave it as Problem 11.1). In a sense, Salukwadze optimum can be considered as a measure of how "fast" the Pareto frontier curve decays.

Using this idea, we can define the scalar partial distance (*Salukwadze distance*) as

$$d_\mathrm{SP}(X, Y) = \inf_{(X', Y') \in \Omega} \|\Phi(X', Y')\|.$$

Here, $\|\cdot\|$ is some norm on $\mathbb{R}^2_+$ that is used to measure the distance from the utopia point. If, for example, we use the weighted L_1-norm, the Salukwadze distance is given as an aggregation of the partiality and the dissimilarity, which is exactly optimization problem (11.3). Changing the value of the multiplier η allows us to control the location of the selected point on the Pareto frontier.

11.4 Fuzzy approximation

A more significant drawback of our multicriterion optimization framework for the computation of partial similarity is that optimization is performed over all possible parts of the objects, $\Sigma_X \times \Sigma_Y$. In a discrete setting, this problem becomes intractable, as the number of possible parts explodes exponentially as the number of samples grows.[9] In order to overcome this difficulty, we need to find a different way to represent the parts. A subset of X can be characterized by a function $m_X : X \to \{0, 1\}$, which indicates whether a point belongs to the subset or not. Our problem is thus posed as optimization over (m_X, m_Y) instead of the parts (X', Y') and is still intractable, as the requirement that m_X and m_Y obtain the values of 0 or 1 only results in a combinatorial optimization problem.

A computationally tractable problem is obtained by relaxing the values of m_X to the entire interval $[0, 1]$. Resorting to such a relaxation, we break the boundaries of the traditional set theory and land in the realm of *fuzzy set theory*, introduced in 1965 by Lotfi Zadeh [407, 408]. In fuzzy set theory, unlike the traditional "crisp" one, instead of saying that a point x belongs to a subset of X, we define the degree of its *membership*. The function m_X is referred to as a *membership function*.

Adapting our problem to the new setting, we need to recall that our partial similarity has three ingredients: sets of parts, partiality, and dissimilarity. The

11.4 Fuzzy approximation

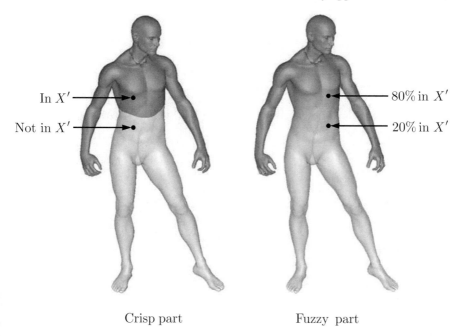

Figure 11.7. Example of a crisp (left) and a fuzzy part (right). Darker shades correspond with larger values of the membership function.

fuzzy versions of these definitions will be an extension of the crisp ones. A *fuzzy part* is defined by its membership function $m_X : X \to [0, 1]$ (see example in Figure 11.7). The complement of a fuzzy part is defined as $m_X^c = 1 - m_X$, which coincides with the standard definition on crisp sets. We say that a membership function m_X is Σ_X-*measurable* if thresholding m_X at a value of $0 \leq \theta \leq 1$ always produces a set $\{x : m_X(x) \leq \theta\}$ belonging to Σ_X. We define the set M_X of all the fuzzy parts as the set of all the Σ_X-measurable membership functions. Obviously, M_X is a superset of Σ_X (the reader is requested to prove this straightforward fact in Problem 11.2).

A fuzzy version of partiality (11.1) can be obtained in the following way. First, we define the *fuzzy measure* as

$$\tilde{\mu}_X(m_X) = \int_X m_X(x) d\mu_X,$$

for all $m_X \in M_X$. For crisp parts, the fuzzy measure $\tilde{\mu}_X$ boils down to the standard measure μ_X. Using this definition, the *fuzzy partiality* is given as

$$\tilde{\lambda}(m_Y, m_Y) = \tilde{\mu}_X(1 - m_X) + \tilde{\mu}_Y(1 - m_Y),$$

and it also coincides with λ on crisp sets.

---
1 **repeat**
2   Fix the fuzzy parts m_X, m_Y and find the correspondences φ and ψ,

$$(\varphi, \psi) = \underset{\substack{\varphi: X \to Y \\ \psi: Y \to X}}{\operatorname{argmin}} \sigma(m_X, m_Y, \varphi, \psi).$$

3   Fix the correspondences φ and ψ and find the fuzzy parts m_X, m_Y,

$$(m_X, m_Y) = \min_{(m_X, m_Y) \in \tilde{\Omega}} \sigma(m_X, m_Y, \varphi, \psi) \quad \text{s.t.} \quad \tilde{\lambda}(m_X, m_Y) \leq \lambda_0.$$

4 **until** *convergence*

Algorithm 11.1. Fuzzy Pareto distance computation.

The definition of the fuzzy dissimilarity is trickier and depends on the type of full similarity used. For the following discussion, we assume that the fuzzy dissimilarity can be written in the form

$$\tilde{d}_F(m_X, m_Y) = \min_{\substack{\varphi: X \to Y \\ \psi: Y \to X}} \sigma(m_X, m_Y, \varphi, \psi),$$

where $\sigma(m_X, m_Y, \varphi, \psi)$ is a fuzzy version of the stress function, measuring an aggregate of some local dissimilarity between the corresponding points. We will see a few specific examples in the following sections.

The multicriterion optimization problem is similar to the crisp case with all the crisp quantities replaced by fuzzy ones: we minimize the objective function $\tilde{\Phi}(m_X, m_Y) = (\tilde{\lambda}(m_X, m_Y), \tilde{d}_F(m_X, m_Y))$ over $\tilde{\Omega} = M_X \times M_Y$. The Pareto optimal set $\tilde{\Omega}^*$ and the Pareto frontier $\tilde{\Phi}(\tilde{\Omega}^*)$ are defined exactly as before. We call the set-valued distance $\tilde{d}_P(X, Y) = \tilde{\Phi}(\tilde{\Omega}^*)$ the *fuzzy Pareto distance*.

A single point on $\tilde{d}_P(X, Y)$ is computed as a scalar partial distance by fixing a value of partiality $\tilde{\lambda}(m_X, m_Y) \leq \lambda_0$ and minimizing $\tilde{d}_F(m_X, m_Y)$ with respect to m_X, m_Y subject to this constraint,

$$\begin{aligned}\tilde{d}_{SP}(X, Y) &= \min_{(m_X, m_Y) \in \tilde{\Omega}} \tilde{d}_F(m_X, m_Y) \quad \text{s.t.} \quad \tilde{\lambda}(m_X, m_Y) \leq \lambda_0, \\ &= \min_{(m_X, m_Y) \in \tilde{\Omega}} \min_{\substack{\varphi: X \to Y \\ \psi: Y \to X}} \sigma(m_X, m_Y, \varphi, \psi) \quad \text{s.t.} \quad \tilde{\lambda}(m_X, m_Y) \leq \lambda_0.\end{aligned}$$

One can notice that this problem involves two sets of variables: the fuzzy parts m_X, m_Y and the correspondences φ and ψ. These two sets of variables can be decoupled, resulting in the following two-stage iterative scheme shown in Algorithm 11.1.

By gradually varying the value of λ_0 from zero to $\mu_X(X) + \mu_Y(Y)$, the entire Pareto frontier can be obtained. We should emphasize that Algorithm 11.1

is generic: depending on the definition of σ (expressing the full similarity criterion), we can obtain different ways to define the partial similarity. In the next sections, we show two specific examples: a partial version of extrinsic and intrinsic similarity criteria.

11.5 Extrinsic partial similarity

A reader remembering our discussion on rigid shape matching in Chapter 6 will find a resemblance between ICP methods and our Algorithm 11.1. In ICP algorithms, we had an alternating minimization over correspondence and rigid motion, which minimized extrinsic dissimilarity (for example, a non-symmetric L_2 version of the Hausdorff distance). Here, we also need to optimize over the parts and can formulate *partial extrinsic similarity* as finding parts of given size, such that the ICP distance between them is minimized.

Recall from Chapter 6 that if we know the closest point correspondences

$$x^*(y) = \min_{x \in X} \|x - y\|_2$$
$$y^*(x) = \min_{y \in Y} \|y - x\|_2$$

between the shapes, the symmetric L_2 version of the ICP distance can be written as

$$d_{\mathrm{ICP}}(X, Y) = \min_{R,t} \int_X \|x - (Ry^*(x) + t)\|_2^2 \, dx$$
$$+ \int_Y \|(Ry + t) - x^*(y)\|_2^2 \, dy,$$

where R and t express the rotation and translation transformation. Once the optimal rigid motion R and t is found, the correspondences $x^*(y)$ and $y^*(x)$ are recomputed, and the process is iterated until convergence.

If we wish to compare parts of X and Y rather than the entire shapes, we should limit the integration domain to those parts only. In the fuzzy setting, this can be done by using the membership functions m_X and m_Y as the weights,

$$\tilde{d}_{\mathrm{ICP}}(m_X, m_Y) = \min_{R,t} \int_X \|x - (Ry^*(x) + t)\|_2^2 \, m_X(x) dx \quad (11.4)$$
$$+ \int_Y \|(Ry + t) - x^*(y)\|_2^2 \, m_Y(y) dy.$$

In Step 2 of Algorithm 11.1, the parts m_X and m_Y are fixed. By solving problem (11.4), we find the optimal weighted rigid alignment between the shapes, realized by the transformation (R^*, t^*). This, in turn, allows us to define the optimal correspondences,

$$\varphi(x) = \underset{y \in Y}{\operatorname{argmin}} \, \|x - (R^*y + t^*)\|_2^2; \qquad (11.5)$$

$$\psi(y) = \underset{x \in X}{\operatorname{argmin}} \, \|(R^*y + t^*) - x\|_2^2.$$

In Step 3, we fix the correspondences according to (11.5) and perform optimization over the parts,

$$(m_X, m_Y) = \underset{m_X, m_Y}{\operatorname{argmin}} \int_X \|x - \varphi(x)\|_2^2 \, m_X(x) dx$$
$$+ \int_Y \|y - \psi(y)\|_2^2 \, m_Y(y) dy$$
$$\text{s.t.} \quad \tilde{\lambda}(m_X, m_Y) \leq \lambda_0.$$

We should note that numerically, Step 2 is nothing else but a weighted version of the ICP algorithm, carried out by optimization schemes we have described in Chapter 6. The weights in the discretized problem are represented as vectors containing the point-wise values of the membership functions. Step 2 involves constrained optimization over the discretized membership functions. This appears to be a well-structured optimization problem, which can be solved efficiently (we refer the reader to [55] for additional details).

To summarize our discussion of partial rigid similarity, we show the following example illustrating the method.

Example 11.2 (extrinsic partial similarity of rigid shapes). As a demonstration of extrinsic partial similarity computation, we reproduce a numerical experiment from [55]. In this experiment, male and female figures in similar poses are compared as rigid shapes using the described approach with partiality set to half of the sum of the shape areas. Figure 11.8 depicts the corresponding parts in the two shapes. Observe that the computed parts suffer from irregularity. We will address this issue in the conclusion of this chapter.

11.6 Intrinsic partial similarity

The derivation of *intrinsic partial similarity* is very similar in its spirit to the derivation of the extrinsic partial similarity above. For simplicity, we start with an L_2 formulation. Given the correspondences φ and ψ between the shapes X and Y, we can use the generalized L_2-stress to measure their intrinsic similarity,

$$\sigma_2(\varphi, \psi) = \int_{X \times X} |d_X(x, x') - d_Y(\varphi(x), \varphi(x'))|^2 \, da \times da \qquad (11.6)$$
$$+ \int_{Y \times Y} |d_Y(y, y') - d_X(\psi(y), \psi(y'))|^2 \, da \times da.$$

Figure 11.8. Extrinsic partial similarity of rigid shapes. Highlighted are the corresponding parts. See insert for image in color.

Expression (11.6) can be regarded as a symmetric L_2 version of the embedding distance we have seen in Chapter 10 (here, we embed X into Y and vice versa). When we wish to compare parts of the shapes, we use a continuous equivalent of the weighted stress,

$$\sigma_2(m_X, m_Y, \varphi, \psi) =$$
$$\int_{X \times X} |d_X(x, x') - d_Y(\varphi(x), \varphi(x'))|^2 \, m_X(x) m_X(x') da \times da$$
$$+ \int_{Y \times Y} |d_Y(y, y') - d_X(\psi(y), \psi(y'))|^2 \, m_Y(y) m_Y(y') da \times da, \quad (11.7)$$

like we did in the case of the ICP algorithm. Note that the weight is *pointwise* rather than *pair-wise*. This means that if we remove a point (by setting its membership to zero), all the distances from this point are not taken into account in the stress computation.

From this point on, Algorithm 11.1 goes straightforwardly. In Step 2, we fix the fuzzy parts and find the best correspondences that minimize the stress $\sigma_2(m_X, m_Y, \varphi, \psi)$. Numerically, this is carried out using a weighted version of GMDS. Note that the variables in the two terms of the stress are decoupled: we can embed X into Y and Y into X independently, or in other words, find φ independently of ψ. Having fixed the correspondences, Step 3 is performed by minimizing $\sigma_2(m_X, m_Y, \varphi, \psi)$ over the fuzzy parts m_X and m_Y under the constraint $\tilde{\lambda}(m_X, m_Y) \leq \lambda_0$.

If one wishes to use the Gromov-Hausdorff distance rather than an L_2 criterion for the computation of partial intrinsic similarity, the derivation is a bit trickier. In the crisp case, the dissimilarity of the parts $d_{\text{GH}}(X', Y')$ is expressed in terms of the distortions of the maps between the parts ($\varphi : X' \to Y'$ and $\psi : Y' \to X'$). In the fuzzy setting, we have to rewrite the Gromov-Hausdorff distance using maps defined on the entire objects ($\varphi : X \to Y$ and

$\psi : Y \to X$). It can be done in the following way,

$$\tilde{d}_{\text{GH}}(m_X, m_Y) = \frac{1}{2} \inf_{\substack{\varphi: X \to Y \\ \psi: Y \to X}} \sigma_{\text{GH}}(m_X, m_Y, \varphi, \psi), \qquad (11.8)$$

where

$$\sigma_{\text{GH}}(m_X, m_Y, \varphi, \psi) = \max \left\{ \begin{array}{l} \sup_{x,x' \in X} m_X(x) m_X(x') |d_X(x,x') - d_Y(\varphi(x), \varphi(x'))| \\ \sup_{y,y' \in Y} m_Y(y) m_Y(y') |d_Y(y,y') - d_X(\psi(y), \psi(y'))| \\ \sup_{\substack{x \in X \\ y \in Y}} m_X(x) m_Y(y) |d_X(x, \psi(y)) - d_Y(\varphi(x), y)| \\ D \sup_{x \in X} (1 - m_Y(\varphi(x))) \, m_X(x) \\ D \sup_{y \in Y} (1 - m_X(\psi(y))) \, m_Y(y) \end{array} \right\}, \qquad (11.9)$$

and $D \geq \max\{\text{diam}(X), \text{diam}(Y)\}$. When m_X and m_Y are crisp parts, such a definition is just an equivalent way to write the Gromov-Hausdorff distance (see Problem 11.3). The computation of the partial similarity is performed in the same way, with $\sigma_2(m_X, m_Y, \varphi, \psi)$ replaced by $\sigma_{\text{GH}}(m_X, m_Y, \varphi, \psi)$.

Example 11.3 (intrinsic partial similarity of mythological creatures). As an example of intrinsic partial similarity computation, we show a numerical experiment performed on the set of mythological creatures. The set contains five different objects (horse, centaur, seahorse, male, and female). Each object appears in five instances, representing its nearly-isometric deformations (Figure 11.9). We compare two criteria: full intrinsic dissimilarity (the Gromov-Hausdorff distance d_{GH}) and partial intrinsic dissimilarity (scalar partial dissimilarity $\tilde{d}_{\text{SP}}$, computed using the fuzzy approximation scheme described above). Both criteria are computed using 50 samples on the shapes. The matching results are visualized in Figure 11.10 as dissimilarity matrices (the color of each element in the matrix represents the dissimilarity; the darker the smaller). Being an intrinsic criterion of similarity, the Gromov-Hausdorff distance captures the intra-class similarity of shapes (i.e., that different instances of the same objects are similar). However, it fails to adequately capture the inter-class similarity: the centaur, horse, and seahorse appear as dissimilar. On the other hand, the partial similarity captures correctly both the intra- and inter-class similarity (the centaur, horse, and seahorse are similar).

11.7 Not only size matters

Thus far, in defining the significance of parts we considered only their size, trying to find the largest and most similar parts, without saying anything

11.7 Not only size matters   233

Figure 11.9. Three-dimensional mythological creatures data set.

about what these parts look like. It appears that in some cases, such a formulation may tend toward preferring to select a large number of disconnected parts, small in size and similar on one hand and summing up to a large area on the other (for instance, this is a behavior observed in Example 11.2). The conclusion is that in the problem of partial similarity, not only size matters: the definition of significance should account not only for the area but also for

234   11 Partial Similarity

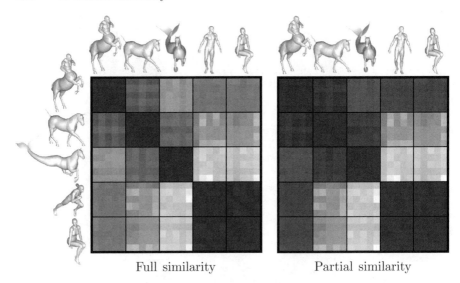

Figure 11.10. Intrinsic similarity of mythological creatures computed using d_{GH} and $\tilde{d}_{\mathrm{SP}}$. Shown are dissimilarity matrices (darker colors mean smaller dissimilarity).

the "quality" of the parts [55]. For example, we would probably prefer a single part than multiple parts even with larger area. To quantitatively express the "quality" of the parts, we can define an *irregularity* function $r(X')$ and add it as the third criterion into our vector-valued objective function. This new multicriterion optimization problem requires simultaneous minimization of dissimilarity, partiality, and irregularity $r(X') + r(Y')$. The Pareto frontier in this case becomes a surface in $\mathbb{R}^3$ (Figure 11.11).

The most straightforward definition of the irregularity is the boundary length: given a part X' with boundary $\partial X'$, we define irregularity as

$$r(X') = \int_{\partial X'} d\ell. \tag{11.10}$$

In case of two-dimensional shapes, fixing the area of the part and minimizing the boundary length will produce a circle, which is the shape with the smallest possible perimeter to area ratio and has the highest regularity.[10]

Unfortunately, this result cannot be immediately generalized to non-Euclidean manifolds, such that small boundary length of a part of a manifold does not necessarily guarantee that the part is "good." It may appear, for example, that a part consisting of a single connected component and another one consisting of multiple disconnected components have equal area and boundary length. In this case, we can use a topological regularity criterion, defined as the genus of X', which, as we have seen in Chapter 2, can be expressed as

$$r(X') = 1 - \frac{1}{2}\int_{X'} K(x)dx - \frac{1}{2}\int_{\partial X'} \kappa_g(x)d\ell,$$

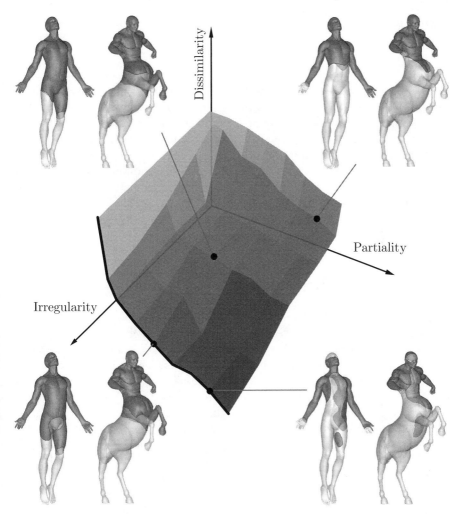

Figure 11.11. Three-dimensional Pareto frontier in the space of the dissimilarity, partiality, and irregularity criteria.

according to the Gauss-Bonnet theorem. Such a definition penalizes parts having holes or multiple disconnected components (and, thus, larger genus), and provide the desired topological regularity.

In the fuzzy case, there is no "boundary" in the strict sense. However, we can replace integration along the boundary by integration of the band in which the membership function changes from small to large values. The regularity criterion (11.10) can be approximated as

$$\tilde{r}(m_X) = \int_X \delta(m_X(x) - 0.5) \, \|\nabla_X m_X(x)\| \, dx, \qquad (11.11)$$

where δ denotes the Dirac delta function, in practice approximated by

$$\delta(t) \approx \frac{\epsilon}{\pi(\epsilon^2 + t^2)},$$

and $\nabla_X m_X(x)$ is the intrinsic gradient of m_X at the point x. The quantity $\|\nabla_X m_X(x)\|$ can be thought of as the length of the extrinsic gradient vector $\nabla_{\mathbb{R}^3} m_X$ projected on the tangent space of X at a point x.

Regularization of the form (11.11) was introduced by Mumford and Shah [288] and used for segmentation problems in computer vision [96]. In our formulation, it can be thought of as an extension of the Mumford-Shah functional to general non-Euclidean manifolds. The influence of such regularization on the selected part shape is depicted in Figure 11.12. Concluding this chapter, we should note that the definition of significance of parts still remains an open question.

Suggested reading

A good review of early shape similarity methods, including the problem of shape partitioning, is presented in the paper of Basri et al. [19]. For a discussion on partial shape similarity, we refer the reader to Latecki et al. [241]. A generic framework of partial similarity of objects, going beyond geometric shapes, is presented in [58]. For additional details on numerical implementation and examples, the reader is referred to [57] and [56, 59] (dealing exclusively with two-dimensional object recognition). A more extensive discussion of regularized partial similarity can be found in [55]. For classic references on multicriterion optimization, see [333, 303]; a more recent reference is the book by Miettinen [274]. Boyd and Vandenberghe [46] present the notion of Pareto optimality briefly but comprehensively. As an introduction to fuzzy logic and fuzzy set theory, we refer to [229, 416]. An interesting view on fuzzy sets from the perspective of metric geometry is presented in [136].

Problems

11.1. Show that a Salukwadze optimum is also a Pareto optimum.

11.2. Show that characteristic functions representing crisp parts belong to M_X.

11.3. Show that the definition (11.8) is equivalent to the Gromov-Hausdorff distance for crisp parts.

11.4. What will be the consequence of choosing large values of D in the computation of the fuzzy Pareto distance? Explain.

11.7 Not only size matters 237

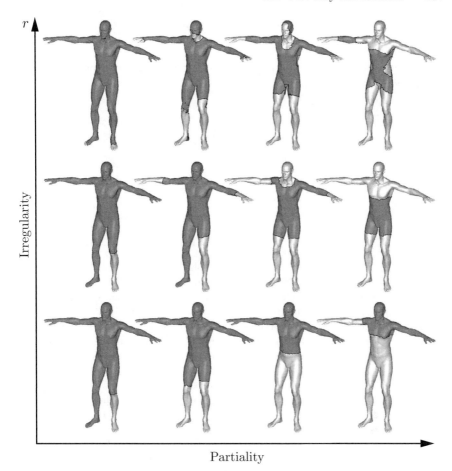

Figure 11.12. The selected part in Example 11.2 for different values of partiality and irregularity. Observe that for stronger regularization (or, equivalently, smaller irregularity), the part moves to decrease the boundary length at the expense of dissimilarity.

11.5. Show that if $D = \max\{\operatorname{diam}(X), \operatorname{diam}(Y)\}/\theta(1-\theta)$, where $0 < \theta < 1$ is a parameter, the following relation between d_{P} and $\tilde{d}_{\mathrm{P}}$ holds:

$$\tilde{d}_{\mathrm{P}}(X, Y) \leq \bigl((1-\theta), \theta^{-2}\bigr) \cdot d_{\mathrm{P}}(X, Y),$$

where the inequality is understood in the vector sense.

11.6. In lossy image and video compression, a fundamental problem is the trade-off between the amount of information used to describe the data (rate) and the amount of error introduced by the compression process (distortion). Describe the rate-distortion optimization problem from the perspective of partial similarity. Prove that the Pareto frontier (the rate-distortion curve) is convex.

11.7. Show a closed form solution for the best rigid motion between the two sets of corresponding points $\{x_i\}_{i=1}^N$ and $\{y_i\}_{i=1}^N$, minimizing the weighted point-to-point distance

$$\sum_{i=1}^N w_i \|x_i - (Ry_i + t)\|_2^2,$$

where w_i are non-negative weights.

Notes

[1] The example of centaurs was given by Jacobs et al. [213].

[2] Distances not satisfying the triangle inequality are generally referred to as *prametrics*. Psychological research (see, e.g., [385]) shows that such type of distances can be used to describe human reasoning when judging similarity.

[3] The paper by Latecki et al. [241] discusses two-dimensional shapes represented as polylinear curves and uses the similarities defined respectively, though their concepts can be generalized to generic shape representation and other similarities, as we do here.

[4] Technically, we require Σ_X to be a σ-algebra (a subset of 2^X closed under complement and countable union) in order to guarantee that the shape simplification is well defined. This requirement guarantees that X is a part of itself and that X'^c is also a part.

[5] In [59], we used a frivolous cartoon *Temptation* by the Danish cartoonist Herluf Bidstrup in order to exemplify an everyday situation in which partial similarity could be misleading. It was removed from the book due to copyright considerations.

[6] Multicriterion optimization problems are widely known in information theory. For example, the bias and the variance of an estimator in statistical estimation or distortion and bitrate in lossy signal compression [130] can be considered as two competing criteria that should be simultaneously minimized.

[7] In some references, Pareto optimum is also referred to as *Edgeworth-Pareto optimum*, crediting the Irish economist Francis Ysidro Edgeworth (1845–1926), whose mathematical formalism presented in the book *Mathematical Psychics* [145] was later adopted by Pareto.

[8] Sometimes alternatively transliterated as Salukvadze.

[9] The number of possible subsets of a discrete surface sampled at N points is 2^N.

[10] This result is known as *isoperimetric inequality*, stating that $4\pi a \leq p^2$, where p is the perimeter of a closed curve and a is the area enclosed by it. Equality $4\pi a = p^2$ is realized on a circle.

12

Non-rigid Correspondence and Calculus of Shapes

> Of bodies changed to various forms, I sing.
>
> OVID, *Metamorphoses*

In Chapter 1, we mentioned two main problems in the analysis of non-rigid shapes: similarity and correspondence. Thus far, our main focus was on finding similarity between objects. Now, we finally arrive to correspondence problems. Automatic computation of correspondence is an important active research field in computer graphics and computational geometry. In simple terms, the correspondence problem consists of finding a mapping between two shapes that copies similar features to similar features. The term "similar features" in this context has a semantic rather than geometric meaning. For example, we have no doubt how a "natural" correspondence between a cat and a dog should look like, as both have two ears, two eyes, a nose, four legs, and a tail. At the same time, it would probably be more difficult to consent about a natural correspondence between a dog and a flamingo – because the bird has only two legs whereas the dog has four, it is not clear, for example, which part of the dog's body corresponds with the bird's wings. In computer graphics, aesthetic considerations are often applied in such cases to judge the quality of the correspondence.

Though the general correspondence problem is not well-defined mathematically, restricting the surfaces of interest to a sufficiently narrow class of similar objects increases the chances to express the problem in purely geometric terms. In Chapter 6, we have already encountered the rigid correspondence problem when discussing iterative closest point algorithms. Here, we are going to explore techniques to construct correspondences between non-rigid objects and study several applications that rely on them. We will also see that correspondence provides a tool to synthesize new objects and construct a calculus of non-rigid shapes, which allows us to apply signal processing-like operations to shapes.

12.1 Intrinsic parameterization

As a motivating example, we first go to a lower dimension and start our discussion from correspondence between curves (one-dimensional manifolds). Let us be given two planar curves $\alpha : [0, a] \to \mathbb{R}^2$ and $\beta : [0, b] \to \mathbb{R}^2$. Our goal is to establish a bijective mapping between α and β copying corresponding points on the two curves. Recall from Chapter 2 that a curve can be parameterized so that

$$\int_{t_0}^{t_1} \|\dot{\alpha}(t)\| dt = t_1 - t_0,$$

for any $t_0, t_1 \in [0, L]$, where L is the length of α. Such a parameterization is called the *arclength* parameterization, as the parameter t measures the distance traveled along the curve. A handy property of arclength parameterization is that it is unique for open curves and unique up to selection of the starting point and direction for closed curves. For this reason, it can be considered as a *canonical parameterization*. Therefore, by reparameterizing α and β using the arclength parameterization, the correspondence problem boils down to finding an isometry $\xi : [0, L] \to [0, L]$ between the two parameterization intervals.

The reader may protest that whereas the example of curves may be insightful in visualizing rigid correspondence, it is over-simplistic and has little sense in the non-rigid setting. Non-rigid deformations of curves are tremendously boring, as all curves are isometric either to a straight line or to a circle of corresponding length. Following the analogy of curves, we could try extending our reasoning to surfaces, attempting to find a canonical parameterization for X and Y. Unfortunately, such a parameterization does not exist for surfaces. The arclength parameterization of curves was possible thanks to the existence of a total order relation between real numbers. Because no such order exists in $\mathbb{R}^2$, there is no analogy of arclength parameterization for surfaces or higher-dimensional manifolds.

Nevertheless, we can still construct an *intrinsic* parameterization of the surface, which will be independent of the extrinsic geometry and, consequently, isometry-invariant. Formally, given a surface X and a parameterization domain U, our goal is to devise a procedure for constructing a bijective mapping $\pi_X : U \to X$, such that for every isometric deformation f applied to X,

$$f \circ \pi_X = \pi_{f(X)}, \tag{12.1}$$

where $(f \circ \pi_X) = f(\pi_X(u))$ denotes function composition. In their study on texture mapping problems, Zigelman et al. [415], motivated by Schwartz et al. [345], proposed to embed the surface into $\mathbb{R}^2$, creating a minimum-distortion parameterization. This approach can be expressed as the familiar minimization problem,

$$\psi = \arg \min_{\psi : X \to \mathbb{R}^2} \text{dis } \psi,$$

and solved using an MDS algorithm.

The embedding ψ can be interpreted as an inverse parameterization π_X^{-1}, and its image $U = \psi(X)$ as the parameterization domain. The process resembles the computation of the canonical form discussed in Chapter 7, with the only exception that here the embedding is performed into a two-dimensional rather than a three-dimensional Euclidean space. Because the MDS problem is expressed only in terms of geodesic distances on X, the parameterization created this way is intrinsic.

As in the case of canonical forms, the minimum distortion parameterization is defined up to the isometry group in $\mathbb{R}^2$, namely, translations, planar rotations, and reflections. Therefore, in order to find correspondence between X and Y, their intrinsic parameterizations have to be aligned by finding a congruence (i.e., a Euclidean isometry) $\xi : \mathbb{R}^2 \to \mathbb{R}^2$, which brings the two parameterizations into correspondence. The mapping $\varphi : X \to Y$ is then expressed as $\varphi = \pi_Y \circ \xi \circ \pi_X^{-1}$ [214].

12.2 An image processing approach

A potential difficulty of using planar embedding for surface correspondence stems from the fact that generally the surfaces are not isometric to the plane, and like in the case of canonical forms, the embedding introduces inevitable distortion. As a consequence, relation (12.1) holds only approximately, and the computed correspondence may be inaccurate, as generally there exists no congruence ξ bringing the two intrinsic parameterizations into a perfect alignment.

The problem can be overcome by allowing ξ to be an arbitrary bijective mapping between the two parameterization domains of X and Y. This can be thought of as creating a low-distortion parameterization of X and Y, followed by translating the correspondence problem into the parameterization domain. Apparently, this reformulation does not make the problem easier, yet, now the correspondence can be computed in the Euclidean parameterization domain, which is usually easier to handle. Many correspondence techniques follow this strategy, differing mainly in the way the map ξ is found [3].

Because the computation is performed in the two-dimensional parameterization domain rather than in $\mathbb{R}^3$, the advantage of such approaches becomes especially pronounced when the surfaces are represented as geometry images, that is, are parameterized and sampled on a regular Cartesian grid. In their 2005 paper, Litke et al. [254] observed that well-established matching and registration methods from image processing can be used to find the correspondence between two shapes. It is common to find the correspondence between two grayscale images I and J by minimization of the least-squares error between I and the warped version of J,

$$\min_{\xi} \int (I(u) - J \circ \xi(u))^2 du$$

where the integration is performed in the image domain. Problems of this type arise, for example, in the computation of *optical flow* in video sequences [209] and *disparity estimation* in stereoscopic imaging [396, 261]. Following the image processing spirit, the pursuit for the best correspondence between two surfaces can be formulated as minimization of some *local matching error* $e(u,\xi)$ between $\pi_X^{-1}(X)$ and $(\pi_Y \circ \xi)^{-1}(Y)$, which gives rise to the following minimization problem,

$$\min_{\xi:U_X \to U_Y} \int_{U_X} e(u,\xi) du. \qquad (12.2)$$

Here, $U_X = \pi_X^{-1}(X)$ and $U_Y = \pi_Y^{-1}(Y)$ denote the parameterization domains of X and Y, respectively. Note that integration over U_X is merely an example; other error criteria such as the L_∞ norm can be used instead. The local match error $e(u,\xi)$ measures the degree of dissimilarity between the point $\pi_X(u)$ on X and the point $\pi_Y \circ \xi(u)$ on Y, based on some local features.

Note that the functional (12.2) and its minimizer depend on the parameterization. While in image processing this is less an issue as the parameterization is fixed, in our setting we have the freedom of choosing the surface parameterizations (ideally, but not necessarily, the parameterizations should have as low distortion as possible). In order to make the minimization problem independent of the choice of π_X and π_Y, we have to substitute all coordinate-dependent quantities by their intrinsic counterparts. For example, the Euclidean differential element du has to be replaced by the differential area element $da = \sqrt{\det G_X} du$ on X, where G_X denotes the first fundamental form of X in the coordinates of the parameterization. This yields the following minimization problem,

$$\min_{\xi:U_X \to U_Y} \int_{U_X} e(u,\xi) \sqrt{\det G_X} du. \qquad (12.3)$$

Clearly, the local match error e should also be parameterization-independent.

Various local matching errors exist in the literature. For example, Surazhsky and Elber [368] proposed to use the inner product between the unit surface normals,

$$e(u,\xi) = 1 - \langle N_X(\pi_X(u)), N_Y(\pi_Y \circ \xi(u)) \rangle_{\mathbb{R}^3}, \qquad (12.4)$$

as a degree of surface mismatch. However, while being invariant to translation and scaling, normals are sensitive to non-rigid deformations. In order to obtain an isometry-invariant local match error, an intrinsic geometric quantity has to be used. Recall that in Chapter 2 we have encountered one such quantity – the Gaussian curvature. A Gaussian curvature-based local match error,

$$e(u,\xi) = (K_X(\pi_X(u)) - K_Y(\pi_Y \circ \xi(u)))^2, \qquad (12.5)$$

is truly invariant to isometries. Its disadvantage is the fact that because the curvature is a second-order quantity, it is more sensitive to noise. In general,

$e(u, \xi)$ is not necessarily limited to geometric features; for example, when X and Y are endowed with photometric properties, those as well could be incorporated into the matching error. A remarkable use of photometric information for registration of human faces is made in the studies by Vetter and Blanz [36, 37, 35].

It is a well-known fact that for most reasonable types of $e(u, \xi)$, the problem (12.3) is *ill-posed* due to the richness of the space of maps $\xi : U_X \to U_Y$. This irregularity is manifested in the possibility to find a very irregular map ξ giving a good local match, yet a poor global match. In such cases, we would prefer to slightly compromise the local match in order to have a more regular ξ. A "healthy" solution to ill-posed problems is generally found by introducing a *regularization* into the cost function, whose goal is to penalize for irregular solutions. Minimization problem (12.3) can be therefore rendered well-posed by solving the regularized problem,

$$\min_{\xi:U_X \to U_Y} \int_{U_X} e(u,\xi)\sqrt{\det G_X}\,du + R(\xi). \tag{12.6}$$

The term $R(\xi)$ is scaled to control the degree of regularization.

The regularizer $R(\xi)$ conceals our prior knowledge on what a good correspondence ξ should look like. Because there exists no universal recipe to select it, different regularizations are possible. In image processing, one of the most common choices for $R(\xi)$ is

$$R(\xi) = \int \|\nabla \xi\|_F^2 du = \int (\|\nabla \xi^1\|^2 + \|\nabla \xi^2\|^2)du. \tag{12.7}$$

Here, $\xi = (\xi^1, \xi^2)^T$, the 2×2 matrix

$$\nabla \xi = \begin{pmatrix} \frac{\partial \xi^1}{\partial u^1} & \frac{\partial \xi^2}{\partial u^1} \\ \frac{\partial \xi^1}{\partial u^2} & \frac{\partial \xi^2}{\partial u^2} \end{pmatrix},$$

is the Jacobian of ξ, and

$$\|\nabla \xi\|_F^2 = \mathrm{trace}((\nabla \xi)^T \nabla \xi) = \|\nabla \xi^1\|_2^2 + \|\nabla \xi^2\|_2^2,$$

is its Frobenius norm.

As before, in order to make this regularizer parameterization-invariant, the differential element du has to be replaced by $\sqrt{\det G_X}\,du$, and the Frobenius norm $\|\nabla \xi\|_F^2$ has to be replaced by its intrinsic counterpart,

$$\|\nabla \xi\|_{\mathrm{HS}}^2 = \mathrm{trace}(G_X^{-1}(\nabla \xi)^T (G_Y \circ \xi) \nabla \xi),$$

called the *Hilbert-Schmidt norm*.[1] Informally, this norm can be interpreted as replacing the gradient $\nabla \xi$ by the intrinsic gradient and the standard Frobenius norm by the Frobenius norm in the tangent space of Y at $\xi(u)$. The obtained regularization functional,

$$R(\xi) = \int_{U_X} \|\nabla \xi\|_{\mathrm{HS}}^2 \sqrt{\det G_X} \, du, \tag{12.8}$$

is independent of the choice of parameterization, and can be rewritten in terms of $\varphi : X \to Y$ as

$$R(\varphi) = \int_X \|\nabla_X \varphi\|_Y^2 \, da. \tag{12.9}$$

All quantities are now intrinsic, meaning that $R(\varphi)$ is also invariant to isometries of X and Y.

Litke et al. [254] give a physical interpretation to this regularizer. In elasticity theory, the term $G_X^{-1}(\nabla \xi)^{\mathrm{T}}(G_Y \circ \xi)\nabla \xi$ is known as the *Cauchy-Green deformation tensor*, which expresses the square of local changes in distances due to an elastic deformation. Its trace, expressed by $\|\nabla \xi\|_{\mathrm{HS}}^2$, measures the square of the average local change of length due to the deformation created by pressing a thin rubber shell X into a mold having the form of Y. R itself is called the *Dirichlet energy* functional and measures (up to a factor of $\frac{1}{2}$) the elastic energy of the deformation.

12.3 Minimum distortion correspondence

The choice of an intrinsic local match error, like the one based on the Gaussian curvature, and an intrinsic regularizer like (12.9), make the minimization problem (12.6) invariant to isometries. By combining the two terms of the minimized functional, we can formulate a general isometry-invariant correspondence problem as

$$\min_{\varphi:X \to Y} \int_X e(x, \varphi) \, da,$$

where $e(s, \varphi)$ can be expressed as

$$e(x, \varphi) = \int_X \sigma(d_X(x, x'), d_Y(\varphi(x), \varphi(x'))) \, da,$$

with $\sigma(d_X, d_Y)$ depending only on the metrics d_X and d_Y. Consequently, our correspondence problem assumes the form

$$\min_{\varphi:X \to Y} \int_{X \times X} \sigma(d_X(x, x'), d_Y(\varphi(x), \varphi(x'))) \, da \times da. \tag{12.10}$$

This minimization problem can be recognized as a GMDS problem with some generalized distortion expressed as the integral over $X \times X$, in which σ can be interpreted as a local stress. For the particular choice of $\sigma(d_X, d_Y) = (d_X - d_Y)^2$, the familiar least-squares version of GMDS is obtained.

Figure 12.1. A minimum distortion correspondence between different postures of a male figure computed using GMDS. Corresponding regions on the meshes are denoted with the same colors (see insert for image in color). Note that the correspondence is defined up to a self-isometry (symmetry) of both shapes; for example, the correspondence between the two left-most and the two right-most shapes is reflected.

If previously we were looking for the distortion dis φ, which quantified the intrinsic dissimilarity of the two objects, we are now interested in φ itself, or formally,

$$\varphi = \arg \min_{\varphi: X \to Y} \text{dis } \varphi. \quad (12.11)$$

We call this map the *minimum-distortion correspondence* between X and Y. An example is shown in Figure 12.1. Note that the computation of φ can be performed either directly on the surfaces themselves or in the parameterization space if X and Y admit a global parameterization. The map φ can also be interpreted as a parameterization of Y in X.

Minimum distortion correspondence is not limited to isometric surfaces. Many classes of objects though not isometric share common (intrinsic) geometric properties – for example, two different faces have nose, eyes, mouth, and so forth. Finding the minimum distortion correspondence between two faces can be visualized as the problem of putting a flexible rubber mask over one face, trying to minimize its stretch. Obviously, in most cases we will place the mask in such a way that the facial features coincide, at least roughly, so that the correspondence is semantically correct. Practice shows that even when dealing with two substantially different objects, the minimum distortion map usually gives a reasonable correspondence (Figure 12.2). The approach can also be applied to partially similar objects using the scheme described in Chapter 11. An example is shown in Figure 12.3.

It is worthwhile noting that any intrinsic correspondence, including the one computed using GMDS, is defined up to the isometry groups of the two

Figure 12.2. A minimum distortion correspondence between male, female, and gorilla figures computed using GMDS. Corresponding regions on the meshes are denoted with the same colors (see insert for image in color). The male-female correspondence (left) appears accurate (up to a symmetry reflecting the left and the right) despite the differences in the intrinsic geometry. On the other hand, the minimum distortion correspondence between the male and the gorilla figures fails, as the human legs are mapped to the elongated gorilla's hands (second from the right). Manually fixing four semantically correct landmarks in GMDS initialization fixes the correspondence (right).

surfaces, meaning that if one or both surfaces have symmetries, there will be multiple mappings $\varphi : X \rightarrow Y$ achieving the minimum distortion. In such a case, there is no way to give any preference to one of these correspondences based on the intrinsic geometry only, and we have to introduce extrinsic or other (e.g., photometric) information in order to make the right choice. For example, our face is known to have an approximate reflection symmetry with respect to the vertical axis. Consequently, there may exist two minimum distortion correspondences, one mapping the left eye of X to the left eye of Y, and another mapping the left eye of X to the right eye of Y (Figure 12.1). Adding an extrinsic constraint demanding that the correspondence does not change the surface *orientation* would exclude the second possibility.

12.4 Texture mapping and transfer

Imagine that an animation studio wishes to create a synthetic animation character for its new movie. Developing an animation system that will allow the character's face to reproduce basic human expressions is a Sisyphean work,

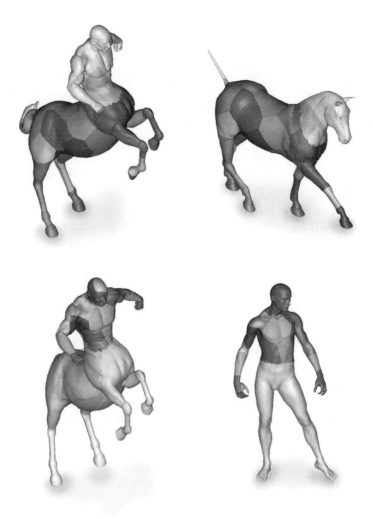

Figure 12.3. Partial minimum distortion correspondence. See insert for image in color.

requiring months of manual labor of a group of artists. Because of budget considerations, the producer may choose a much cheaper approach: scan a human actor using a real-time three-dimensional scanner[2] and create the synthetic character by drawing his features on the actor's face. Such a "virtual makeup" is achieved by means of *texture mapping* and can create a realistic face.

Consider a three-dimensional video (see an example in Figure 12.4) as a sequence of surfaces X_t, where t represents the time index. Our goal is to map

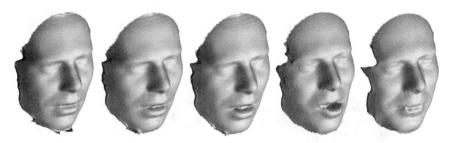

Figure 12.4. An example of a three-dimensional video acquired with a coded light range scanner at 3 frames per second. Photometric information (texture) is not shown.

a single texture image onto all X_t, which essentially consists of drawing a high resolution RGB image containing the "texture" of the personage and finding a set of correspondences that maps every point on the scanned surfaces into image coordinates.[3] Practically, this can be thought of as attaching a set of vector fields $\rho_t : X_t \to \mathbb{R}^3$ to the surfaces X_t.

Despite some automation provided by modern 3D modeling tools, this process is a manual work performed by an artist, usually taking a non-negligible amount of time. Although it is feasible to manually map the texture onto a single surface, it is clearly impractical to do such a work for each frame in the scanned sequence, which may easily contain thousands of frames. However, if we were able to automatically compute a correspondence $\varphi_t : X_t \to X_0$ between a frame X_t and the first frame X_0, to which a texture $\rho_0 : X_0 \to \mathbb{R}^3$ was mapped manually, we could generate a sequence of textures $\rho_t = \rho_0 \circ \varphi_t$ for all the frames X_t. The correspondence allows one to *transfer* the texture from the manually created reference surface to the entire sequence (Figure 12.5). Because the expressions of the human face are approximately isometries, the minimum distortion correspondence generated by GMDS accurately maps the features throughout the entire sequence.

The same technique can be applied to other objects, for example the human body, which fits the isometric model even better than does the human face. Imagine that we would like a human actor to be used as a character in a movie. The actor is first scanned in several poses, then, an artist draws the texture that should be mapped onto the character. In order to avoid drawing a different texture for each pose, the texture from some reference pose has to be transferred to the rest of the poses of the character (Figure 12.6). Following the "virtual makeup" analogy, such texture transfer can be thought of as a computer-age version of *body painting*, a contemporary stream of art of drawing clothes directly on the skin, in order to create an illusion of a genuine dress. When the person moves, the drawn picture deforms naturally with the skin, thus looking realistic practically in every pose of the body.

Texture transfer can also be used for the reverse problem of mapping the texture of a human actor onto a synthetic animated personage, or "trans-

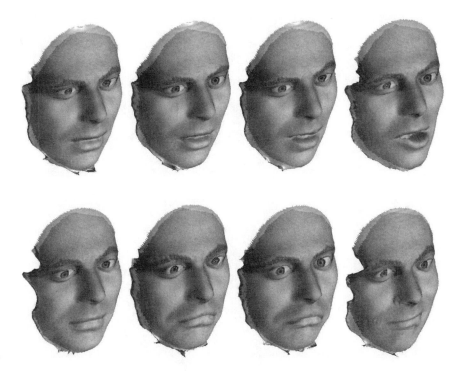

Figure 12.5. A virtual makeup experiment from [72]. A green texture image (inspired by the DreamWorks Shrek movie) is automatically transferred from a manually created reference frame using the minimum distortion correspondence established by GMDS. See insert for image in color.

plantation" of photometric properties from one face to another (Figure 12.7). Other surface attributes such as geometric details expressed as normal displacement maps can can be transferred as well [195].

12.5 Morphing

Another application heavily relying on the ability to find correspondence between objects is *morphing* or *metamorphosis* (from Greek μετά, "beyond," and μόρφως, "shape"), a seamless transformation of one object into another (Figure 12.8). Morphing techniques have become widespread in computer graphics, animation, and modeling, achieving spectacular results. One of such first impressive uses of digital morphing appeared in Michael Jackson's 1991 music video *"Black or white."* Although much research has been conducted in the field since then, the morphing process still requires a significant amount of manual work to achieve aesthetically pleasant results. Automation of digital morph remains a major research challenge.

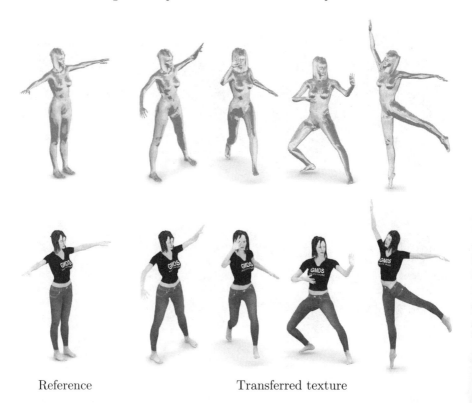

Reference   Transferred texture

Figure 12.6. A virtual body painting experiment from [72]. The texture is transferred from a reference pose of the human body (left column, outlined in gray) to its different poses. The correspondence between the objects is established by embedding 200 points on the reference object into its poses using the GMDS algorithm. See insert for image in color.

In our formulation, given the *source* and the *target* surfaces X and Y, a morphing process aims at finding a temporal sequence of *intermediate* surfaces, X_t ($t \in [0,1]$) such that $X_0 = X$ and $X_1 = Y$. Typically, morphing is performed in three steps. The first step consists of finding a *correspondence* relating points on X to points on Y. We have already seen that this can be done using GMDS, which finds the minimum distortion correspondence between the source and the target. Practically, we select a subset of N points $\{x_i\}$ on X and find a corresponding set of points $\{y_i\}$ on Y. Neither x_i, nor y_i have necessarily to be restricted to mesh vertices. Once the correspondence is found, we can work with two clouds of corresponding points $\{x_1, \ldots, x_N\}$ and $\{y_1, \ldots, y_N\}$ in $\mathbb{R}^3$. From this point on, we do not have to deal anymore with the intrinsic geometry of X and Y and will work only with their extrinsic geometries.

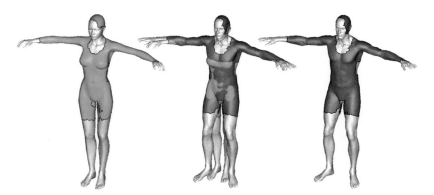

Figure 11.8. Extrinsic partial similarity of rigid shapes.

Figure 12.1. Minimum distortion correspondence.

Figure 12.2. Minimum distortion correspondence.

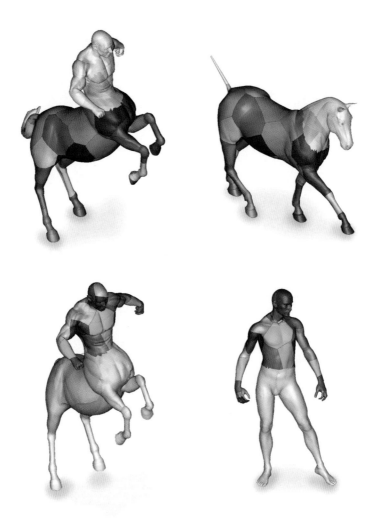

Figure 12.3. Partial minimum distortion correspondence.

Figure 12.8. Morphing of faces.

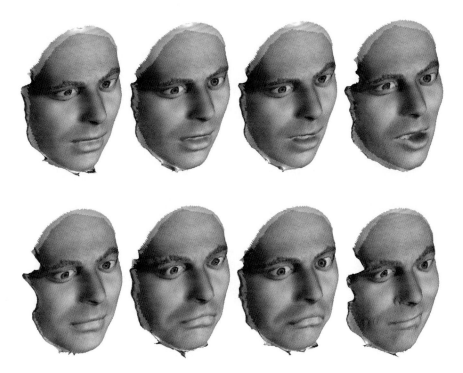

Figure 12.5. Virtual makeup.

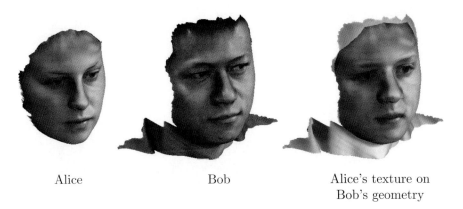

Alice Bob Alice's texture on Bob's geometry

Figure 12.7. Face texture substitution.

Reference   Transferred texture

Figure 12.6. Virtual body painting.

Figure 13.3. A face with different illuminations, head positions, and expressions.

Figure 13.4. Is our eye more sensitive to texture or geometry?

Alice Bob Alice's texture on Bob's geometry

Figure 12.7. In the texture substitution application, GMDS is used to find the minumum distortion correspondence between two faces. Using this mapping, the texture is transferred from one face to the other. See insert for image in color.

In the discrete setting where the surfaces are represented as triangular meshes, X and Y often have different number of vertices and different connectivity. Consequently, morphing should involve computations not only on the extrinsic geometry of the meshes but also on their topology. Given the source mesh $T_X(\{x_i\})$ and the target mesh $T_Y(\{y_i\})$, our second step consists of creating a *common mesh connectivity* T such that $T(\{x_i\})$ and $T(\{y_i\})$ are two valid meshes. Such meshes are usually referred to as *compatible* or *isomorphic*. In order to allow creation of compatible meshes, $T(\{x_i\})$ and $T(\{y_i\})$ have to be topologically equivalent (homeomorphic). For example, a morph between a sphere and a cube (which are topologically equivalent) is simple and intuitive, whereas a morph between a sphere and a torus (which have different topologies) is not well-defined. As most morphing techniques assume homeomorphic source and target shapes, we will stick to this setting, which appears to be still sufficiently wide.

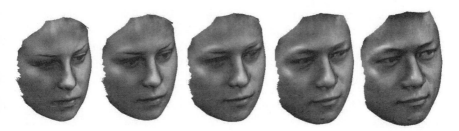

Figure 12.8. Morphing. In the example from Figure 12.7, the correspondence is used to transform the texture and the extrinsic geometry of the source (left) into the corresponding texture and extrinsic geometry of the target (right), creating a morphing effect. See insert for image in color.

One of the most frequently used approaches for finding a common connectivity of two homeomorphic meshes is *mesh overlay*, generating a supergraph of the two mesh connectivities [3]. We construct a new connectivity T, containing all edges and faces of T_X and T_Y, and add new vertices if some edges intersect. Mesh overlay guarantees that if a point lies on the original mesh $T_X(\{x_i\})$, the new mesh $T(\{x_i\})$ will also necessarily contain it. This does not take into account the fact that in most practical cases, a mesh is only a piecewise linear approximation of some smooth surface. As there is nothing special in this specific approximation, the construction by which the new mesh has to reproduce the given mesh may be disadvantageous. We can therefore reformulate the goal of our second step as finding a common connectivity T, such that the meshes $T(\{x_i\})$ and $T(\{y_i\})$ approximate respectively the surfaces X and Y themselves. This problem is known as *compatible meshing* (or *compatible re-meshing* if X and Y are already given as meshes). The exact description of mesh overlay and compatible meshing techniques is beyond the scope of this book. For an overview, the reader is referred to [3].

Once compatible meshes are created, we have to actually define the set of intermediate surfaces X_t. Therefore, the third step of a morphing process consists of creating a set of smooth *trajectories* $x_i(t)$ for $t \in [0,1]$, such that $x_i(0) = x_i$ and $x_i(1) = y_i$ for $i = 1, \ldots, N$. Intermediate surfaces are defined as the set of meshes $T(\{x_i(t)\})$. Computation of trajectories is where morphing techniques differ the most. As in the case of correspondence, there exists no formal definition of what a "good" trajectory should look like, and aesthetic criteria are often applied. The simplest solution to the trajectory problem is to displace each source point toward the corresponding target point with constant velocity along a straight line, creating the following linear trajectories,

$$x_i(t) = (1-t)x_i + ty_i. \qquad (12.12)$$

A morph generated in this manner is referred to as *linear*. It is optimal in the sense that linear trajectories are the shortest among all possible trajectories connecting x_i and y_i.

Unfortunately, linear morph often leads to undesirable results. For example, trajectories may intersect at some $t \in (0,1)$, leading to an invalid self-intersecting intermediate mesh (Figure 12.9, first row). One may argue that we can try to maintain the linear trajectories while relaxing the constant velocity constraint and allowing x_i to move with some variable velocity $v_i(t) > 0$, obeying

$$\int_0^1 v_i(\tau) d\tau = 1.$$

This yields straight trajectories of the following form,

$$x_i(t) = \left(1 - \int_0^t v_i(\tau) d\tau\right) x_i + \int_0^t v_i(\tau) d\tau \; y_i. \qquad (12.13)$$

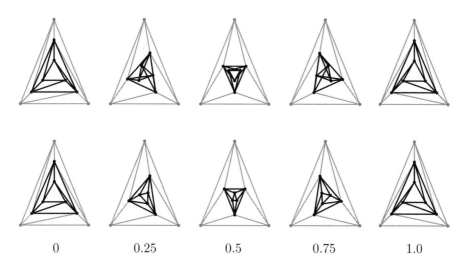

0	0.25	0.5	0.75	1.0

Figure 12.9. First row: linear morph produces a sequence of invalid self-intersecting intermediate meshes. Second row: guaranteed self-intersection free morph obtained by linear interpolation of barycentric coordinates.

However, it appears that this more general setting cannot completely avoid self-intersections of the intermediate meshes [161]. Even when no such catastrophes happen, linear morph often produces intermediate surfaces, which differ significantly from the source and the target.

Example 12.1. We show an example when linear morph fails, following the scenario described in Lemma 4.3 in [161]. Let $\{x_1, \ldots, x_N\}$ and $\{y_1, \ldots, y_N\}$ be two clouds of points with $x_1 = y_2$ and $x_2 = y_1$. Then, the trajectories

$$x_1(t) = (1-t)x_1 + ty_1 = (1-t)x_1 + tx_2;$$
$$x_2(t) = (1-t)x_2 + ty_2 = (1-t)x_2 + tx_1$$

intersect at $t = 0.5$, yielding an invalid morph. Considering a more general linear morph of the form (12.13), we define the weight functions,

$$w_i(t) = \int_0^t v_i(\tau) d\tau.$$

Here $w_i(t)$ are increasing continuous functions, obeying $w_i(0) = 0$ and $w_i(1) = 1$. Hence, $w(t) = w_1(t) + w_2(t)$ is an increasing continuous function, obeying $w(0) = 0$ and $w(1) = 2$. This implies that there exists $t_0 \in (0, 1)$, for which $w(t_0) = 1$. Consequently, the trajectories

$$x_1(t) = (1 - w_1(t))x_1 + w_1(t)y_1 = (1 - w_1(t))x_1 + w_1(t)x_2;$$
$$x_2(t) = (1 - w_2(t))x_2 + w_2(t)y_2 = (1 - w_2(t))x_2 + w_2(t)x_1$$

intersect at t_0.

12.6* Guaranteed self-intersection free morph

The fact that the linear morph as well as the majority of other morphing techniques are unable to guarantee the validity of intermediate shapes was probably the main consideration driving Floater and Gotsman to propose in their 1999 paper [161] a *guaranteed self-intersection free* morphing method. In their study, Floater and Gotsman considered the particular case of morphing between two compatible planar triangulations having a common convex boundary. In follow-up works by Gotsman and Surazhsky [182, 369, 370], the method was extended to arbitrary shapes.

The method requires both $\{x_i\}_{i=1}^N$ and $\{y_i\}_{i=1}^N$ to be bounded by a common convex shape. Let us assume that a set of points $\{x_{N+1}, \ldots, x_{N'}\}$ is found such that their *convex hull*

$$\mathrm{conv}(\{x_{N+1}, \ldots, x_{N'}\}) = \left\{ \sum_{i=N+1}^{N'} \lambda_i x_i : \sum_{i=N+1}^{N'} \lambda_i = 1 \right\} \quad (12.14)$$

contains both $\{x_i\}_{i=1}^N$ and $\{y_i\}_{i=1}^N$. We can think of two larger clouds of points $\{x_i\}_{i=1}^{N'}$ and $\{y_i\}_{i=1}^{N'}$ as of two sets of *interior* vertices $\{x_i\}_{i=1}^N$ and $\{y_i\}_{i=1}^N$ and a common convex boundary $\{x_i\}_{i=N+1}^{N'} = \{y_i\}_{i=N+1}^{N'}$. This allows us to express each interior vertex x_i as a strictly convex combination of its neighbors $\mathcal{N}_i$, that is, there exist some $\lambda_{ij} > 0$ (not necessarily unique) for all $i = 1, \ldots, N$ and $j \in \mathcal{N}_i$ such that $\sum_j \lambda_{ij} = 1$ and

$$x_i = \sum_{j \in \mathcal{N}_i} \lambda_{ij} x_j.$$

Using matrix notation, we can write

$$X_\mathrm{i} = \Lambda \begin{pmatrix} X_\mathrm{i} \\ X_\mathrm{b} \end{pmatrix} = (\Lambda_\mathrm{i}\ \Lambda_\mathrm{b}) \begin{pmatrix} X_\mathrm{i} \\ X_\mathrm{b} \end{pmatrix} = \Lambda_\mathrm{i} X_\mathrm{i} + \Lambda_\mathrm{b} X_\mathrm{b}, \quad (12.15)$$

where X_i and X_b are two $N \times 3$ and $(N' - N) \times 3$ matrices containing the interior and the boundary vertices, respectively, as the rows, and Λ is an $N \times N'$ matrix with non-negative elements, whose rows sum to one. Gotsman and Surazhsky refer to Λ as the *neighborhood matrix*, as it describes both the connectivity of the mesh and the position of the interior vertices in barycentric coordinates relative to their neighbors [182, 369].

Because the barycentric representation of the interior vertices is not unique, there exist many ways to construct a valid neighborhood matrix for a given mesh. On the other hand, it appears that $\Lambda_\mathrm{i} - I$ is non-singular, which implies that the equation

$$(I - \Lambda_{\text{i}})X_{\text{i}} = \Lambda_{\text{b}}X_{\text{b}} \tag{12.16}$$

has a unique solution [159, 160]. In other words, given the neighborhood matrix Λ and the boundary points X_{b}, the interior points X_{i} can be uniquely determined. Moreover, in [159], Floater proved that given *any* neighborhood matrix of *any* valid mesh with triangulation T, the interior vertices constructed by solving (12.16) form a valid mesh with the same triangulation.

This result allows us to create a self-intersection free morph. Given two clouds of corresponding points $\{x_i\}_{i=1}^{N}$ and $\{y_i\}_{i=1}^{N}$ with a compatible triangulation T and a common convex boundary $\{x_i\}_{i=N+1}^{N'}$, we first construct two neighborhood matrices Λ^0 and Λ^1 for $\{x_i\}$ and $\{y_i\}$, respectively. Next, instead of interpolating between x_i and y_i directly, we interpolate between their barycentric coordinates, creating a sequence of matrices $\Lambda(t)$. The simplest way to do so is by using linear interpolation,

$$\Lambda(t) = (1-t)\Lambda^0 + t\Lambda^1. \tag{12.17}$$

Observe that if Λ^0 and Λ^1 are valid neighborhood matrices of T, then $\Lambda(t)$ is also a valid neighborhood matrix of T. The sequence $\Lambda(t)$ uniquely defines a set of trajectories $x_i(t)$, obtained by solving the linear system (12.16). The morph is constructed from intermediate meshes $T(\{x_i(t)\})$, which are guaranteed to be valid triangulations (Figure 12.9, second row).

Currently, this appears to be the only mesh morphing technique with theoretical guarantee of non self-intersecting intermediate meshes. However, there exist other algorithms and heuristics that *practically* avoid self-intersections. For more details, the reader is referred to [353, 176, 334, 4]. The reader might be wondering about what happens to the intrinsic geometry of the shape undergoing a metamorphosis. Even if X and Y are isometric, generally there is no guarantee that the intermediate shapes will be isometric, as a continuous bending of X might not exist. In [224], Kilian *et al.* address this important question by proposing an *as isometric as possible* morphing technique. For details, the reader is referred to the original paper.

12.7 Calculus of shapes

By morphing between source and target shapes, intermediate shapes are created. This gives us a tool to *synthesize* new non-rigid objects. In a broader perspective, we can think of non-rigid objects as points in some infinite-dimensional space. Let X be a shape, and let $\mathbb{M}$ denote the shape space created by the deformations of X. It is known empirically that the dimensionality of $\mathbb{M}$ is usually low, and it can be represented approximately as some abstract manifold [377]. Let X_0 and X_1 be two deformations of X. If the difference between X_0 and X_1 is sufficiently small, we can linearize the manifold $\mathbb{M}$ around the point X_0, approximating its generally non-Euclidean structure

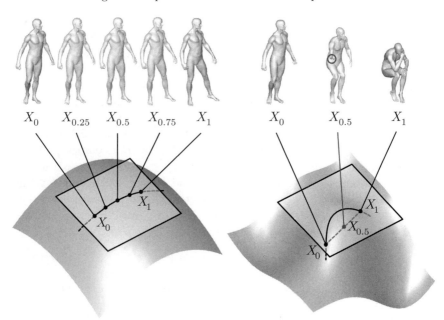

Figure 12.10. Geometric illustration of calculus of shapes. Surfaces can be represented as points in the shape space $\mathbb{M}$, visualized here as a two-dimensional surface. Left: by linearization around some point, an analog of a locally linear tangent space is created. Right: when X_0 and X_1 are significantly distinct, the convex combination $X_t = tX_1 + (1-t)X_0$ defined using linear morph may produce an invalid surface (point outside $\mathbb{M}$). A self-intersection free morph guarantees that X_t lies on $\mathbb{M}$.

by a Euclidean one (Figure 12.10, left). This construction resembles the notion of *tangent space* in Riemannian geometry, which allows us to consider surfaces as vectors in a locally linear space.[4] In other words, it provides us with a *calculus*, i.e., the ability to "add," "subtract," and "scale" surfaces.

The knowledge of a correspondence $\varphi : X_0 \to X_1$ between X_0 and X_1 is crucial in order to define operations on their extrinsic coordinates in $\mathbb{R}^3$. Once the correspondence is available, we can express the "difference" $dX = X_1 - X_0$ as

$$dx(x) = \varphi(x) - x. \qquad (12.18)$$

for all $x \in X_0$. Moving from X_0 to X_1 along dX can be expressed as the convex combination $X_t = X_0 + t\, dX$, or

$$x_t = x + t\, dx = t\,\varphi(x) + (1-t)x, \qquad (12.19)$$

which is nothing but a linear morph between X_0 and X_1. If in addition the surfaces are endowed with textures represented as vector fields $\rho_0 : X_0 \to \mathbb{R}^3$ and $\rho_1 : X_1 \to \mathbb{R}^3$, we can similarly construct the texture ρ_t by blending between the corresponding texture intensities, e.g.,

$$\rho_t(x) = t\rho_1 \circ \varphi(x) + (1-t)\rho_0(x). \qquad (12.20)$$

Again, the term "corresponding" is well-defined as soon as a correspondence has been established.

We have already seen that when the difference between X_0 and X_1 is large (i.e., dX is large), $X_0 + t\,dX$ defined as a linear morph can cease from being a valid surface. Geometrically, this can be interpreted as follows: because the manifold $\mathbb{M}$ is locally Euclidean, if X_0 and X_1 are sufficiently close, the straight line connecting them lies on $\mathbb{M}$. Yet, when X_0 and X_1 are distant, some points on that line may lie outside the manifold. The use of a self-intersection free morphing technique replaces the linear trajectory $X_0 + t\,dX$ by a non-linear one, guaranteeing that X_t is a valid surface. Geometrically, this is equivalent to traveling between X_0 and X_1 *on* the manifold $\mathbb{M}$ (Figure 12.10, right).

As an example visualizing the calculus of surfaces, consider a temporal sequence of deformations of X, which can be represented as a smooth trajectory X_t on $\mathbb{M}$. As in our previous examples, this can be a 3D video of a human actor acquired by a range scanner. In practice, only a discrete set of frames $X_{n\Delta t}$ can be acquired with finite sampling rate Δt, rarely exceeding 30 frames per second. Without loss of generality, we assume $\Delta t = 1$ and denote this sequence by X_n. Because our eyes are very sensitive to motion discontinuities, low frame rate sequences usually appear unpleasant to a human observer. A way to improve the visual experience is by creating a higher frame rate sequence by interpolating the "missing" frames. An intermediate frame $X_{n+t} : t \in (0,1)$ can be created as a convex combination $(1-t)X_n + tX_{n+1}$ between the two adjacent frames X_n and X_{n+1}, using either linear or non-linear morphing (see Figure 12.10 for a geometric illustration). The trajectory connecting two corresponding points on X_n and X_{n+1} describes the *motion* of a point on X_n, and interpolation along that trajectory creates an illusion of a smoothly moving surface.

In video processing, such an interpolation is often referred to as *frame rate up conversion* or *temporal super-resolution*, as it increases the sampling rate (or resolution) along the time axis.[5] Interpolating along the trajectory of a point is similar in its spirit to *motion-compensated* frame rate conversion techniques widely used for conventional two-dimensional video sequences. There, correspondence between points in adjacent frames is computed by means of a *motion estimation* algorithm, based essentially on finding the highest correlation between displaced regions in the two images [238, 100, 103, 243]. In this terminology, the use of non-rigid correspondence produces a *deformation-compensated* interpolation. Figure 12.11 shows an interpolated sequence between two frames in a 3D video.

The combination of two surfaces X_0 and X_1 does not necessarily have to be convex. Allowing for $t < 0$ or $t > 1$, we can *extrapolate* X_t beyond X_0 or X_1. As a particular example in the facial animation problem, if X_0 is a neutral posture of the face and X_1 is an expression, we can exaggerate this

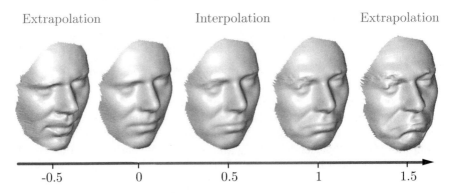

Figure 12.11. Expression interpolation between two frames in the video sequence is obtained by taking $t \in (0, 1)$. Allowing for $t < 0$ or $t > 1$, it is possible to extrapolate the expression, creating a caricaturization effect as shown on the right.

expression by constructing the non-convex combination $(1-t)X_0 + t X_1$ with $t > 1$ (Figure 12.11). The opposite process allows us to undo the expression.

Suggested reading

A survey of surface parameterization methods is available in Michael Floater's review papers [163, 162]. The book *Warping and Morphing of Graphical Objects* by Gomes [178] presents perhaps the widest perspective on morphing techniques in various computer graphics applications. A comprehensive overview of mesh correspondence and morphing techniques is presented in Alexa's review paper [3]. For more details on the guaranteed self-intersection free morphing technique discussed in this chapter, the reader is referred to the original paper by Floater and Gotsman [161] and follow-up works by Surazhsky and Gotsman [182, 370]. The "as isometric as possible" morphing technique is introduced in [224]. Theoretical foundation of shape spaces is presented systematically in [273]. For details on the texture transfer and calculus of shapes experiments, the reader is referred to [72]. An interesting point of view on geometry manipulation through the prism of signal and image processing was pioneered by Taubin [374, 375] and Schröder and Sweldens [344] under the name of *geometric signal processing* or *digital geometry processing*.

Software

Intrinsic correspondence computation by means of GMDS is available as a part of the TOSCA toolbox. The toolbox also contains Euclidean MDS routines, which can be used to create planar minimum distortion parameterization. The C++ GoTools parameterization library distributed by SINTEF contains an implementation of Floater's surface parameterization algorithms.

Problems

12.1 (Research question). Elaborate a way to transfer bump maps and normal displacement maps using the correspondence-based texture transfer technique.

12.2 (Research question). Devise a self-intersection free morphing technique, which given two ϵ-isometric surfaces X and Y, guarantees that all intermediate meshes are $c\epsilon$-isometric to X or Y.

12.3. Does the guaranteed self-intersection free morph guarantee valid shapes for $t < 0$ or $t > 1$?

Notes

[1] Formally, the Hilbert-Schmidt norm is defined for a linear operator $L : U \to V$ from a Hilbert space U to a Hilbert space V, as

$$\|L\|^2_{\text{HS}} = \sum_i \|Le_i\|^2_V,$$

where $\{e_i\}$ is an orthonormal basis in U. The norm is independent on the choice of this basis.

[2] Current technology allows the acquisition of three-dimensional geometry at video frame rates, see, e.g., [211].

[3] The texture image coordinate system (also known as the *UV coordinates*) is usually normalized between 0 and 1.

[4] More rigorously, we construct a locally *affine* space with the origin X_0.

[5] Clearly, temporal super-resolution creates new information from nowhere. Under the assumption of smooth motion, interpolated frames are usually close to the true "continuous" version of the sequence. However, if the trajectories are irregular between the samples, peculiar *aliasing* effects are likely to arise.

13
Three-dimensional Face Recognition

> "And in stature he is small, chest broad, one arm
> shorter than the other, blue eyes, red hair, a wart
> on his cheek, another on his forehead." Then is it
> not you, my friend?
>
> A. S. PUSHKIN, *Boris Godunov*

In this chapter, the theoretical and algorithmic apparatus we have developed so far will be used for an important practical application: biometric authentication. The term *biometric* refers to methods of identifying a person's identity according to individual characteristics of the human body, like fingerprints, iris texture, the vein pattern in the eye retina, or the face structure. Just a decade ago, biometrics were considered almost science-fiction technologies, seen in the James Bond movies. Today, many laptop computers are equipped with a fingerprint scanner, which is used as an alternative to typing a password.

Being apparently the fruit of the hi-tech era, biometric technologies have been exploited from the dawn of the human civilization [78], long before the famed Agent 007 was born. One of the oldest written testimonies of a biometric technology and the first identity theft dates back to the age of the Patriarchs, when Jacob fraudulently used the identity of his twin brother Esau to benefit from their father's blessing. The book of Genesis describes a multimodal biometric test comprising hand scan and voice recognition that Isaac, already blind at his age, used in an attempt to verify his son's identity, without knowing that the smooth-skinned Jacob had wrapped his hands in kidskin: "And Jacob went near unto Isaac his father; and he felt him, and said, 'The voice is Jacob's voice, but the hands are the hands of Esau.' And he discerned him not, because his hands were hairy, as his brother Esau's hands." [171]. The recognition error, as we know today, had consequences of historic proportions.

In our everyday routine, we use biometric technology without paying attention. From the very first years of infancy, our brain develops the ability to distinguish between faces of different people. A grown person is able to recognize thousands of individuals encountered during his life. It is not a surprise that law enforcement forces were among the first to take advantage of this most natural kind of biometrics. A legend accounts that the runaway French king Louis XVI was arrested after a clerk recognized his face on a gold coin (a nineteenth century engraving in Figure 13.2 shows this dramatic moment in

Figure 13.1. "And Jacob went near unto Isaac his father..." Gustave Doré's illustration of the biblical blessing scene, in which one of the earliest identity thefts known to human civilization occurred.

Figure 13.2. An engraving from Camille Pelletan's book *Les guerres de la Révolution* [309], showing the arrest of the French king Louis XVI.

the history of the French Revolution). In our days, in most countries citizens are obliged by law to carry a photo-ID, in order to allow the respective authorities to confirm the document bearer's identity by comparing the photograph with his or her face.

13.1 Some terminology

Though a routine task for humans, face recognition is thus far a great challenge for the machine. From the point of view of pattern recognition, face recognition is a classification problem. A generic face recognition algorithm tries to find similarity between the face of an enrolled person (referred to as *probe*) and a database of faces of known identities (*gallery*). For every entry in the gallery, a distance from the probe is computed. The best match (i.e., one with the smallest distance) below some threshold is then identified with the probe. This scenario is referred to as a *one-to-many* or *identification* problem. It can be applicable, for example, if face recognition is used to track employees of a large corporation. The recognition accuracy in this setting is measured as the fraction of correct identifications within the closest k matches (referred to as *rank-k recognition rate*). The plot of the recognition rate versus the rank is called the *cumulative match characteristic* (CMC).

A simpler setting is the *one-to-one* or *authentication* problem, in which the supposed identity of the subject is known in advance, and the comparison is performed only with one gallery entry corresponding with the claimed identity. The output in this case is binary: identity confirmed or not confirmed. As an application, one may think of an automated teller machine (ATM) that uses face recognition instead of typing a PIN code. The algorithm performance in this setting is quantified in terms of the *false acceptance* and *false rejection* rates (FAR and FRR, respectively). The first refers to the fraction of different subjects wrongly recognized as being the same, and the latter refers to the fraction of instances of the same subject wrongly recognized as being different. Obviously, the two criteria are competing. Defining a threshold on the distance value below which the probe is accepted, it is possible to trade off between the false acceptance and the false rejection rates. This trade-off can be represented graphically as a curve, referred to as the *receiver operating characteristic* (ROC) – a term dating back to the early 1940s in relation to radar detection efficiency characterization. A point on the ROC curve at which the false acceptance and the false rejection rates coincide is called the *equal error rate* (EER).

13.2 A retrospective

First insights on face recognition came from the psychological community [87] and eventually penetrated into the realm of engineering [38]. As the inception of the field of automatic face recognition, the landmark papers of Kelly [222] and Kanade [216] are usually cited. These and other early publications on face recognition were motivated by the studies of human vision and perception, which found that the human eye concentrates its attention on local features of the face (such as the mouth, nose, eyes, etc.) [356, 86]. Attempting to imitate this property of the human visual system, algorithms based on measuring geometric relations between feature points were proposed [38, 222, 216, 399] in the 1960s.

Another landmark in face recognition research was the paper of Sirovich and Kirby [365] followed by Turk and Pentland [384], who proposed an algebraic approach to the face recognition problem. They considered facial images as vectors in a linear space and, by applying principal component analysis, created bases of faces (Turk and Pentland coined the famous term *eigenfaces*

Figure 13.3. A face shown with different illuminations, head positions, and facial expressions. The variability of the image due to these factors is very high, which makes automatic two-dimensional face recognition a great challenge. (This image was rendered using data from the Notre Dame database [97]; see insert for a color version).

referring to such bases) in which the recognition was performed. This led to the birth of the so-called *holistic* algorithms (from the Greek $o\lambda o\varsigma$, meaning "entire"), which, as opposed to feature-based ("local") methods, use the whole facial image as the input [152, 20, 412].

It appeared that one of the main obstacles complicating the face recognition problem is the large variability of the facial image due to external factors, for example, variations in illumination and head pose (see Figure 13.3). Both feature matching and holistic approaches are sensitive to such factors: in feature-based algorithms, illumination and head pose tamper with the feature detection accuracy [117], and holistic methods work well only when a face has been observed beforehand in similar conditions.

These limitations stem from the fact that the face recognition problem is approached as a two-dimensional pattern recognition one, whereas in practice, the human face is a three-dimensional object. If we had a rough estimation of the three-dimensional face geometry forming the two-dimensional image, we could compensate for the pose and the illumination, or generate new images, in previously unseen conditions. This approach, based on a generic three-dimensional model as an intermediate stage for a two-dimensional face recognition algorithm, was introduced and popularized by Blanz and Vetter [37] and Gheorghiades *et al.* [173].[1]

Taking one step forward, the problem of face recognition can be considered as a three-dimensional pattern recognition problem, i.e., as a problem of surface comparison. Here, the technology deviates from Nature, as it appears that the human visual perception does not attribute much attention to the three-dimensional structure of faces. We demonstrate this phenomenon in a simple experiment shown in the following example.

Example 13.1 (2D vs. 3D in human perception of faces). In this example, we created the faces of George Bush and Osama Bin Laden by simply mapping the respective textures onto a three-dimensional surface of Prof.

Figure 13.4. Is our eye more sensitive to texture or geometry? Simple texture mapping makes a three-dimensional surface look like George Bush or Osama Bin Laden (third and fourth from the left, respectively), though the person from whom the facial geometry has been taken looks completely different (first and second from left). This proves that texture information is more important for human face recognition. (This image was rendered using data from the Notre Dame database [97] and photos of Bush and Bin Laden from the World Wide Web; see insert for a color version).

Kevin Bowyer, who has resemblance to none of the above individuals (Figure 13.4). Our eye recognizes the synthetic faces as the forty-third president of the United States and the most wanted international terrorist, though the underlying geometry belongs to Prof. Bowyer.

Example 13.1 teaches us about an inherent sensitivity of two-dimensional approaches to variations in the facial texture, which, in practical situations, may result from the use of cosmetics. Theoretically, had it been possible to draw any texture on our faces, we could fool any two-dimensional face recognition algorithm.

Unlike two-dimensional images, three-dimensional geometric information is less sensitive to illumination and head orientation, and therefore adds new information that can be employed for more accurate face recognition. Cartoux *et al.* [93] were among the first to realize this potential of the three-dimensional data. They still did not take advantage of the whole facial surface and used a one-dimensional section (profile) for recognition. In subsequent research, their approach was extended to multiple sections of the face [289, 32].

As a historical comment, we should note that these ideas bear an amazing resemblance to the results of Sir Francis Galton (1822–1911). In his 1888 paper [167], Galton proposed to compare profiles in order to identify a suspected criminal and even designed a mechanical device to carry out this task. Today, Galton's ideas would probably be deemed as politically incorrect, being heavily motivated by eugenics, a controversial field of science aimed at "improving the quality of human stock." Yet, together with Alphonse Bertillon [28], he may be considered the father of biometric methods that revolutionized the criminology of the late nineteenth century. The methods of Galton

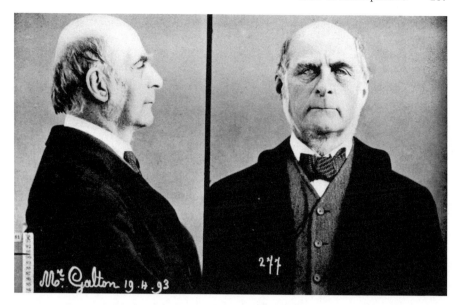

Figure 13.5. Identity recognition in the nineteenth century: a mugshot of Sir Francis Galton, taken during his visit to Bertillon's laboratory in 1893.

and Bertillon belong to the field called *anthropometry* (from Greek $\alpha\nu\theta\rho\omega\pi o\varsigma$ meaning "man" and $\mu\epsilon\tau\rho o\nu$, "measure"), a science concerning the measurement of physical parameters of the human body. The breakthrough of Galton and Bertillon was the systematization of anthropometric data and its application for the recognition of recidivists. Their methods bear much resemblance with modern biometrics. Previous anthropometric studies in the field of criminology, like the works of Cesare Lombroso (1835-1909), attempted mainly to predict an inherent criminal predisposition of an individual, based on the theory that a "born criminal" could be identified by certain characteristic physical features [257], which modern science has rejected.

In the late 1990s and the beginning of the 2000s, the face recognition community has turned to three-dimensional methods in pursuit of better accuracy and robustness. Along with attempts to generalize two-dimensional PCA-like approaches to the three-dimensional problem [2, 262, 203, 306, 305], new recognition methods based on geometric features like curvatures were proposed [242, 181, 372, 104, 282, 245]. Another class of popular geometric approaches is based on the Hausdorff distance and variants of the ICP algorithm [1, 267, 244, 259, 305].

The disadvantage of the majority of these methods is rooted in treating faces as rigid surfaces, an assumption that fails to hold when considering facial expressions. The richness of human expressions and the fact that they involve changes of many facial features has been noticed already by Charles Darwin

[122]. Facial expression may result in significant deformations of the three-dimensional structure of the face, and consequently, we have to think of the face as of a non-rigid surface. This links the face recognition problem to the scope of our book.

13.3 Isometric model of facial expressions

Describing expressions as deformations of the facial surface, we can formulate the problem of *expression-invariant face recognition* as the comparison of facial surfaces invariant to such a class of deformations. The question is whether it is actually possible to model facial expressions. Synthesizing the deformations of the face as the result of expressions appears to be too complicated a task: in the animated movie industry, months are spent on rendering realistic faces of three-dimensional characters. Yet, in our problem, we just need to characterize the class of deformations that result from natural expressions. Here, we have a surprisingly simple result: facial expressions can be approximated by isometries!

In order to convince ourselves in this model, we conducted an experiment, in which a set of flexible markers was attached to the face of a person (Eyal Gordon, then the engineer of the Geometric Image Processing Laboratory, volunteered to act as the "guinea pig" in this experiment). The subject was asked to demonstrate different expressions, ranging from mild to extreme (see some examples in Figure 13.6). By tracking the markers, we could measure the geodesic distances between a set of fixed points on his facial surface. It appeared that the geodesic distances remain approximately invariant to facial expressions. On the other hand, the Euclidean counterparts (describing the extrinsic geometry of the face) show a notably higher variance (Figure 13.7). Later, Mpiperis *et al.* [286, 287] repeated our experiments and independently confirmed the isometric model of facial expressions. Another validation was performed by Gupta *et al.* [194].

The isometric model is of course an approximation; deviations from the model may be attributed to the fact that facial tissues may stretch and therefore are not precisely isometric. For example, expressions with open mouth

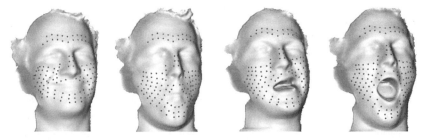

Figure 13.6. Empirical verification of the isometric model of facial expressions. Shown are the points that were tracked.

13.4 Expression-invariant face recognition

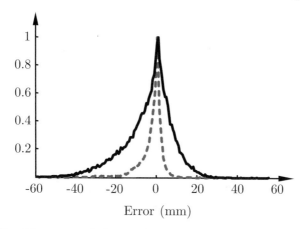

Figure 13.7. Histogram of the geodesic distance deviation from the isometric model (solid); for comparison, a histogram for the Euclidean distances is shown (dashed).

deviate from the isometric model, being topologically different from those with closed mouth. Thus, a geodesic between two points on the upper and the lower lip that crossed the lips when the mouth was closed will circumflex the lip contour when the mouth is open. However, if we introduce a topological constraint into the model, we can easily handle both types of expressions. The easiest way to enforce such a constraint is by cropping out the lips region [73].

13.4 Expression-invariant face recognition

Stated concisely, the isometric model can be formulated as follows: the identity is described by the intrinsic geometry of the face and the expression by the extrinsic one. This means that under the assumption of the isometric model, expression-invariant face recognition is formulated as isometry-invariant comparison of surfaces.

We realized this fact in 2002, thinking of an interesting application for Asi Elad and R.K's canonical forms algorithm. Our first experiments were based on the canonical form distance [63]. A face was first scanned by a coded-light range scanner designed for this purpose, and the input data (in the form of a geometry image) were cropped and sub-sampled to about 2500 points. The matrix of all pair-wise geodesic distances between these points was computed using the fast marching method. Next, the sampled surfaces were embedded into $\mathbb{R}^3$ using the SMACOF algorithm. The obtained canonical form (see examples in Figure 13.8) was aligned by canceling the first-order and the mixed second-order moments, and a signature of moments was computed. The probe signature was then compared with the signatures from the gallery using

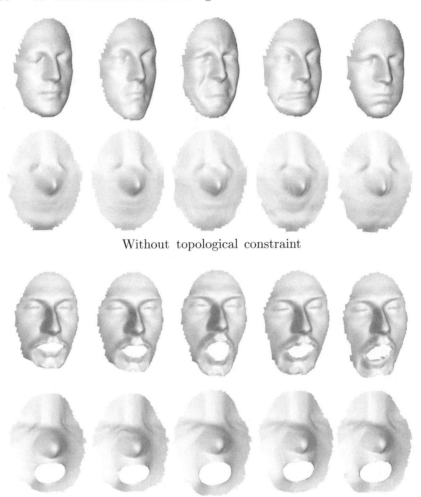

Figure 13.8. Examples of canonical forms of faces obtained by embedding into $\mathbb{R}^3$ using the L_2-stress criterion. Top row shows expressions with closed mouth and the corresponding canonical forms without any pre-processing. Bottom row shows expressions with both open and closed mouth, which were pre-processed by removing the lips region.

the Euclidean distance. Such a comparison procedure allowed quick matching of the probe to a gallery of almost arbitrarily large size.

Because the canonical form distance does not allow comparison of partially overlapping surfaces, it is crucial that every canonical form contain the same parts of the face. The extraction of the facial contour has to be invariant to facial expressions and hence based on the intrinsic geometry only. For this

13.4 Expression-invariant face recognition

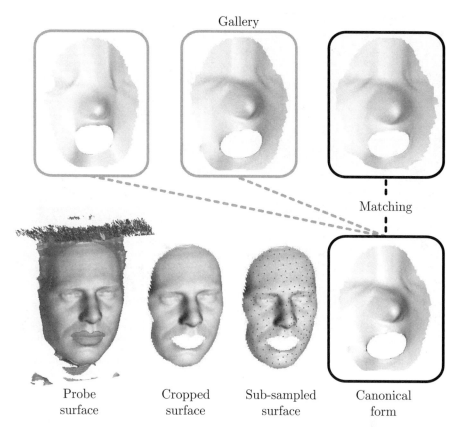

Figure 13.9. Block diagram of the expression-invariant face recognition algorithm.

purpose, we located two invariant fiducial points on the face, the nose tip and the upper point of the nose bridge located between the eyes (referred to as *sellion* in anthropometry), and measured a geodesically equidistant contour around it. This procedure was termed the *geodesic mask* and allowed to crop approximately the same region of the face regardless of the expression.

Figure 13.9 depicts the stages in the three-dimensional face recognition system. In our first experiments, the entire processing took about five minutes per face; this time was gradually reduced to less than a minute. Further improvements lowered the end-to-end processing time to about five seconds on a standard PC, using the parallel raster-scan version of fast marching described in Chapter 4. For demonstration purposes, the face recognition was also equipped with a graphics interface shown in Figure 13.10.

The data for our first experiments were acquired during the Science Festival organized by the Ministry of Science at the Hebrew University campus in Jerusalem in September 2002. The 3D face scanner we used at the time fea-

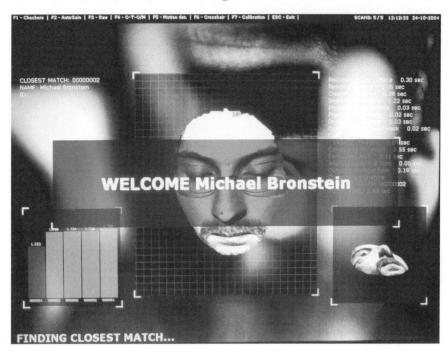

Figure 13.10. A screen shot of the three-dimensional face recognition system user's interface.

tured a menacingly-looking wooden headrest, having certain resemblance to a guillotine. For this reason, people initially felt reluctant to scan their faces. The first feasibility tests of the face recognition algorithm gave encouraging results: even in the presence of extreme facial expressions, the canonical form distance yielded less than 2% EER recognition error, about three times better compared with rigid surface matching of the same data [66]. In addition, it exhibited sufficient accuracy to distinguish between the subtle differences in the geometry of the faces of two identical twins, co-authors of this book. This curious fact attracted the media and gave rise to the weird title "Twins crack the face recognition puzzle" of a CNN news item reporting about our research, afterwards repeated in dozens of newspapers in various languages. Some time later, we repeated the "twins experiment" with other pairs of identical twins. In all cases, the canonical form distance was able to tell apart the visually indistinguishable siblings.

A few improvements followed. First, it became clear that in order for the algorithm to be insensitive to expressions with open mouth, comparison of surfaces with different topology is required. In [73], we used an improved geodesic mask that also included a cut through the lips, thus imposing the same topology for all expressions, both with open and closed mouth. Alternatively, in

[71] we proposed removing geodesic distances that become inconsistent as the result of facial expressions. More recently, Li and Zhang [252] extended this idea, proposing weighting different geodesic distances according to their consistency under facial expression. By means of training, they selected the weights in such a way that gives a description with the highest discrimination between different faces and the smallest between different expressions of the same face.

Secondly, attempting to combat the embedding error inherent to the canonical forms approach, we tried different embedding spaces. In [65, 71], we showed that the representation accuracy of faces (and as a result, the face recognition accuracy) can be improved by embedding into a space with spherical geometry, using methods discussed in Chapter 9. In [71], we showed that using GMDS instead of canonical forms, much higher accuracy could be achieved. In addition, GMDS naturally allowed handling partial comparison of faces, for example, in situations where acquisition imperfections results in parts of the facial surfaces missing or noisy.

13.5 Comparison of photometric properties

Thus far, we focused our attention on the recognition of the intrinsic facial geometry, which appeared to be insensitive to expressions. However, our face is also endowed with rich photometric characteristics that carry additional information useful for recognition. A somewhat simplified model incorporating both geometric and photometric properties of the face consists of a non-rigid surface $\mathcal{S}$ and a texture represented as a scalar field $\rho : \mathcal{S} \to [0,1]$ (or a vector field $\rho : \mathcal{S} \to [0,1] \times [0,1] \times [0,1]$ if a color sensor is used). The texture represents the *reflectance coefficient* or *albedo* at each point on the surface, that is, the fraction of the incident light reflected by the surface.

Unfortunately, the albedo cannot be directly measured by a camera; what we observe is the *brightness* $\beta : \mathcal{S} \to \mathbb{R}$ (or $\beta : \mathcal{S} \to \mathbb{R}^3$ in the color case, respectively) of the surface, or, in simple words, the amount of radiation scattered from it in the direction of the camera. It appears that our skin behaves approximately like a diffusive reflector, which means that its apparent brightness is roughly the same regardless of the observer's angle of view. Formally, this can be expressed as

$$\beta(s) = \max\{\langle \hat{n}(s), \hat{\ell} \rangle, 0\}, \qquad (13.1)$$

where $\hat{n}(s)$ denotes the unit normal vector to the surface at the point s, and $\hat{\ell}$ stands for the illumination direction. This formula is known as the *Lambert cosine law*, named after the German mathematician Johann Heinrich Lambert, who discovered it in 1760 [239]. The knowledge of the three-dimensional surface geometry gives us the surface normal field, which, under controlled illumination conditions, allows us to estimate the reflectance coefficient ρ. Note that such information is unavailable in purely two-dimensional face recognition methods, which are all based essentially on the brightness image of the face.

In this setting, the problem of expression-invariant face recognition fits into the context of correspondence problems we discussed in Chapter 12. Our goal is to measure the similarity of two faces, based on the similarity of their geometries $(\mathcal{S}, d_\mathcal{S})$ and $(\mathcal{Q}, d_\mathcal{Q})$, as well as their photometric properties, $\rho_\mathcal{S}$ and $\rho_\mathcal{Q}$. In order to be able to compare between $\rho_\mathcal{S}$ and $\rho_\mathcal{Q}$, we have to bring them first to some common system of coordinates, in which corresponding facial features coincide. This can be achieved by finding a common parameterization, $\pi_\mathcal{S} : U \to \mathcal{S}$ and $\pi_\mathcal{Q} : U \to \mathcal{Q}$, where $U \subset \mathbb{R}^2$. Such a parameterization should be invariant to facial expressions, which in terms of the isometric model means that π should depend on the intrinsic geometry of the surface only. After the surfaces $\mathcal{S}$ and $\mathcal{Q}$ are re-parameterized, $\rho_\mathcal{S}$ and $\rho_\mathcal{Q}$ can be represented as $\rho_\mathcal{S} \circ \pi_\mathcal{S}$ and $\rho_\mathcal{Q} \circ \pi_\mathcal{Q}$ on the common parameterization domain U, which makes the comparison trivial using standard image comparison techniques. Expression-invariant comparison of the photometric properties of the faces therefore reduces to finding an isometry-invariant "canonical" parameterization of the facial surfaces.

As we mentioned in Chapter 12, the simplest way to construct an isometry-invariant parameterization of a surface is by embedding it into $\mathbb{R}^2$ using an MDS algorithm. The problem is very similar to the computation of the canonical form, except that now the embedding space, serving as the parameterization domain, has to be two-dimensional. Because the embedding is based on the intrinsic geometry only, such a parameterization will be invariant to isometries, and consequently, the albedo image in the embedding space will be insensitive to facial expressions. We term such an image as the *canonical image* of the face. However, recall that the embedding into $\mathbb{R}^2$ is defined up to a Euclidean isometry, implying that the canonical images are defined up to planar rotation, translation and reflection, which has to be undone in order to enable the comparison. Also, the inevitable distortion of the metric introduced by the embedding into the plane makes the canonical image only approximately invariant.

As we have already seen in Chapter 9, a partial remedy to the latter problem comes from non-Euclidean embedding, for example into the two-dimensional sphere $\mathbb{S}^2$ in our case. Because a face is more similar to a sphere than to a plane, spherical embedding produces more accurate canonical images with lower distortion. The minimum distortion, better by about a factor of two compared to the Euclidean embedding, is typically achieved for a sphere radius ranging between 75 and 100 millimeters. In addition, a clear correlation between the representation error and the recognition error is observed: the recognition error closely follows the embedding distortion [64]. Another advantage of the spherical embedding is that the obtained spherical canonical images (Figure 13.11) can be represented using a signature of the spherical harmonic coefficients, which are known to be invariant to rigid isometries on

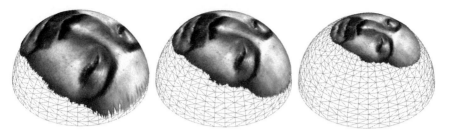

Figure 13.11. Canonical images of a face in $\mathbb{S}^2$ with different radii.

$\mathbb{S}^2$. This property is analogous to the translation invariance of the Fourier transform [219].

Nevertheless, canonical images give only an approximately invariant representation, as a fixed embedding space implies necessarily embedding distortion. Exactly as in the case of isometry-invariant comparison of non-rigid surfaces, an alternative approach is to resort to GMDS for embedding $\mathcal{S}$ into $\mathcal{Q}$. In addition to quantifying the similarity of the two intrinsic geometries, the embedding $\varphi : \mathcal{S} \to \mathcal{Q}$ would also bring $\rho_\mathcal{S}$ and $\rho_\mathcal{Q}$ to the same system of coordinates in $\mathcal{Q}$. The photometric distance between $\rho_\mathcal{Q}$ and $\rho_\mathcal{S} \circ \varphi$ measured either locally or globally provides additional information about the similarity of the two faces. Such an approach is inherently distortion-free and naturally allows for partial comparison of photometric and geometric information.

Suggested reading

For a general overview of biometric technologies, the reader is referred to [12] and [302]. The most complete and up-to-date review of face recognition methods is presented in the paper of Zhao *et al.* [413]. A review with the emphasis on three-dimensional methods can be found in the paper of Bowyer *et al.* [45]. Details on expression-invariant face recognition algorithms based on the isometric model are presented in [66, 77, 65, 71] (geometry only) and [63, 76, 64, 73] (geometry and texture). Face recognition using spherical harmonics is detailed in [64, 73]. A less technical presentation can be found in [75, 68]. For a reader interested in biological and psychological studies on the nature of human facial expressions, we recommend the book by Paul Ekman [146].

Notes

[1] Two-dimensional face recognition methods based on a three-dimensional model to produce facial images in new, previously unseen conditions, are often referred to as *generative* methods.

14
Epilogue

> I hope that posterity will judge me kindly, not only as to the things which I have explained, but also to those which I have intentionally omitted so as to leave to others the pleasure of discovery.
>
> R. DESCARTES, *La géométrie*.

Here comes the end of our journey in the non-rigid world, a new and fascinating field of research on the boundary of computer vision, pattern recognition, computer graphics, and numerical geometry. However, the exploration does not end here, as there is still much to say and much to discover. Some parts of the non-rigid world still remain a *terra incognita* (unknown land), or, like self-confident cartographers of the great era of exploration preferred to write on their incomplete maps, *terra nondum cognita* (not yet known land), to optimistically say that those coming after them would discover this territory.

This book, to the best of our knowledge, is the first attempt to consistently present the map of the non-rigid world and bring together a wide spectrum of problems and approaches. As any first attempt, it is prone to critique. Anticipating the question why a certain work is not mentioned, we should say that comprehensive overview of all the research related to non-rigid shape analysis was not our main goal. Instead, we rather tried to present a common denominator of numerous apparently unrelated fields, problems, solutions, and algorithms. We are aware of the fact that a large number of works has been left outside the scope of our discussion and hope to extend our discussion in future editions to address new problems and new methods.

The decision to finish this book at the face recognition chapter leaves us with a slight sensation of dissatisfaction of a mission not completely accomplished. At the same time, we believe in the value of timely publication, and however imperfect and incomplete, our treatise hopefully presents new points of view and gives some food for thought. We hope this book will catalyze the interest in non-rigid shapes and generate research leading to new results and approaches that will eventually remove the uncharted spots from our map of the non-rigid world.

Solutions of Selected Problems

Chapter 2

2.1 Let $f : X \to Y$ be a bi-Lipschitz function. For any $x \neq x' \in X$, $d_X(x, x') > 0$ and hence $d_Y(f(x), f(x')) \geq C^{-1}d_X(x, x') > 0$, from where it follows that $f(x) \neq f(x')$. Hence, f is injective, implying that there exists $f^{-1} : f(X) \to X$. Let us now select some $y \neq y' \in f(X)$, and denote $x = f^{-1}(y), x' = f^{-1}(y')$. Because f is bi-Lipschitz, $d_Y(y, y') = d_Y(f(x), f(x')) \geq C^{-1}d_X(x, x') = C^{-1}d_X(f^{-1}(x), f^{-1}(x'))$, implying that f^{-1} is Lipschitz with the constant C^{-1}.

2.4 See proof of Proposition 2.4.1 and the relevant discussion in Burago et al. [88].

2.12 Taking the derivatives of $(x, y, z(x, y))$ with respect to x and y yields $x_1 = (1, 0, z_x)^{\mathrm{T}}$ and $x_2 = (0, 1, z_y)^{\mathrm{T}}$, where z_x and z_y denote the partial derivatives of z. Expressing the inner products of x_i, we obtain the first fundamental form matrix

$$G = \begin{pmatrix} \langle x_1, x_1 \rangle & \langle x_1, x_2 \rangle \\ \langle x_1, x_2 \rangle & \langle x_2, x_2 \rangle \end{pmatrix} = \begin{pmatrix} 1 + z_x^2 & z_x z_y \\ z_x z_y & 1 + z_y^2 \end{pmatrix}.$$

The normal to the surface is given by the cross-product

$$N = \pm \frac{x_1 \wedge x_2}{\|x_1 \wedge x_2\|_2} = \pm \frac{(z_x, z_y, -1)^{\mathrm{T}}}{\sqrt{1 + z_x^2 + z_y^2}}.$$

The second fundamental form can be expressed as

$$B = \begin{pmatrix} \langle N, \partial_x x_1 \rangle & \langle N, \partial_y x_1 \rangle \\ \langle N, \partial_x x_2 \rangle & \langle N, \partial_y x_2 \rangle \end{pmatrix} = \frac{1}{\sqrt{1 + z_x^2 + z_y^2}} \begin{pmatrix} z_{xx} & z_{yx} \\ z_{xy} & z_{yy} \end{pmatrix}.$$

The shape operator is given by

$$S = BG^{-1} = \frac{1}{(1+z_x^2+z_y^2)^{3/2}} \begin{pmatrix} z_{xx} & z_{xy} \\ z_{xy} & z_{yy} \end{pmatrix} \begin{pmatrix} 1+z_y^2 & -z_x z_y \\ -z_x z_y & 1+z_x^2 \end{pmatrix}$$

$$= \frac{1}{(1+z_x^2+z_y^2)^{3/2}} \begin{pmatrix} z_{xx} + z_{xx}z_y^2 - z_x z_y z_{xy} & -z_{xx}z_x z_y + z_{xy} + z_{xy}z_x^2 \\ -z_{xx}z_x z_y + z_{xy} + z_{xy}z_x^2 & z_{yy} + z_{yy}z_x^2 - z_x z_y z_{xy} \end{pmatrix},$$

from where one can compute the Gaussian and the mean curvatures

$$K = \det(S) = \frac{z_{xx}z_{yy} - z_{xy}^2}{(1+z_x^2+x_y^2)^2}$$

$$H = \operatorname{trace}(S) = \frac{z_{xx} + z_{xx}z_y^2 - 2z_{xy}z_x z_y + z_{yy} + z_{yy}z_x^2}{(1+z_x^2+x_y^2)^{3/2}}.$$

Chapter 3

3.3 Let us assume that the sampling $X'_n = \{x_1, \ldots, x_n\}$ is created according to the recursion formula

$$\begin{cases} X'_1 = \{x_1\} \\ X'_n = X'_{n-1} \cup \{\arg\max_{x \in X} \min_{x_i \in X'_{n-1}} d_X(x, x_i)\} \end{cases}$$

and let us denote

$$r_n = \max_{x \in X} \min_{x_i \in X'_{n-1}} d_X(x, x_i) = \min_{x_i \in X'_{n-1}} d_X(x_n, x_i).$$

Clearly, as $X'_{n-1} \subset X'_n$,

$$\min_{x_i \in X'_{n-2}} d_X(x, x_i) \geq \min_{x_i \in X'_{n-1}} d_X(x, x_i),$$

and hence $r_{n-1} \geq r_n$. Let us denote $X' = X'_n$ and $r = r_n$. Let us pick any x_i and x_j with $j > i$. Then,

$$d_X(x_i, x_j) \geq \min_{k=1,\ldots,j-1} d_X(x_k, x_j) = r_j \geq r,$$

meaning that X' is r-separated. To show that X' is an r-covering, let us pick a point $x \in X$. Then,

$$d_X(x, X') = \min_{x_i \in X'_n} d_X(x, x_i) \leq \max_{x \in X} \min_{x_i \in X'_n} d_X(x, x_i) = r_{n+1} \leq r.$$

Chapter 4

4.5 Our argument closely follows [61]. In order to show that the update formula is numerically stable, we assume that an input d_i, $i = 1, 2$, is affected by a small error ϵ, which, in turn, influences the computed time of arrival d_3. Denoting by $\tilde{d}_3$ the perturbed value of d_3, our goal is to establish the relationship between the magnitude of the perturbation ϵ and $\|\tilde{d}_3 - d_3\|$. Using first-order Taylor expansion, we have

$$\tilde{d}_3 \approx d_3 + \epsilon \cdot \frac{\partial d_3}{\partial d_i} \leq d_3 + \epsilon \cdot \left(\left|\frac{\partial d_3}{\partial d_1}\right| + \left|\frac{\partial d_3}{\partial d_2}\right| \right).$$

Under the monotonicity condition $\nabla_d d_3 > 0$, and we can therefore write

$$\tilde{d}_3 \approx d_3 + \epsilon \cdot 1_{2\times 1}^T \nabla_d d_3 = d_3 + \epsilon \cdot \frac{1_{2\times 1}^T Q (d - p \cdot 1_{2\times 1})}{1_{2\times 1}^T Q (d - p \cdot 1_{2\times 1})} = d_3 + \epsilon.$$

In the case of the Dijkstra-type update from either d_1 or d_2, we also trivially get

$$\tilde{d}_3 \approx d_3 + \epsilon.$$

This implies that regardless of the value of d_i, $\|\tilde{d}_3 - d_3\| \approx \epsilon$, meaning that update formula stable.

4.7 This alternative update scheme based on a circular wavefront model was proposed by Novotni and Klein in [299]. Here, we show a derivation following [61]. The wavefront is modeled as a circular wave propagating from a virtual point source x. Demanding that the supporting vertices x_1, x_2 of the triangle being updated lie at distances d_1 and d_2, respectively, from the source, we obtain for $i = 1, 2$

$$d_i^2 = (x_i - x)^T (x_i - x) = x_i^T x_i - 2 x_i^T x + x^T x. \tag{S.1}$$

The time of arrival of the wavefront to the updated vertex x_3 is given by its distance from the point source,

$$d_3^2 = (x_3 - x)^T (x_3 - x) = x^T x$$

(the last transition is due to the fact that $x_3 = 0$). Denoting $s_i = d_i^2$, and $q = (s_1 - x_1^T x_1, s_2 - x_2^T x_2)^T$, we obtain

$$s_3 \cdot 1_{2\times 1} - 2 V^T x = q.$$

Assuming the mesh to be non-degenerate,

$$x = \frac{1}{2} V (V^T V)^{-1} (s_3 \cdot 1_{2\times 1} - q) = \frac{1}{2} V Q (s_3 \cdot 1_{2\times 1} - q).$$

Plugging the later result into the expression for d_i^2 in (S.1), we have

$$s_3 = x^T x = \frac{1}{4} (s_3 \cdot 1_{2\times 1} - q)^T Q (s_3 \cdot 1_{2\times 1} - q)$$
$$= \frac{1}{4} \left(s_3^2 \cdot 1_{2\times 1}^T Q 1_{2\times 1} - 2 s_3 \cdot 1_{2\times 1}^T Q q + q^T Q q \right).$$

Consequently, d_3 is given as the largest positive solution of the following bi-quadratic equation

$$d_3^4 \cdot 1_{2\times 1}^T Q 1_{2\times 1} - 2d_3^2 \cdot (1_{2\times 1}^T Qq + 2) + q^T Qq = 0.$$

The reader is referred to [61], where an analysis similar to that in Chapter 4 is presented. It appears that for the circular wavefront model, there is no analogy of the non-obtuse angle condition we had in the planar wavefront case, which guarantees that all update directions coming from within the triangle, the update will be monotonous and consistent. Furthermore, repeating the stability analysis from the solution of Problem 4.5 shows that the circular wavefront scheme may amplify noise. The amplification factor depends on the values of d_1 and d_2 and in some cases may even grow unbounded, making the scheme numerically unstable.

Chapter 5

5.2 Commutativity of matrix multiplication under the trace operator is easy to verify. Let A and B be two $N \times M$ matrices, and let $C = A^T B$ and $D = B^T A$ be the $M \times M$ product matrices. The elements of C are given by

$$c_{ij} = \sum_{k=1}^{N} a_{ki} b_{kj},$$

so that

$$\text{trace}(C) = \sum_{i=1}^{M} c_{ii} = \sum_{i=1}^{M} \sum_{k=1}^{N} a_{ki} b_{ki}.$$

In the same manner, the elements of D are given by

$$d_{ij} = \sum_{k=1}^{N} b_{ki} a_{kj},$$

so that

$$\text{trace}(D) = \sum_{i=1}^{M} d_{ii} = \sum_{i=1}^{M} \sum_{k=1}^{N} b_{ki} a_{ki}.$$

5.5 Let $A, B \subseteq \mathbb{X}$ be two convex sets. Given any $x, x' \in A \cap B$, the combination $\lambda x + (1-\lambda)x'$ belongs both to A and to B for all $\lambda \in [0, 1]$. Hence, $A \cap B$ is convex.

5.7 Let $A \subseteq \mathbb{X}$ be a convex set, $f : A \to \mathbb{R}$ a convex function, and $X_c = \{x \in A : f(x) \le c\}$ its c sub-level set. Let $x, x' \in X_c$. Because $f(x), f(x') \le c$ and

f is convex, $f(\lambda x + (1-\lambda)x') \leq \lambda f(x) + (1-\lambda)f(x') \leq c$ for every $\lambda \in [0,1]$. Hence, $\lambda x + (1-\lambda)x' \in X_c$, which proves that X_c is a convex set.

5.8 Let $X \subseteq \mathbb{X}$ be a convex set, $f : \mathbb{X} \to \mathbb{R}$ a strictly convex function, and $x^* \in \mathbb{X}$ its local minimizer. Therefore, there exists an open neighborhood U of x where $f(x^*) \leq f(x)$ for all $x \in U$. Let $y \in X$ be an arbitrary point; by continuity of scalar multiplication and addition in the vector space $\mathbb{X}$, the combination $x(\lambda) = \lambda y + (1-\lambda)x^*$ approached x^* as λ approaches 0. Therefore, for a sufficiently small λ, $x(\lambda) \in U$. Then, $f(x) \leq f(x(\lambda)) \leq f(\lambda y + (1-\lambda)x^*) \leq \lambda f(y) + (1-\lambda)f(x^*)$. Rearranging terms, we get $f(x^*) \leq f(x)$, which proves that x^* is a global minimizer. For a strictly convex function, the latter inequality is strong, implying that x^* is also unique.

5.15 The constraints are two circles of radii 1 and 2 tangent at the origin. Because $x^* = (0,0)$ is the only point where both constraints are satisfied, it is the minimizer of the constrained optimization problem (objective value 0). However, because the gradients to both constraints are aligned at the same direction $(1,0)$ at the origin, there exist infinitely many Lagrange multipliers satisfying $\mu_1 \nabla h_1(x^*) + \mu_2 \nabla h_2(x^*) + \nabla f(x^*) = 0$. This does not contradict the KKT conditions, because x^* is not a *regular* point, as ∇h_i are not linearly independent.

5.16 The objective $x^1 + x^2$ is a linear function, whose gradient is constant and directed in $(1,1)$. The constraint $(x^1)^2 + (x^2)^2 = 2$ is a circle of radius 2 centered at the origin. At a constrained minimizer, the normal to the circle has to be aligned with the gradient direction. This happens exactly at two points: $(1,1)$ and $(-1,-1)$. The former one is a constrained maximum (objective value 2), whereas the latter one is a constrained minimum (objective value -2).

Chapter 6

6.3 The main issue in consistently discretizing the geometric moment integral

$$m_{pqr} = \int_X (x^1)^p (x^2)^q (x^3)^r da$$

on a triangular mesh $T(\{x_1, \ldots, x_N\})$ is to correctly account for the possibly non-uniform triangle areas. A reasonable accuracy can be obtained by computing the centroid $\overline{x}_t = \frac{1}{3}(x_{t1} + x_{t2} + x_{t3})$ for each of the mesh faces and replacing the integral with the weighted sum,

$$\hat{m}_{pqr} = \sum_{t \in T} (\overline{x}_t^1)^p (\overline{x}_t^2)^q (\overline{x}_t^3)^r a_t,$$

where a_t is the area of the triangle t.

6.5 Let x be a real number and $y = x \pm \epsilon$ its finite precision approximation. Raising y to the p-th power yields $y^p \leq (x+\epsilon)^p$, which by first-order Taylor expansion can be approximated as

$$y^p \approx x^p + p\, x^{p-1}\epsilon.$$

Hence, the relative error of the approximation of x^p by y^p is

$$\frac{y^p - x^p}{x^p} \approx \frac{p\, x^{p-1}\epsilon}{x^p} = p\frac{\epsilon}{x}.$$

Because $\frac{\epsilon}{x}$ is fixed for given data, the error grows linearly with the increase of the p, which complicates the practical use of high-order geometric moments.

6.10 Let us be given two sets of corresponding points $\{x_1, \ldots, x_n\}$ and $\{y_1, \ldots, y_n\}$ in $\mathbb{R}^3$. Our goal is to find the rotation matrix R and translation vector t minimizing

$$d^2 = \sum_{i=1}^{n} \|x_i - (Ry_i + t)\|_2^2.$$

Let us fix R and find the t minimizing d^2. Denoting by $z_i = Ry_i$, we have

$$d^2 = \sum_{i=1}^{n} \|x_i - z_i + t\|_2^2 = \sum_{i=1}^{n} (x_i - z_i + t)^{\mathrm{T}}(x_i - z_i + t)$$
$$= \sum_{i=1}^{n} (x_i - z_i)^{\mathrm{T}}(x_i - z_i) + 2(x_i - z_i)^{\mathrm{T}} t + t^{\mathrm{T}} t.$$

Taking the derivative with respect to t yields

$$\nabla_t d^2 = 2 \sum_{i=1}^{n} (x_i - z_i + t).$$

Requiring $\nabla_t d^2 = 0$, we obtain

$$t = \frac{1}{n}\sum_{i=1}^{n}(x_i - z_i) = \frac{1}{n}\sum_{i=1}^{n} x_i - \frac{1}{n}\sum_{i=1}^{n} z_i,$$

which is nothing but the difference between the centroids of $\{x_i\}$ and $\{z_i\}$. In other words, $t = \bar{x} - R\bar{y}$ minimizing d^2 translates the centroid $\bar{x}$ of $\{x_i\}$ to the centroid $R\bar{y}$ of $\{Ry_i\}$, for any rotation matrix R. We will therefore assume for the following discussion that both $\{x_i\}$ and $\{y_i\}$ are zero-centered, i.e., have

$$\bar{x} = \frac{1}{n}\sum_{i=1}^{n} x_i = 0, \bar{y} = \frac{1}{n}\sum_{i=1}^{n} y_i = 0.$$

In this case, the optimal t will always be zero, and our problem reduces to

$$d^2 = \sum_{i=1}^{n} \|x_i - Ry_i\|_2^2.$$

Arranging the x_i into the columns of a $3 \times n$ matrix X, and the y_i into the columns of a $3 \times n$ matrix Y, the latter can be rewritten in terms of the Frobenius norm as

$$d^2 = \|X - RY\|_F^2 = \text{trace}((X - RY)^T(X - RY))$$
$$= \text{trace}(X^T X) - 2\,\text{trace}(X^T RY) + \text{trace}(Y^T R^T RY).$$

Because R is a rotation matrix, $R^T R = I$. Furthermore, due to the commutativity of multiplication under the trace operator (see solution to Problem 5.2), $\text{trace}(X^T RY) = \text{trace}(RYX^T)$. This allows us to write

$$d^2 = \text{trace}(X^T X + Y^T Y) - 2\,\text{trace}(RYX^T) = \text{const} - 2\,\text{trace}(RH),$$

where H is the "covariance matrix," defined as

$$H = YX^T = \sum_{i=1}^{n} y_i x_i^T.$$

Therefore, in order to minimize d^2, R has to maximize $\text{trace}(RH) = \langle R, H^T \rangle$. We leave to the reader to prove the simple algebraic fact that the latter inner product is maximized by $R = VU^T$, where U and V are the unitary matrices obtained by the singular value decomposition $H = U \Lambda V^T$ of H.

Chapter 7

7.2 The gradient of the first term of $\sigma_2(Z)$ is immediate using the results of Example 5.1 in Chapter 5. Differentiating the second term is slightly more complicated, as it involves a non-linear matrix function $B(Z; D_X)$.

To derive the gradient, we start writing the stress as

$$\sigma_2^2(Z) = \sum_{i>j} (d_{ij}(Z) - d_X(x_i, x_j))^2$$
$$= \sum_{i>j} d_{ij}^2(Z) - 2d_{ij}(Z)d_X(x_i, x_j) + d_X^2(x_i, x_j).$$

and do some acrobatics with the formulae. The first term can be written as

$$\sum_{i>j} d_{ij}^2(Z) = \sum_{i>j}\sum_{k=1}^{m}(z_i^k - z_j^k)^2 = \sum_{i>j}\sum_{k=1}^{m}(z_i^k)^2 - 2z_i^k z_j^k + (z_j^k)^2$$

$$= \sum_{i>j}\langle z_i, z_i\rangle + \langle z_j, z_j\rangle - 2\langle z_i, z_j\rangle$$

$$= (N-1)\sum_{i=1}^{N}\langle z_i, z_i\rangle - \left(\sum_{i,j=1}^{N}\langle z_i, z_j\rangle - \sum_{i=1}^{N}\langle z_i, z_i\rangle\right)$$

$$= N\sum_{i=1}^{N}\langle z_i, z_i\rangle - \sum_{i,j=1}^{N}\langle z_i, z_j\rangle, \qquad (S.2)$$

where $\langle z_i, z_i\rangle$ are the inner products of the vectors z_i, z_j in $\mathbb{R}^m$. Note that what we got is an expression in $\langle z_i, z_j\rangle$, i.e., elements of the *Gram matrix* ZZ^T. Using this observation, we can express equation (S.2) as trace(VZZ^T), where V is an $N \times N$ matrix with elements

$$v_{ij} = \begin{cases} -1 & i \neq j \\ N-1 & i = j. \end{cases}$$

Using the property of matrix commutativity under trace, we obtain

$$\sum_{i>j} d_{ij}^2(Z) = \text{trace}(Z^T V Z).$$

The second term is written in a slightly more complicated but similar way,

$$\sum_{i>j} d_{ij}(Z) d_X(x_i, x_j) = \sum_{i>j} d_X(x_i, x_j) \left(\sum_{k=1}^{m}(z_i^k - z_j^k)^2\right)^{1/2}$$

$$= \sum_{i>j} d_X(x_i, x_j) d_{ij}^{-1}(Z) \sum_{k=1}^{m}(z_i^k - z_j^k)^2$$

$$= \sum_{i>j} d_X(x_i, x_j) d_{ij}^{-1}(Z)(\langle z_i, z_i\rangle + \langle z_j, z_j\rangle - 2\langle z_i, z_j\rangle)$$

$$= \sum_{i,j=1}^{N} d_X(x_i, x_j) d_{ij}^{-1}(Z)(\langle z_i, z_i\rangle - \langle z_i, z_j\rangle)$$

$$= \text{trace}(B(Z)ZZ^T) = \text{trace}(Z^T B(Z) Z), \qquad (S.3)$$

where $B(Z)$ is an $N \times N$ matrix with elements,

$$b_{ij}(Z) = \begin{cases} -d_X(x_i, x_j) d_{ij}^{-1}(Z) & i \neq j \text{ and } d_{ij}(Z) \neq 0 \\ 0 & i \neq j \text{ and } d_{ij}(Z) = 0, \\ -\sum_{k \neq i} b_{ik} & i = j. \end{cases}$$

The last term $\sum_{i>j} d_X^2(x_i, x_j)$ does not depend on Z, and therefore, we leave it as is. Arranging all the intermediate results, we finally arrive at the matrix expression of the stress

$$\sigma_2(Z) = \text{trace}(Z^\mathrm{T} V Z) - 2\text{trace}(Z^\mathrm{T} B(Z)Z) + \sum_{i>j} d_X^2(x_i, x_j).$$

7.3 In order to prove the majorizing inequality (7.7), it is sufficient to show that $\text{trace}(Z^\mathrm{T} B(Q)Q) \leq \text{trace}(Z^\mathrm{T} B(Z)Z)$ for all $Q \in \mathbb{R}^{N\times m}$. Our proof is similar to one given by Borg and Groenen [44]. First, from the Cauchy-Schwarz inequality, it follows that

$$\sum_{k=1}^m (z_i^k - z_j^k)(q_i^k - q_j^k) \leq \left(\sum_{k=1}^m (z_i^k - z_j^k)^2\right)^{1/2} \left(\sum_{k=1}^m (q_i^k - q_j^k)^2\right)^{1/2}$$
$$= d_{ij}(Z) d_{ij}(Q),$$

where an equality is achieved for $Q = Z$. Using this result, we have

$$\text{trace}(Z^\mathrm{T} B(Z)Z) = \sum_{i>j} d_{ij}(Z) d_X(x_i, x_j)$$
$$\geq \sum_{i>j} \sum_{k=1}^m (z_i^k - z_j^k)(q_i^k - q_j^k) d_{ij}^{-1}(Q) d_X(x_i, x_j).$$

Recognizing the elements of the matrix $B(Z)$ in the last term, we can rewrite it as

$$\text{trace}(Z^\mathrm{T} B(Z)Z) \geq \text{trace}(Z^\mathrm{T} B(Q)Q),$$

similarly to (S.3), which completes the proof.

7.10 The derivation of (7.16) is based on the chain rule for matrix functions, which we show here. Let us be given a function $f : \mathbb{R}^{N\times m} \to \mathbb{R}$, a matrix $A \in \mathbb{R}^{N\times N'}$, and a vector $Y \in \mathbb{R}^{N'\times m}$. By definition of the gradient,

$$f(X + dX) = f(X) + \langle \nabla_X f(X), dX \rangle$$
$$= f(X) + \text{trace}(dX^\mathrm{T} \nabla_X f(X)).$$

On the other hand, denoting $X = AY$ and $dX = AdY$, we have

$$f(A(Y + dY)) = f(AY + AdY) = f(AY) + \langle \nabla_Y f(AY), dY \rangle$$
$$= f(AY) + \langle \nabla_X f(AY), AdY \rangle$$

Rewriting the last equality explicitly,

$$\langle \nabla_Y f(AY), dY \rangle = \text{trace}(dY^\mathrm{T} \nabla_Y f(AY))$$
$$= \text{trace}((AdY)^\mathrm{T} \nabla_X f(AY)) = \text{trace}(dY^\mathrm{T}(A^\mathrm{T} \nabla_X f(AY))),$$

we obtain the chain rule,

$$\nabla_Y f(AY) = A^\mathrm{T} \nabla_X f(AY).$$

7.11 Equation 7.21 amounts to solving the constrained optimization problem,

$$\min_{\gamma} \frac{1}{2}\|A_{K+1}\gamma\|_2^2 \quad \text{s.t.} \quad 1_{(K+1)\times(K+1)}^{\mathrm{T}}\gamma = 1. \tag{S.4}$$

From the optimality conditions for problem (S.4), we obtain

$$A_{(K+1)}^{\mathrm{T}}A_{(K+1)}\gamma - \lambda 1_{(K+1)\times(K+1)} = 0, \tag{S.5}$$

where λ is the Lagrange multiplier (for convenience, we choose it with a negative sign). Taking the inner product with γ, we obtain

$$\gamma^{\mathrm{T}}A_{(K+1)}^{\mathrm{T}}A_{(K+1)}\gamma - \lambda\gamma^{\mathrm{T}}1_{(K+1)\times(K+1)} = 0,$$

and plugging in the constraint $1_{(K+1)\times(K+1)}^{\mathrm{T}}\gamma = 1$, we finally have $\lambda = \gamma^{\mathrm{T}}A_{(K+1)}^{\mathrm{T}}A_{(K+1)}\gamma$. This result allows us to solve (S.5) as follows: first, we solve the system

$$A_{(K+1)}^{\mathrm{T}}A_{(K+1)}\tilde{\gamma} = 1_{(K+1)\times(K+1)},$$

whose solution is related to the solution of (S.5) as

$$\gamma = \lambda\tilde{\gamma}. \tag{S.6}$$

Taking the inner product with $1_{(K+1)\times(K+1)}$ on both sides of equation (S.6), we have

$$1 = 1_{(K+1)\times(K+1)}^{\mathrm{T}}\gamma = \lambda 1_{(K+1)\times(K+1)}^{\mathrm{T}}\tilde{\gamma},$$

from where

$$\lambda = \frac{1}{\tilde{\gamma}_0 + \ldots \tilde{\gamma}_K}.$$

Chapter 8

8.5 Our derivation is similar to one we did in Problem 7.2 to derive the matrix form of the L_2-stress:

$$\begin{aligned}\sum_{i>j} w_{ij}d_{ij}(Z) &= \sum_{i>j} w_{ij}\langle z_i - z_j, z_i - z_j\rangle \\ &= \sum_{i>j} w_{ij}(\langle z_i, z_i\rangle - 2\langle z_i, z_j\rangle + \langle z_j, z_j\rangle) \\ &= \sum_{i,j=1}^{N} w_{ij}(\langle z_i, z_i\rangle - \langle z_i, z_j\rangle) \\ &= \sum_{i=1}^{N}\left(\sum_{j\neq i} w_{ij}\right)\langle z_i, z_i\rangle - \sum_{i\neq j} w_{ij}\langle z_i, z_j\rangle \\ &= \text{trace}(L_X ZZ^{\mathrm{T}}) = \text{trace}(Z^{\mathrm{T}}L_X Z),\end{aligned}$$

where L_X is an $N \times N$ matrix with elements

$$l_{ij} = \begin{cases} -w_{ij} & i \neq j \\ \sum_{k \neq i} w_{ik} & i = j. \end{cases}$$

8.6 The positive semidefiniteness of L_X follows straightforwardly from our previous derivation, as a particular case when we take $z \in \mathbb{R}^N$ instead of Z. The expressions simplify in the following way,

$$z^T L_X z = \sum_{i>j} w_{ij}(z_i - z_j)^2 \geq 0,$$

which lead to $L_X \succeq 0$.

8.7 The geometric intuition of this result is the following: we are looking for the vector x of fixed length, which is shortened the most by the operation of the matrix A. The factor by which this vector is shortened is the smallest eigenvalue $\lambda_{\min}$ of A, and the vector x itself is the corresponding eigenvector.

Chapter 9

9.11 Let us fix some i. Because the stress $\sigma(u_i) = u_i^T A_i u_i + 2 b_i^T u_i + c_i$ is quadratic, in order to show convexity with respect to u_i, it is sufficient to show that the matrix

$$A_i = \sum_{j \neq i} D_Y(t_i, t_j) u_j u_j^T D_Y(t_i, t_j)^T$$

is positive semi-definite. For that purpose, it is further sufficient to show that $D_Y(t_i, t_j) u_j u_j^T D_Y(t_i, t_j)^T \succeq 0$ for each $j \neq i$, as the sum of positive semi-definite matrices is positive semi-definite. Let us fix some j and denote $R = D_Y(t_i, t_j) u_j$. We have to show that $RR^T \succeq 0$. This, however, follows straightforwardly, as for any x, $x^T RR^T x = (R^T x)^T (R^T x) = \|R^T x\|_2^2 \geq 0$.

Chapter 10

10.2 See proof of Theorem 7.3.25 in [89].

10.3 Let us start by evaluating $d_{\text{GH}}(X, X^r)$. We construct $\psi : X^r \to X$ simply as the identity map, copying x_i to x_i. Clearly, dis $\psi = 0$. Because the image $\varphi(X^r) = X^r$ is an r-net in X, the map is r-surjective. Let us now construct another map $\varphi : X \to X^r$, copying a Voronoi region of x_i in X to x_i (points along Voronoi edges can be copied arbitrary to any of the adjacent regions). The map is clearly surjective. To evaluate its distortion, let us take x and x' in X, belonging to the Voronoi regions of some x_i and x_j (not necessarily distinct). Using the triangle inequality on X,

$$d_X(x, x') \le d_X(x, x_i) + d_X(x_i, x') \le d_X(x, x_i) + d_X(x', x_j) + d_X(x_i, x_j),$$

and, similarly,

$$d_X(x_i, x_j) \le d_X(x_i, x) + d_X(x, x_j) \le d_X(x_i, x) + d_X(x_j, x') + d_X(x, x').$$

Combining these results and using the fact that X^r is an r-net yields

$$|d_X(x, x') - d_X(x_i, x_j)| \le d_X(x_i, x) + d_X(x_j, x') \le 2r,$$

from where dis $\varphi \le 2r$. In a very similar way, we obtain dis $(\varphi, \psi) \le r$. Combining the previous results, we have $d_{GH}(X, X^r) \le r$. Now, using the triangle inequality on $\mathbb{M}$, we have

$$d_{GH}(X, Y) \le d_{GH}(X, X^r) + d_{GH}(X^r, Y)$$
$$\le d_{GH}(X, X^r) + d_{GH}(Y, Y^r) + d_{GH}(X^r, Y^r);$$

similarly,

$$d_{GH}(X^r, Y^r) \le d_{GH}(X^r, X) + d_{GH}(X, Y^r)$$
$$\le d_{GH}(X^r, X) + d_{GH}(Y^r, Y) + d_{GH}(X, Y).$$

Combining these results and using the symmetry of d_{GH}, we can write

$$|d_{GH}(X^r, Y^r) - d_{GH}(X, Y)| \le d_{GH}(X, X^r) + d_{GH}(Y, Y^r) \le 2r.$$

Chapter 11

11.1 Let us be given a Salukwadze optimal solution (X'^*, Y^*), on which the minimum of $\|\Phi(X', Y')\|$ on Ω is achieved. Assume that (X'^*, Y^*) is not Pareto optimal. Then, there exists another $(X', Y') \in \Omega$, such that $\Phi(X', Y') < \Phi(X^*, Y^*)$. It therefore follows that $\|\Phi(X', Y')\| < \|\Phi(X^*, Y^*)\|$ by properties of the norm, which contradicts the assumption that (X'^*, Y^*) is Salukwadze optimal. Therefore, (X'^*, Y^*) is necessarily Pareto optimal.

11.3 In order to show the equivalence, we have to show that though the maps are defined as $\varphi : X \to Y$ and $\psi : Y \to X$, their ranges and images are X' and Y'. Given a crisp part X', we denote by

$$\delta_{X'}(x) = \begin{cases} 1 & x \in X' \\ 0 & \text{else} \end{cases}$$

its characteristic function. The characteristic functions in the infima terms restrict the ranges,

$$\frac{1}{2}\inf_{\substack{\varphi:X\to Y\\\psi:Y\to X}}\max\left\{\begin{array}{l}\sup_{x,x'\in X}\delta_{X'}(x)\delta_{X'}(x')|d_X(x,x')-d_Y(\varphi(x),\varphi(x'))|\\ \sup_{y,y'\in Y}\delta_{Y'}(y)\delta_{Y'}(y')|d_Y(y,y')-d_X(\psi(y),\psi(y'))|\\ \sup_{\substack{x\in X\\y\in Y}}\delta_{X'}(x)\delta_{Y'}(y)|d_X(x,\psi(y))-d_Y(\varphi(x),y)|\\ D\sup_{x\in X}(1-\delta_{Y'}(\varphi(x)))\,\delta_{X'}(x)\\ D\sup_{y\in Y}(1-\delta_{X'}(\psi(y)))\,\delta_{Y'}(y)\end{array}\right\}$$

$$=\frac{1}{2}\inf_{\substack{\varphi:X'\to Y\\\psi:Y'\to X}}\max\left\{\begin{array}{l}\sup_{x,x'\in X'}|d_X(x,x')-d_Y(\varphi(x),\varphi(x'))|\\ \sup_{y,y'\in Y'}|d_Y(y,y')-d_X(\psi(y),\psi(y'))|\\ \sup_{\substack{x\in X'\\y\in Y'}}|d_X(x,\psi(y))-d_Y(\varphi(x),y)|\\ D\sup_{x\in X'}(1-\delta_{Y'}(\varphi(x)))\\ D\sup_{y\in Y'}(1-\delta_{X'}(\psi(y)))\end{array}\right\}$$

assuming $D=\max\{\mathrm{diam}(X),\mathrm{diam}(Y)\}$. If $\varphi(X')\not\subseteq Y'$ or $\psi(Y')\not\subseteq X'$, we have $\sup_{x\in X'}(1-\delta_{Y'}(\varphi(x)))=1$ (respectively, $\sup_{y\in Y'}(1-\delta_{X'}(\psi(y)))=1$); hence, the values of the above expression will be at least D. Because the other terms are bounded above by D, it follows that for $\varphi(X')\subseteq Y'$ and $\psi(Y')\subseteq X'$, the above expression will be at most D. As the result, we can rewrite the infimum on the maps $\varphi:X'\to Y'$ and $\psi:Y'\to X'$,

$$=\frac{1}{2}\inf_{\substack{\varphi:X'\to Y'\\\psi:Y'\to X'}}\max\left\{\begin{array}{l}\sup_{x,x'\in X'}|d_X(x,x')-d_Y(\varphi(x),\varphi(x'))|\\ \sup_{y,y'\in Y'}|d_Y(y,y')-d_X(\psi(y),\psi(y'))|\\ \sup_{\substack{x\in X'\\y\in Y'}}|d_X(x,\psi(y))-d_Y(\varphi(x),y)|\end{array}\right\}=d_{\mathrm{GH}}(X',Y').$$

Software

TOSCA Code accompanying *Numerical geometry of nonrigid shapes*. Fast marching algorithms, farthest point sampling, Voronoi diagrams. Includes MATLAB interface to *QSlim*. MDS with vector extrapolation acceleration; Generalized MDS.
http://tosca.cs.technion.ac.il
License: Free
Language: MATLAB/MEX

Visualization Toolkit (VTK) Library for 3D data handling and visualization. Includes everything that can be desired for rendering, basic surface manipulation routines, importers/exporters for many 3D data formats.
http://www.vtk.org
License: Free
Language: C++

QSlim Mesh simplification code by Michael Garland [168].
http://graphics.cs.uiuc.edu/~garland/software/qslim.html
License: Free
Language: C++

Afront Advancing front remeshing; mesh extraction from volumetric data and point clouds. Based on [343, 342, 338].
http://afront.sourceforge.net
License: Free
Language: C++

Qhull One of the fastest codes for the computation of the convex hull and Euclidean Voronoi and Delaunay tesselations.
http://www.qhull.org
License: Free
Language: C

MATLAB Optimization Toolbox The best starting point for MATLAB users. Includes constrained and unconstrained optimization problems; nonlinear and multi-objective optimization; nonlinear least-squares; data fitting; nonlinear equations; quadratic and linear programming; integer programming.
http://www.mathworks.com/products/optimization
License: Commercial
Language: MATLAB

TOMLAB Nonlinear optimization; data fitting; global, non-convex and non-smooth optimization; linear, quadratic, and semidefinite programming. Supports large-scale sparse matrices. Compatible with *MATLAB Optimization Toolbox*.
http://tomopt.com
License: Commercial
Language: MATLAB

Iterative Methods for Optimization Codes accompanying Kelley's book [221]. Line search algorithms; trust region; Newton; conjugate gradients; BFGS.
http://www4.ncsu.edu/~ctk/matlab_darts.html
License: Free
Language: MATLAB

OPT++ Object-oriented nonlinear optimization library. Newton methods; nonlinear interior-point method; parallel direct search; trust region.
http://csmr.ca.sandia.gov/opt++
License: Free
Language: C++

LANCELOT Unconstrained and constrained optimization problems; nonlinear equations; nonlinear least-squares problems. Includes an implementation of Augmented Lagrangian method.
http://www.cse.scitech.ac.uk/nag/lancelot
License: Free
Language: Fortran 77

ANN David Mount's library for approximate nearest neighbor search.
http://www.cs.umd.edu/~mount/ANN
License: Free
Language: C++

MATLAB Statistics Toolbox Supports different types of MDS problems, including classic scaling.
http://www.mathworks.com/products/statistics
License: Commercial
Language: MATLAB

Nonmetric MDS Mark Steyvers' code allows minimization of different variants of stress.
http://www.mathworks.com/products/statistics
License: Free
Language: MATLAB

Isomap Low-dimensional embedding of Euclidean data.
http://isomap.stanford.edu
License: Free
Language: MATLAB

Toolbox for Dimensionality Reduction by Laurens van der Maaten. Includes PCA, LLE, HLLE, LTSA, and diffusion maps.
http://www.cs.unimaas.nl/l.vandermaaten
License: Free
Language: MATLAB

An up-to-date list of relevant software is available on the book website tosca.cs.technion.ac.il/book.

Notation

$\mathbb{R}$	Real numbers
$\mathbb{R}_+$	Non-negative real numbers
$\mathbb{R}^m$	m-dimensional Euclidean space
$\mathbb{R}^m_+$	Non-negative m-dimensional orthant
$\mathbb{R}^{m \times n}$	Space of $m \times n$ matrices
$\mathcal{C}^r$	Class of r-times continuously differentiable maps
$\mathcal{C}^\infty$	Class of smooth maps
$\|\cdot\|$	Norm
$\|\cdot\|_p$	p-norm
$\|\cdot\|_F$	Frobenius matrix norm
$\|\cdot\|_Q$	Q-norm
$\langle \cdot, \cdot \rangle$	Inner product
$\cdot \wedge \cdot$	Wedge (cross) product
2^X	Powerset of X
Σ_X	σ-algebra on X
$\sqcup$	Disjoint union
(X, d_X)	Metric space
$d_X|_{X'}$	Restricted metric
$B_X(x, r)$	Open ball of radius r around the point x with respect to the metric d_X
$\overline{B}_X(x, r)$	Closed ball of radius r around the point x with respect to the metric d_X
$x_n \to x$	Convergence of sequence $\{x_n\}$ to the limit x
$f^{-1}(A)$	Preimage of the set A under the map f
f^{-1}	Inverse of a bijective map f
dil f	Dilation of the map f
dis f	Distortion of the map f
Iso(X)	Isometric group of the metric space X
d_{L}	Length metric
d_{G}	Geodesic metric

Notation

$L(\gamma)$	Length of curve γ
$\dot{\gamma}(t)$	First derivative of $\gamma(t)$ with respect to t (velocity)
$\ddot{\gamma}(t)$	Second derivative of $\gamma(t)$ with respect to t (acceleration)
X	Surface (Riemannian manifold)
X_N	Finite sampling of surface X
X_N^r	Finite r-covering of surface X
∂X	Boundary of X
$\text{int}(X)$	Interior of X
$\text{conv}(X)$	Convex hull of a set X
g	Riemannian metric
TX	Tangent space (plane)
κ	Curvature
κ_n	Normal curvature
κ_g	Geodesic curvature
K	Gaussian curvature
H	Normal curvature
$\mathcal{N}(x)$	Neighborhood of point x
$T(X_N)$	Triangulation of a cloud of points X_N (triangular mesh)
df	Differential of map f
∇f	Gradient of map f
$\nabla^2 f$	Hessian of map f
$\nabla_X f$	Intrinsic gradient of map $f : X \to \mathbb{R}$
A^{T}	Transpose of matrix A
A^{-1}	Inverse of matrix A
$A^\dagger$	Pseudoinverse of matrix A
$\det(A)$	Determinant of matrix A
$\text{trace}(A)$	Trace of matrix A
$\text{rank}(A)$	Rank of matrix A
$\odot$	Hadamard (element-wise) matrix product
$\succ$	Positive definite
$\succeq$	Positive semi-definite
a^i	ith coordinate of vector a
a_{ij}	ijth element of matrix A
$\text{vec}(A)$	Column-stack representation of matrix A
Δ_X	Laplace-Beltrami operator
L_X	Laplacian matrix of a graph

Acronyms

2D	Two-dimensional
3D	Three-dimensional
AL	Augmented Lagrangian
BFGS	Broyden-Fletcher-Goldfarb-Shanno
CMC	Cumulative match characteristic
EER	Equal error rate
FAR	False acceptance rate
FMM	Fast marching method
FPS	Farthest point sampling
FRR	False rejection rate
GMDS	Generalized multidimensional scaling
GPU	Graphics processing unit
GSS	Generalized Shanks-Schmidt transform
HLLE	Hessian locally linear embedding
ICP	Iterative closest point
IRBL	Implicitly restarted block-Lanczos
IRLS	Iteratively reweighted least squares
KKT	Karush-Kuhn-Tucker
KLT	Karhunen-Loéve transform
LLE	Locally linear embedding
LTSA	Local tangent space alignment
MDS	Multidimensional scaling
MG	Multigrid
MPE	Minimal polynomial extrapolation
PBM	Penalty/barrier method
PCA	Principal component analysis
RGB	Red green blue
ROC	Receiver operating characteristic
RRE	Reduced rank extrapolation
SIMD	Single instruction multiple data

SMACOF	Scaling by minimizing a convex function
SSE	Streaming SIMD extensions
TEA	Topological ϵ-algorithm

Glossary

ϵ-**isometry** a map that has *distortion* ϵ and is ϵ-*surjective*.
ϵ-**surjection** a map $f : (X, d_X) \to (Y, d_Y)$, such that, $d_Y(Y, f(X)) \le \epsilon$.
$\mathcal{C}^k$-**function** a function that is k-times continuously differentiable.
σ-**algebra** on X (denoted by Σ_X) is a subset of the *power set* of X closed under complement and countable union, i.e., (i) if $X' \in \Sigma_X$ then $X'^c \in \Sigma_X$; (iii) if $X'_i \in T$ for $i = 1, 2, \ldots$ then $\bigcup_i X_i \in T$.

Armijo rule an algorithm for inexact *line search*.

bending of a surface X is a deformation $f : X \to Y$, satisfying $d_X(x_1, x_2) = d_Y(f(x_1), f(x_2))$ for every $x_1, x_2 \in X$.
bi-Lipschitz function an injective *Lipschitz function* whose inverse is also a Lipschitz function.
bijection a map that is *surjective* and *injective*. Bijective maps have an inverse.

characteristic function of a subset X' of X is a function obtaining the value 1 on X' and zero on $X \setminus X'$.
column stack of an $N \times M$ matrix A (denoted as vec(A)) is a column vector of size NM produced by appending the columns of A, such that vec$(A)_{i+(j-1)N} = A_{ij}$.
condition number of a matrix is the ratio of its maximum and minimum *eigenvalues*.
continuous bending of a surface X is a family $\{f_\lambda\}$ of bendings continuous in λ such that $f_0(X) = X$.
convex function a function whose *epigraph* is a convex set.
convex hull of a subset A of a metric space (X, d_X) is the minimum set containing A that is convex in X.
convex set a subset A of a vector space X, such that for every $x, x' \in X$ and $\lambda \in [0, 1]$, it holds $\lambda x + (1 - \lambda)x' \in A$.

crisp set is a fuzzy set whose membership function obtains discrete values $\{0, 1\}$. Crisp set coincides with the traditional definition of a subset in set theory.

diffeomorphism a *bijective* map between two smooth manifolds that is *differentiable* and its inverse is also differentiable.

differentiable function a function whose derivative exists at every point of its domain.

dilation of a map $f : (X, d_X) \to (Y, d_Y)$ is the measure of the maximum relative change of the metric,

$$\operatorname{dil} f = \sup_{x \neq x' \in X} \frac{d_Y(f(x), f(x'))}{d_X(x, x')}.$$

disjoint union (or *discriminated union*) of two sets A and B is denoted by $A \sqcup B$ and can be thought of as $(A \times \{0\}) \cup (B \times \{1\})$.

eigenvalue or a square matrix A is a scalar λ such that $Au = \lambda u$ for some vector u, referred to as an *eigenvector*.

epigraph of a function f (denoted by $\operatorname{epi}(f)$) is a set of functions lying on or above the graph of f.

equivalence class of $x \in X$ under the *equivalence relation* $\sim$ is the set $[x] = \{y \in X : x \sim y\}$.

equivalence relation a binary relation, which is *reflexive*, *symmetric*, and *transitive*.

Euler characteristic of a polyhedron (denoted by χ_X) is defined as $\chi_X = N_F - N_E + N_V$, where N_F is the number of faces, N_E is the number of edges, and N_V is the number of vertices in the polyhedron. For manifolds, the Euler characteristic is defined through the Gauss-Bonnet theorem.

fuzzy set on X is a generalization of the notion of subset, described by a membership function $m : X \to [0, 1]$ that determines the degree of "belonging" of a point in x to the fuzzy set.

genus of a manifold X is the largest number of cuts along nonintersecting closed simple curves that leave the manifold connected. Genus can be intuitively interpreted as the number of "handles" or "holes" a manifold has.

geodesically convex set a subset A of a length space X, containing the geodesics between all $x, y \in A$.

gradient of a differentiable function $f : X \to \mathbb{R}$ (denoted by ∇f) is an operator on X satisfying $df(x) = \langle \nabla f(x), dx \rangle$ for an infinitesimal dx. In coordinate notation, the gradient is a vector of the first-order partial derivatives of f.

Gram matrix a symmetric matrix of inner products.

group a pair $(G, *)$ of a set G and a binary operator $* : G \to G$, satisfying (i) *closure*: for all $a, b \in G$, $a * b \in G$; (ii) *associativity*: for all $a, b, c \in G$, $(a * b) * c = a * (b * c)$; (iii) there exists an *identity element* $e \in G$ such that for all $a \in G$ $e * a = a * e = a$; and (iv) for all $a \in G$, there exists an *inverse element* $b \in G$ such that $a * b = b * a = e$.

Hadamard product element-wise product of two matrices.
Hausdorff space a *topological space* (X, T), in which for every distinct x, y, there exist disjoint open sets $U, V \in T$ such that $x \in U$ and $y \in V$.
Hessian of a twice-differentiable function $f : X \to \mathbb{R}$ (denoted by $\nabla^2 f$) is a bilinear operator satisfying $d(\nabla f)(x) = \nabla^2 f(x) dx$. In coordinate notation, the Hessian is a symmetric matrix of the second-order partial derivatives of f.
homeomorphism a *bijective* continuous map with a continuous inverse. Homeomorphisms preserve topological properties.

injection (one-to-one map) a map $f : X \to Y$ associating distinct argument to distinct values, such that $f(x_1) = f(x_2)$ implies $x_1 \neq x_2$ for all $x_1, x_2 \in X$.
intrinsic geometry generic name for properties of a Riemannian manifold, expressible in terms of the distance structure.
isometric embedding a distance-preserving map.
isometry *bijective* distance-preserving map.

line search generic name for a procedure for finding the minimum of a one-dimensional function.
Lipschitz function a function whose *dilation* is bounded.
lower triangular matrix a matrix A with elements $a_{ij} = 0$ for all $j > i$.

measurable function (or Σ_S-*measurable* function) is a function $f : X \to \mathbb{R}$ such that $\{x : f(x) \leq \delta\} \in \Sigma_S$ for all δ, where Σ_X is a σ-*algebra* on X.
metric a non-negative function $d : X \to \mathbb{R}$ satisfying for every $x, y, z \in X$ (i) $d(x, y) = 0$ if and only if $x = y$; (2) $d(x, y) = d(y, x)$; and (3) $d(x, y) \leq d(x, z) + d(y, z)$.
metric space (denoted by (X, d_X)) a space X equipped with a *metric* d_X.
minor of an $N \times N$ matrix A (denoted by $M_{ij}(A)$) is the determinant of the $(N-1) \times (N-1)$ matrix, obtained by removing the ith row and jth column from A.
multicriterion optimization optimization problem in which the objective function is vector-valued.

normed vector space a vector space equipped with a *norm*.

over-determined a system of linear equations containing more equations than variables.

Pareto optimum a solution of a multicriterion optimization problem, at which none of the criteria can be improved without compromising the other ones.
positive-definite matrix a matrix A satisfying $Au > 0$ for all vectors $u \neq 0$. This definition can be extended to operators.
positive-semidefinite matrix a matrix A satisfying $Au \geq 0$ for all vectors u. This definition can be extended to operators.
power set of X (denoted by 2^X) is the set of all subsets of X.
preimage of $A \subseteq Y$ under the map $f : X \to Y$ (denoted by $f^{-1}(A)$) is the set of all the arguments mapped by f into A, i.e., $f^{-1}(A) = \{x \in X : f(x) \in A\}$.
pseudoinverse of a matrix A (denoted by $A^\dagger$) is a generalization of inverse for non-invertible matrices. The pseudoinverse is given by $A^\dagger = (A^T A)^{-1} A^T$ and coincides with the standard definition of inverse in the case when A is invertible.

quotient space of a space X under the equivalence relation $\sim$ (denoted by $X \backslash \sim$) is the set X^*, whose members are *equivalence classes*.

rank of a matrix A is the dimensionality spanned by the columns (column rank) or the rows (row rank) of A. For square matrices, the row and column rank is equal to the number of non-zero *eigenvalues*.
reflexive relation a binary relation on X, such that every $x \in X$ is in relation with itself.

Salukwadze optimum the closest of all Pareto optima to the *utopia point*.
smooth function ($\mathcal{C}^\infty$-function) a function that has continuous derivatives of all orders.
sparse matrix a matrix containing mostly zero values.
spectrum set of *eigenvalues* of a matrix or an operator.
suboptimality of an objective function f at the solution x is the difference $f(x) - f(x^*)$.
surjection (onto map) a map $f : X \to Y$, whose range spans the whole codomain, i.e., $f(X) = Y$.
symmetric matrix a square matrix A satisfying $A^T = A$.
symmetric relation a binary relation on X, such that if x is in relation with y, then y is in relation with x for all $x, y \in X$.

topological space denoted by (X, T) is a space X equipped with a *topology* T.
topology on X is a subset T of the *power set* of X closed under union and intersection, i.e., (i) $X, \emptyset \in \Sigma_X$; (ii) if $X_\alpha \in T$, then $\bigcup_\alpha X_\alpha \in T$; (iii) if $X_\alpha \in T$, then $\bigcap_\alpha X_\alpha \in T$. Broadly, topology also refers to topological properties of an object.
transitive relation such that if x is in relation with y, and y is in relation with z, then x is in relation with z for all $x, y, z \in X$.

unitary matrix a square matrix A satisfying $A^\mathrm{T} A = I$.

utopia point an ideal, usually non-achievable solution of a multicriterion optimization problem.

vector space a set that is closed under finite vector addition and scalar multiplication.

References

1. B. Achermann and H. Bunke, *Classifying range images of human faces with Hausdorff distance*, Int'l Conf. Pattern Recognition (ICPR), 2000, pp. 809–813.
2. B. Achermann, X. Jiang, and H. Bunke, *Face recognition using range images*, Proc. Int. Conf. Virtual Systems and Multimedia, 1997, pp. 129–136.
3. M. Alexa, *Recent advances in mesh morphing*, Computer Graphics Forum **21** (2002), no. 2, 173–196.
4. M. Alexa, D. Cohen-Or, and D. Levin, *As-rigid-as-possible polygon morphing*, Proc. SIGGRAPH (2000), 157–164.
5. H. Alt, K. Mehlhorn, H. Wagener, and E. Welzl, *Congruence, similarity, and symmetries of geometric objects*, Discrete Comput. Geom. **3** (1988), 237–256.
6. S. Amari, *Methods of information geometry*, AMS, 2000.
7. N. Amenta, M. Bern, and M. Kamvysselis, *Surface reconstruction by Voronoi filtering*, Discrete and Computational Geometry **22** (1999), no. 4, 481–504.
8. D. Anguelov, *Learning models of shape from 3D range data*, Ph.D. thesis, Stanford University, 2005.
9. W. E. Arnoldi, *The principle of minimized iterations in the solution of the matrix eigenvalue problem*, Quarterly Applied Mathematics **9** (1951), 17–29.
10. D. Arthur and S. Vassilvitskii, *Worst-case and smoothed analysis of the ICP algorithm, with an application to the k-means method*, Proc. Symp. Foundations of Computer Science (2006), 153–164.
11. S. Arya, D. M. Mount, N. S. Netanyahu, R. Silverman, and A. Y. Wu, *An optimal algorithm for approximate nearest neighbor searching*, J. ACM **45** (1998), 891–923.
12. J. Ashbourn, *Biometrics: advanced identity verification*, Springer, 2002.
13. M. J. Atallah, *On symmetry detection*, IEEE Trans. Computers **34** (1985), no. 7, 663–666.
14. F. Aurenhammer, *Voronoi diagrams: A survey of a fundamental geometric data structure*, ACM Computing Surveys **23** (1991), no. 3, 345–405.
15. F. Aurenhammer and R. Klein, *Voronoi diagrams*, Karl-Franzens-Universität Graz, 1996.
16. J. Baglama, D. Calvetti, and L. Reichel, *IRBL: An implicitly restarted block Lanczos method for large-scale hermitian eigenproblems*, SIAM J. Scientific Computing (SISC) **24** (2003), no. 5, 1650–1677.

17. R. Bajcsy and F. Solina, *Three dimensional object representation revisited*, Proc. Int'l Conf. Computer Vision (ICCV), 1987, pp. 231–240.
18. N. S. Bakhvalov, *On the convergence of a relaxation method under natural constraints on an elliptic operator*, USSR Computational Mathematics and Mathematical Physics **6** (1966), 861–883.
19. R. Basri, L. Costa, D. Geiger, and D. Jacobs, *Determining the similarity of deformable shapes*, Vision Research **38** (1998), 2365–2385.
20. P. N. Belhumeur, J. P. Hespanha, and D. J. Kriegman, *Eigenfaces vs Fisherfaces: recognition using class specific linear projection*, IEEE Trans. Pattern Analysis and Machine Intelligence (PAMI) **19** (1997), 711–720.
21. M. Belkin and P. Niyogi, *Laplacian eigenmaps for dimensionality reduction and data representation*, Neural Computation **13** (2003), 1373–1396.
22. _____, *Towards a theoretical foundation for Laplacian-based manifold methods*, Proc. Conf. Learning Theory (COLT), Lecture Notes in Art. Intelligence, no. 3559, Springer, 2005, pp. 486–500.
23. R. E. Bellman, *Dynamic programming*, Dover, 2003.
24. C. Benoit, *Note sur une méthode de résolution des équations normales provenant de l'application de la méthode des moindres carrés à un système d'équations linéaires en nombre inférieur à celui des inconnues (procédé du Commandant Cholesky)*, Bulletin géodésique **2** (1924), 67–77.
25. J. L. Bentley, *Multidimensional binary search trees used for associative searching*, Communications of the ACM **18** (1975), no. 9, 509–517.
26. M. Berger, *A panoramic view of riemannian geometry*, Springer, 2003.
27. M. Berger, P. Gauduchon, and E. Mazet, *Le spectre d'une variété riemannienne*, 1971.
28. A. Bertillon, *Instructions signalétiques pour l'identification anthropométrique*, Melun, 1893.
29. D. P. Bertsekas, *Constrained optimization and Lagrange multiplier methods*, Academic Press, 1982.
30. _____, *Nonlinear programming*, 2nd ed., Athena Scientific, 1999.
31. P. J. Besl and N. D. McKay, *A method for registration of 3D shapes*, IEEE Trans. Pattern Analysis and Machine Intelligence (PAMI) **14** (1992), 239–256.
32. C. Beumier and M. P. Acheroy, *Automatic face authentication from 3D surface*, Proc. British Machine Vision Conf., 1988, pp. 449–458.
33. I. Biderman, *Human image understanding: recent research and a theory*, Computer Graphics, Vision and Image Processing **32** (1985), 29–73.
34. T. O. Binford, *Visual perception by computer*, Proc. IEEE Conf. Systems and Control, 1987, pp. 231–240.
35. V. Blanz, C. Basso, T. Poggio, and T. Vetter, *Reanimating faces in images and video*, Computer Graphics Forum **22** (2003), no. 3, 641–650.
36. V. Blanz and T. Vetter, *A morphable model for the synthesis of 3D faces*, Proc. Conf. Computer Graphics and Interactive Techniques (1999), 187–194.
37. V. Blanz and T. Vetter, *Face recognition based on fitting a 3D morphable model*, IEEE Trans. Pattern Analysis and Machine Intelligence (PAMI) **25** (2003), 1063–1074.
38. W. W. Bledsoe, *Man-machine face recognition*, Technical Report 22, Panoramic Research Inc., 1966.
39. H. Blum, *A transformation for extracting new descriptors of shape*, Models for the Perception of Speech and Visual Form (1967), 362–380.

40. _____, *Biological shape and visual science (Part I)*, J. Theoretical Biology **38** (1973), no. 2, 205–87.
41. H. Blum and R. N. Nagel, *Shape description using weighted symmetric axis features*, Pattern Recognition **10** (1978), no. 3, 167–180.
42. O. Boiman and M. Irani, *Similarity by composition*, Int'l Conf. Neural Information Processing Systems (NIPS), 2006.
43. P. O. Bonnet, *Mémoire sur la théorie génerale des surfaces*, J. École Polytechnique **19** (1848), 1–133.
44. I. Borg and P. Groenen, *Modern multidimensional scaling - theory and applications*, Springer, 1997.
45. K. W. Bowyer, K. Chang, and P. Flynn, *A survey of approaches and challenges in 3D and multi-modal 3D+2D face recognition*, Computer Vision and Image Understanding **101** (2006), 1–15.
46. S. Boyd and L. Vandenberghe, *Convex optimization*, Cambridge University Press, 2003.
47. Y. Boykov and V. Kolmogorov, *Computing geodesics and minimal surfaces via graph cuts*, Proc. Int'l Conf. Computer Vision (ICCV), 2003, pp. 26–33.
48. M. Brand, *Nonrigid embeddings for dimensionality reduction*, Proc. European Conf. Machine Learning (ECML), vol. 3720, 2005.
49. A. Brandt, *Multi-level adaptive technique (MLAT) for fast numerical solution to boundary value problems*, Proc. Int'l Conf. Numerical Methods in Fluid Mechanics, Springer, 1973, pp. 82–89.
50. _____, *Multi-level adaptive solutions to boundary-value problem*, Mathematics of Computation **37** (1977), 333–390.
51. C. Brezinski, *Application de l'ε-algorithme à la résolution des systèmes non linéaires.*, Compte Rendus, Academie des Science **271 A** (1970), 1174–1177.
52. _____, *Accélération de la convergence en analyse numérique*, Lecture notes in mathematics, Springer, 1977.
53. _____, *Extrapolation algorithms and Padé approximations: a historical survey*, Applied Numerical Mathematics **20** (1996), 299–318.
54. W. L. Briggs, *A multigrid tutorial*, SIAM, 1987.
55. A. M. Bronstein and M. M. Bronstein, *Partial matching of rigid shapes*, Tech. Report CIS-2008-02, Dept. of Computer Science, Technion, Israel, 2008.
56. A. M. Bronstein, M. M. Bronstein, A. M. Bruckstein, and R. Kimmel, *Matching two-dimensional articulated shapes using generalized multidimensional scaling*, Proc. Articulated Motions and Deformable Objects (AMDO), 2006, pp. 48–57.
57. _____, *Paretian similarity of non-rigid objects*, Proc. Int'l Conf. Scale Space and Variational Methods in Computer Vision (SSVM) (F. Sgallari, A. Murli, and N. Paragios, eds.), Lecture Notes in Computer Science, no. 4485, Springer, 2007, pp. 264–275.
58. _____, *Partial similarity of objects and text sequences*, Proc. Information Theory and Applications Workshop (ITA), 2007.
59. _____, *Analysis of two-dimensional non-rigid shapes*, Int'l J. Computer Vision (IJCV) (2008).
60. A. M. Bronstein, M. M. Bronstein, Y. S. Devir, R. Kimmel, and O. Weber, *Parallel algorithms for approximation of distance maps on parametric surfaces*, ACM Trans. Graphics, submitted.
61. _____, *Parallel algorithms for approximation of distance maps on parametric surfaces*, Tech. Report CIS-2007-03, Dept. of Computer Science, Technion, Israel, 2007.

62. A. M. Bronstein, M. M. Bronstein, E. Gordon, and R. Kimmel, *High-resolution structured light range scanner with automatic calibration*, Tech. Report CIS-2003-06, Dept. of Computer Science, Technion, Israel, 2003.
63. A. M. Bronstein, M. M. Bronstein, and R. Kimmel, *Expression-invariant 3D face recognition*, Proc. Audio and Video-based Biometric Person Authentication (AVBPA) (J. Kittler and M. S. Nixon, eds.), Lecture Notes in Computer Science, no. 2688, Springer, 2003, pp. 62–69.
64. _____, *Expression-invariant face recognition via spherical embedding*, Proc. Int'l Conf. Image Processing (ICIP), vol. 3, 2005, pp. 756–759.
65. _____, *On isometric embedding of facial surfaces into $\mathbb{S}^3$*, Proc. Int'l Conf. Scale Space and PDE Methods in Computer Vision, Lecture Notes in Computer Science, no. 3459, Springer, 2005, pp. 622–631.
66. _____, *Three-dimensional face recognition*, Int'l J. Computer Vision (IJCV) **64** (2005), no. 1, 5–30.
67. _____, *Efficient computation of isometry-invariant distances between surfaces*, SIAM J. Scientific Computing **28** (2006), no. 5, 1812–1836.
68. _____, *Expression invariant face recognition: faces as isometric surfaces*, Face Processing: Advanced Modeling and Methods (R. Chellappa and W. Zhao, eds.), Academic Press, 2006.
69. _____, *Expression-invariant representations of faces*, IEEE Trans. Image Processing (2006).
70. _____, *Generalized multidimensional scaling: a framework for isometry-invariant partial surface matching*, Proc. National Academy of Science (PNAS) **103** (2006), no. 5, 1168–1172.
71. _____, *Robust expression-invariant face recognition from partially missing data*, Proc. European Conf. Computer Vision (ECCV), 2006, pp. 396–408.
72. _____, *Calculus of non-rigid surfaces for geometry and texture manipulation*, IEEE Trans. on Visualization and Computer Graphics **13** (2007), no. 5, 903–913.
73. _____, *Expression-invariant representation of faces*, IEEE Trans. Image Processing **16** (2007), no. 1, 188–197.
74. A. M. Bronstein, M. M. Bronstein, and R. Kimmel, *Rock, Paper, and Scissors: extrinsic vs. intrinsic similarity of non-rigid shapes*, Proc. Int'l Conf. Computer Vision (ICCV), 2007.
75. A. M. Bronstein, M. M. Bronstein, and R. Kimmel, *Story of Cinderella: biometrics and isometry-invariant distances*, 3D Imaging for Safety and Security (A. Koschan, M. Pollefeys, and M. Abidi, eds.), Springer, 2007.
76. A. M. Bronstein, M. M. Bronstein, R. Kimmel, and E. Gordon, *Fusion of 2D and 3D data in three-dimensional face recognition*, Proc. Int'l Conf. Image Processing (ICIP), 2004.
77. A. M. Bronstein, M. M. Bronstein, R. Kimmel, and A. Spira, *Face recognition from facial surface metric*, Proc. European Conf. Computer Vision (ECCV), Lecture Notes in Computer Science, no. 3023, Springer, 2004, pp. 225–237.
78. M. M. Bronstein and A. M. Bronstein, *Biometrics was no match for hair-rising tricks*, Nature **420** (2002), 739.
79. M. M. Bronstein, A. M. Bronstein, R. Kimmel, and I. Yavneh, *A multigrid approach for multidimensional scaling*, Proc. Copper Mountain Conf. Multigrid Methods, 2005.
80. _____, *Multigrid multidimensional scaling*, Numerical Linear Algebra with Applications (NLAA) **13** (2006), 149–171.

81. R. Brooks, *Symbolic reasoning among 3-D models and 2-D images*, Artificial Intelligence **17** (1981), 285–348.
82. _____, *On manifolds of negative curvature with isospectral potentials*, Topology **26** (1987), 63–66.
83. R. Brooks and R. Tse, *Isospectral surfaces of small genus*, Nagoya Mathematical J. **107** (1987), 13–24.
84. C. G. Broyden, *A class of methods for solving nonlinear simultaneous equations*, Mathematics of Computation **19** (1965), 577–593.
85. C. G. Broyden, J. E. Dennis, and J. J. Moré, *On the local and superlinear convergence of quasi-newton methods*, J. Institute of Mathematical Applications **12** (1973), 223–246.
86. V. Bruce, *Recognizing faces*, Lawrence Erlbaum Associates, 1988.
87. I. S. Bruner and R. Tagiuri, *The perception of people*, Handbook of social psychology (G. Lindzey, ed.), vol. 2, Addison-Wesley, 1954, pp. 634–654.
88. D. Burago, Y. Burago, and S. Ivanov, *A course in metric geometry*, Graduate studies in mathematics, vol. 33, AMS, 2001.
89. _____, *A course in metric geometry*, Graduate studies in mathematics, vol. 33, AMS, 2001.
90. P. Buser, *Isospectral Riemann surfaces*, Annales de l'Institut Fourier **36** (1986), 167–192.
91. S. Cabay and L. Jackson, *Polynomial extrapolation method for finding limits and antilimits of vector sequences*, SIAM J. Numerical Analysis (SINUM) **13** (1976), no. 5, 734–752.
92. D. Calvetti, L. Reichel, and D. C. Sorensen, *An implicitly restarted Lanczos method for large symmetric eigenvalue problems*, Electronic Trans. Numerical Analysis (ETNA) **2** (1994), 1–21.
93. J. Y. Cartoux, J. T. Lapreste, and M. Richetin, *Face authentication or recognition by profile extraction from range images*, Proc. Workshop on Interpretation of 3D Scenes, 1989, pp. 194–199.
94. A. L. Cauchy, *Sur les polygones et polyedres, second memoire*, J. Ecole Polytechnique **19** (1813), 87–90.
95. M. Chalmers, *A linear iteration time layout algorithm for visualizing high dimensional data*, Proc. IEEE Visualization, 1996, pp. 127–132.
96. T. F Chan and L. A. Vese, *Active contours without edges*, Trans. Image Processing **10** (2001), no. 2, 266–277.
97. K. Chang, K. W. Bowyer, and P. Flynn, *Face recognition using 2D and 3D facial data*, Workshop on Multimodal User Authentication, 2003, pp. 25–32.
98. G. Charpiat, O. Faugeras, and R. Keriven, *Approximations of shape metrics and application to shape warping and empirical shape statistics*, Foundations of Computational Mathematics **5** (2005), no. 1, 1–58.
99. Y. Chen and G. Medioni, *Object modeling by registration of multiple range images*, Proc. Conf. Robotics and Automation, 1991.
100. Y. K. Chen, A. Vetro, H. Sun, and S. Y. Kung, *Frame-rate upconversion using transmitted true motion vectors*, Proc. Workshop on Multimedia Signal Processing, 1998, pp. 622–627.
101. S. W. Cheng, T. K. Dey, E. A. Ramos, and T. Ray, *Sampling and meshing a surface with guaranteed topology and geometry*, Proc. Symp. Computational Geometry (2004), 280–289.
102. E. F. F. Chladni, *Entdeckungen über die theorie des klanges*, Leipzig, 1787.

103. B. T. Choi, S. H. Lee, and S. J. Ko, *New frame rate up-conversion using bi-directional motion estimation*, IEEE Trans. Consumer Electronics **46** (2000), no. 3, 603–609.
104. C. Chua, F. Han, and Y. K. Ho, *3D human face recognition using point signature*, IEEE Int'l Conf. Automatic Face and Gesture Recognition, 2000, pp. 233–238.
105. F. R. K. Chung, *Spectral graph theory*, AMS, 1997.
106. L. Cohen and R. Kimmel, *Global minimum for active contour models: A minimal path approach*, Int'l J. Computer Vision (IJCV) **24** (1997), no. 1, 57–78.
107. D. Cohen-Steiner, P. Alliez, and M. Desbrun, *Variational shape approximation*, ACM Trans. Graphics **23** (2004), no. 3, 905–914.
108. S. E. Cohn-Vossen, *Nonrigid closed surfaces*, Annals of Mathematics **102** (1929), 10–29.
109. R. R. Coifman, S. Lafon, A. B. Lee, M. Maggioni, B. Nadler, F. Warner, and S. W. Zucker, *Geometric diffusions as a tool for harmonic analysis and structure definition of data: Diffusion maps*, Proc. National Academy of Sciences (PNAS) **102** (2005), no. 21, 7426–7431.
110. A. R. Conn, N. I. M. Gould, and P. Toint, *A globally convergent augmented lagrangian algorithm for optimization with general constraints and simple bounds*, SIAM J. Numerical Analysis (SINUM) **28** (1991), no. 2, 545–572.
111. A. R. Conn, N. I. M. Gould, and Ph. L. Toint, *LANCELOT: a Fortran package for large-scale nonlinear optimization (release A)*, Computational Mathematics, vol. 17, Springer, 1992.
112. J. H. Connell and M. Brady, *Generating and generalizing models of visual objects*, Artificial intelligence **31** (1987), 159–183.
113. R. Connelly, *A flexible sphere*, Math. Intell **1** (1978), no. 3, 130–131.
114. _____, *The rigidity of polyhedral surfaces*, Mathematics Magazine **52** (1979), no. 5, 275–283.
115. _____, *Rigidity*, Handbook of Convex Geometry (1993), 223–271.
116. W. J. Cook, W. H. Cunningham, W. R. Pulleyblank, and A. Schrijver, *Combinatorial optimization*, Wiley, 1997.
117. I. Cox, J. Ghosn, and P. Yianilos, *Feature-based face recognition using mixture distance*, Proc. Computer Vision and Pattern Recognition (CVPR), 1996, pp. 209–216.
118. T. F. Cox and M. A. A. Cox, *Multidimensional scaling on a sphere*, Communications in Statistics: Theory and Methods **20** (1991), 2943–2953.
119. _____, *Multidimensional scaling*, Chapman and Hall, London, 1994.
120. M. G. Crandal and P.-L. Lions, *Viscosity solutions of Hamilton–Jacobi equations*, Trans. AMS **277** (1983), 1–43.
121. P. Danielsson, *Euclidean distance mapping*, Computer Graphics and Image Processing **14** (1980), 227–248.
122. C. Darwin, *The expression of the emotions in man and animals*, Murray, 1872.
123. J. Dattorro, *Convex optimization and euclidean distance geometry*, Meboo Publishing, 2006.
124. E. F. D'Azevedo and R. B. Simpson, *On optimal interpolation triangle incidences*, SIAM J. Scientific Computing (SISC) **10** (2006), no. 6, 1063–1075.
125. J. de Leeuw, *Applications of convex analysis to multidimensional scaling*, Recent developments in statistics (J. R. Barra, F. Brodeau, G. Romier, and B. van Cutsem, eds.), North-Holland, 1977, pp. 133–145.

126. _____, *Convergence of the majorization method for multidimensional scaling*, J. Classification **5** (1988), 163–180.
127. _____, *Accelerated least squares multidimensional scaling*, Preprint 493, Department of Statistics, UCLA, 2006.
128. J. de Leeuw and W. Heiser, *Multidimensional scaling with restrictions on the configuration*, Multivariate Analysis (P. R. Krishnaiah, ed.), vol. 5, North-Holland Publishing Company, Amsterdam, 1980, pp. 501–522.
129. J. de Leeuw and I. Stoop, *Upper bounds on Kruskal's stress*, Psychometrika **49** (1984), 391–402.
130. S. de Rooij and P. Vitanyi, *Approximating rate-distortion graphs of individual data: Experiments in lossy compression and denoising*, CoRR **abs/cs/0609121** (2006).
131. A. de Saint-Exupéry, *Le petit prince*, 1943.
132. V. de Silva and J. B. Tenenbaum, *Sparse multidimensional scaling using landmark points*, Technical report, Stanford University, 2004.
133. B. Delaunay, *Sur la sphére vide*, Izvestia Akademii Nauk SSSR, Otdelenie Matematicheskikh i Estestvennykh Nauk **7** (1934), 793–800.
134. M. Desbrun, M. Meyer, P. Schröder, and A. H. Barr, *Implicit fairing of irregular meshes using diffusion and curvature flow*, Proc. SIGGRAPH, 1999, pp. 317–324.
135. P. Deuflhard and M. Weiser, *Local inexact Newton multilevel FEM for nonlinear elliptic problems*, Computational science for the 21st century (M.-O. Bristeu et al., ed.), Wiley, 1997, pp. 129–138.
136. P. Diamond, *Metric spaces of fuzzy sets: Theory and applications*, World Scientific, 1994.
137. E. W. Dijkstra, *A note on two problems in connexion with graphs*, Numerische Mathematik **1** (1959), 269–271.
138. G. L. Dirichlet, *Über die reduktion der positiven quadratischen Formen mit drei unbestimmten ganzen Zahlen*, J. für die Reine und Angewandte Mathematik **40** (1850), 209–227.
139. M. P. do Carmo, *Differential geometry of curves and surfaces*, Prentice-Hall, 1976.
140. Q. Du, V. Faber, and M. Gunzburger, *Centroidal Voronoi tessellations: Applications and algorithms*, SIAM Review **41** (2006), no. 4, 637–676.
141. R. O. Duda, P. E. Hart, and D. G. Stork, *Pattern classification*, Wiley, 2000.
142. P. Dupuis and J. Oliensis, *Shape from shading: Provably convergent algorithms and uniqueness results*, Proc. European Conf. Computer Vision (ECCV), 1994, pp. 259–268.
143. C. Eckart and G. Young, *Approximation of one matrix by another of lower rank*, Psychometrika **1** (1936), 211–218.
144. R. P. Eddy, *Extrapolationg to the limit of a vector sequence*, Information Linkage between Applied Mathematics and Industry (P. C. Wang, ed.), Academic Press, 1979, pp. 387–396.
145. F. Y. Edgeworth, *Mathematical psychics: An essay on the application of mathematics to the moral sciences*, 1881.
146. P. Ekman, *Darwin and facial expression; a century of research in review*, Academic Press, 1973.
147. A. Elad and R. Kimmel, *Bending invariant representations for surfaces*, Proc. Computer Vision and Pattern Recognition (CVPR), 2001, pp. 168–174.

148. _____, *Spherical flattening of the cortex surface*, Geometric methods in biomedical image processing (R. Malladi, ed.), vol. 2191, Springer, 2002, pp. 77–89.
149. _____, *On bending invariant signatures for surfaces*, IEEE Trans. Pattern Analysis and Machine Intelligence (PAMI) **25** (2003), no. 10, 1285–1295.
150. M. Elad, P. Milanfar, and G. H. Golub, *Shape from moments-an estimation theory perspective*, IEEE Trans. Signal Processing **52** (2004), no. 7, 1814–1829.
151. Y. Eldar, M. Lindenbaum, M. Porat, and Y. Y. Zeevi, *The farthest point strategy for progressive image sampling*, IEEE Trans. Image Processing **6** (1997), no. 9, 1305–1315.
152. K. Etemad and R. Chellappa, *Discriminant analysis for recognition of human face images*, J. Optical Society of America (JOSA) A **14** (1997), 1724–1733.
153. L. Euler, *Elementa doctrinæ solidorum*, Novi commentarii academiæ scientiarum Petropolitanæ **4** (1758), no. 1752/3, 109–140.
154. E. Ezra, M. Sharir, and A. Efrat, *On the ICP algorithm*, Proc. Symp. Computational Geometry (2006), 95–104.
155. C. Faloutsos and K.-I. Lin, *FastMap: a fast algorithm for indexing, datamining, and visualization*, Proc. Special Interest Group on Management of Data (SIGMOD), 1995, pp. 163–174.
156. H. Federer, *Curvature measures*, Trans. AMS **93** (1959), no. 3, 418–491.
157. R. P. Fedorenko, *On the speed of convergence of an iterative process*, USSR Computational Mathematics and Mathematical Physics **4** (1964), 559–564.
158. R. Fletcher, *A new approach to variable metric methods*, Computer J. **13** (1970), 317–322.
159. M. S. Floater, *Parametrization and smooth approximation of surface triangulations*, Computer Aided Geometric Design **14** (1997), no. 3, 231–250.
160. _____, *Parametric tilings and scattered data approximation*, Int'l J. Shape Modeling **4** (1998), no. 1, 165–182.
161. M. S. Floater and C. Gotsman, *How to morph tilings injectively*, J. Computational and Applied Mathematics **101** (1999), no. 1-2, 117–129.
162. M. S. Floater and K. Hormann, *Recent advances in surface parameterization*, Multiresolution in Geometric Modelling (2003), 259–284.
163. _____, *Surface parameterization: a tutorial and survey*, Advances in Multiresolution for Geometric Modelling **1** (2005).
164. D. A. Forsyth and J. Ponce, *Computer vision: A modern approach*, Prentice Hall, 2003.
165. R. Fourer, *Nonlinear programming frequently asked questions*, online, http://www-unix.mcs.anl.gov/otc/Guide/faq/ nonlinear-programming-faq.html, 2000.
166. J. Fu, *Convergence of curvatures in secant approximations*, Differential Geometry **37** (1993), 177–190.
167. F. Galton, *Personal identification and description*, Nature (1888), 173–188.
168. M. Garland and P. S. Heckbert, *Surface simplification using quadric error metrics*, 1997, pp. 209–216.
169. C. F. Gauss, *Disquisitiones generales circa superficies curvas*, Dietrich, Göttingen, 1828.
170. N. Gelfand, N. J. Mitra, L. Guibas, and H. Pottmann, *Robust global registration*, Proc. Symp. Geometry Processing (SGP) (2005).
171. *The Bible: Authorized King James version with Apocrypha*, Oxford University Press, 1998.

172. A. Gersho and R. M. Gray, *Vector quantization and signal compression*, Kluwer, 1992.
173. A. S. Gheorghiades, P. N. Belhumeur, and D. J. Kriegman, *From few to many: illumination cone models for face recognition under variable lighting and pose*, IEEE Trans. Pattern Analysis and Machine Intelligence (PAMI) **23** (2001), no. 6, 643–660.
174. H. Gluck, *Almost all simply connected closed surfaces are rigid*, Proc. Conf. Geometric Topology (1974).
175. D. Goldfarb, *A family of variable metric methods derived by variational means*, Mathematics of Computation **24** (1970), 23–26.
176. E. Goldstein and C. Gotsman, *Polygon morphing using a multiresolution representation*, Proc. Graphics Interface, 1995.
177. G. H. Golub and C. F. van Loan, *Matrix computations*, John Hopkins University Press, 1996.
178. J. Gomes, *Warping & morphing of graphical objects*, Morgan Kaufmann, 1998.
179. T. F. Gonzalez, *Clustering to minimize the maximum intercluster distance*, Theoretical Computer Science **38** (1985), no. 2, 293–306.
180. C. Gordon, D. L. Webb, and S. Wolpert, *One cannot hear the shape of the drum*, Bulletin AMS **27** (1992), no. 1, 134–138.
181. G. Gordon, *Face recognition based on depth and curvature features*, Proc. Computer Vision and Pattern Recognition (CVPR), 1992, pp. 108–110.
182. C. Gotsman and V. Surazhsky, *Guaranteed intersection-free polygon morphing*, Computers and Graphics **25** (2001), no. 1, 67–75.
183. J. C. Gower, *Some distance properties of latent root and vector methods used in multivariate analysis*, Biometrika **53** (1966), 325–338.
184. _____, *Euclidean distance geometry*, Mathematical Scientist **7** (1982), 1–14.
185. R. M. Gray, *Vector quantization*, IEEE ASSP Magazine **1** (1984), 4–29.
186. G. Green, *The elegant universe – superstrings and hidden dimensions*, W. Norton, New York (1999).
187. C. Grimes and D. L. Donoho, *Hessian eigenmaps: Locally linear embedding techniques for high-dimensional data*, Proc. National Academy of Sciences (PNAS) **100** (2003), no. 10, 5591–5596.
188. H. Groemer, *Geometric applications of Fourier series and spherical harmonics*, Cambridge University Press, New York, 1996.
189. P. Groenen and M. van de Velden, *Multidimensional scaling*, Econometric Institute Report 15, Erasmus University, 2004.
190. M. Gromov, *Structures métriques pour les variétés riemanniennes*, Textes Mathématiques, no. 1, 1981.
191. P. M. Gruber, *Optimal arrangement of finite point sets in Riemannian 2-manifolds*, Proc. Steklov Institute of Mathematics, 1999.
192. X. Gu, S. Gortler, and H. Hoppe, *Geometry images*, Proc. SIGGRAPH (2002), 355–361.
193. A. Gueziec, F. Bossen, G. Taubin, and C. T. Silva, *Efficient compression of non-manifold polygonal meshes*, Computational Geometry **14** (1999), no. 1-3, 137–166.
194. S. Gupta, M. K. Markey, J. K. Aggarwal, and A. C. Bovik, *3D face recognition based on geodesic distances*, Proc. SPIE Electronic Imaging, 2007.
195. I. Guskov, K. Vidimče, W. Sweldens, and P. Schröder, *Normal meshes*, Proc. Conf. Computer Graphics and Interactive Techniques (2000), 95–102.

196. L. Guttman, *A general nonmetric technique for finding the smallest coordinate space for a configuration of points*, Psychometrika (1968), 469–506.
197. W. Hackbusch, *A multi-grid method applied to a boundary problem with variable coefficients in a rectangle*, Tech. Report 77-17, Institut für Angewandte Mathematik, Universität Köln, 1977.
198. J. Ham, D. D. Lee, S. Mika, and B. Schölkopf, *A kernel view of the dimensionality reduction of manifolds*, Technical Report TR-110, Max Planck Institute for Biological Cybernectics, 2003.
199. F. R. Hampel, E. M. Ronchetti, P. J. Rousseeuw, and W. A. Stahel, *Robust statistics: The approach based on influence functions*, New York (1986).
200. F. Hausdorff, *Grundzüge der Mengenlehre*, Verlag Veit & Co, 1914.
201. M. Hein, J.-Y. Audibert, and U. von Luxburg, *From graphs to manifolds - weak and strong pointwise consistency of graph Laplacians*, Proc. Conf. Learning Theory (COLT), Lecture Notes in Art. Intelligence, no. 3559, Springer, 2005, pp. 470–485.
202. Y. Hel-Or and M. Werman, *Constraint-fusion for interpretation of articulated objects*, Proc. Computer Vision and Pattern Recognition (CVPR), 1994, pp. 39–45.
203. C. Hesher, A. Srivastava, and G. Erlebacher, *A novel technique for face recognition using range images*, Int'l Symp. Signal Processing and Its Applications, vol. 2, 2003, pp. 201–204.
204. M. R. Hestenes, *Multiplier and gradient methods*, J. Optimization Theory and Applications **4** (1969), 303–320.
205. K. Hildebrandt, K. Polthier, and M. Wardetzky, *On the convergence of metric and geometric properties of polyhedral surfaces*, Geometricae Dedicata **123** (2006), 89–112.
206. D. Hoffman and W. Richards, *Parts of recognition*, Visual cognition (S. Pinker, ed.), MIT Press, 1984.
207. D. Hoffman and M. Singh, *Salience of visual parts*, Cognition **63** (1997), 29–78.
208. B. K. P. Horn, *Closed-form solution of absolute orientation using unit quaternions*, J. Optical Society of America (JOSA) A **4** (1987), 629–642.
209. B. K. P. Horn and B. G. Schunck, *Determining optical flow*, Artificial Intelligence **17** (1981), no. 1-3, 185–203.
210. H. Hotelling, *Analysis of a complex of statistical variables into principal components*, (1933).
211. P. S. Huang, C. P. Zhang, and F. P. Chiang, *High speed 3-D shape measurement based on digital fringe projection*, Optical Engineering **42** (2003), no. 1, 163–168.
212. P. J. Huber, *Robust statistics*, Wiley, 2004.
213. D. Jacobs, D. Weinshall, and Y. Gdalyahu, *Class representation and image retrieval with non-metric distances*, IEEE Trans. Pattern Analysis and Machine Intelligence (PAMI) **22** (2000), no. 6, 583–600.
214. V. Jain, H. Zhang, and O. van Kaick, *Non-rigid spectral correspondence of triangle meshes*, Int'l J. Shape Modeling (2006).
215. M. Kac, *Can one hear the shape of a drum?*, American Mathematical Monthly **73** (1966), 1–23.
216. T. Kanade, *Computer recognition of human faces*, Birkhauser, 1973.
217. C. Kao, S. Osher, and Y. Tsai, *Fast sweeping methods for static Hamilton-Jacobi equations*, Tech. report, Department of Mathematics, UCLA, 2002.

218. Z. Karni and C. Gotsman, *Spectral compression of mesh geometry*, Proc. Conf. Computer Graphics and Interactive Techniques, 2000, pp. 279–286.
219. M. Kazhdan, T. Funkhouser, and S. Rusinkiewicz, *Rotation invariant spherical harmonic representation of 3D shape descriptors*, Proc. Symp. Geometry Processing (SGP), 2004, pp. 156–164.
220. A. Kearsley, R. Tapia, and M. Trosset, *The solution of the metric STRESS and SSTRESS problems in multidimensional scaling using Newton's method*, Computational Statistics **13** (1998), no. 3, 369–396.
221. C. T. Kelley, *Iterative methods for optimization*, Frontiers in Applied Mathematics, SIAM, 1999.
222. M. D. Kelly, *Visual identification of people by computer*, Technical Report AI-130, Stanford University, 1970.
223. J. Kepler, *Strena seu de nive sexangula*, 1611.
224. M. Kilian, N. J. Mitra, and H. Pottmann, *Geometric modeling in shape space*, ACM Trans. Graphics **26** (2007), no. 3.
225. R. Kimmel, *Numerical geometry of images*, Springer, 2004.
226. R. Kimmel and J. A. Sethian, *Computing geodesic paths on manifolds*, Proc. National Academy of Sciences (PNAS) **95** (1998), no. 15, 8431–8435.
227. R. Kimmel and J. A. Sethian, *Optimal algorithm for shape from shading and path planning*, J. Mathematical Imaging and Vision **14** (2001), no. 3, 237–244.
228. R. Kimmel, D. Shaked, N. Kiryati, and A. M. Bruckstein, *Skeletonization via distance maps and level sets*, Computer Vision and Image Understanding **62** (1995), no. 3, 382–391.
229. G. J. Klir and B. Yuan, *Fuzzy sets and fuzzy logic: theory and applications*, Prentice Hall, 1994.
230. J. J. Koenderink and A. J. van Doorn, *The shape of smooth objects and the way contours end*, Perception **11** (1981), 129–137.
231. T. Kohonen, *Self-organizing maps*, Springer, 2001.
232. J. B. Kruskal, *Multidimensional scaling to a nonmetric hypothesis*, Psychometrika **29** (1964), 1–27.
233. _____, *Nonmetric multidimensional method*, Psychometrika **29** (1964), 115–129.
234. H. W. Kuhn and A. W. Tucker, *Nonlinear programming*, Proc. Berkeley Symp. Mathematical Statistics and Probability (J. Neyman, ed.), 481–491, University of California Press, 1957.
235. R. Kunze, F. E. Wolter, and T. Rausch, *Geodesic Voronoi diagrams on parametric surfaces*, Proc. Computer Graphics International **97** (1997), 230–237.
236. _____, *Geodesic Voronoi diagrams on parametric surfaces*, Tech. report, Welfen Laboratory, Hannover University, 1997.
237. K. Kupeev and H. Wolfson, *On shape similarity*, Proc. Int'l Conf. Pattern Recognition (ICPR), 1994, pp. 227–237.
238. R. L. Lagendijk and M. I. Sezan, *Motion compensated frame rate conversion of motion pictures*, Proc. Int'l Conf. Acoustics, Speech, and Signal Processing (ICASSP), vol. 3, 1992.
239. J. H. Lambert, *Photometria sive de mensura et gradibus luminis, colorum et umbra*, Augustæ Vindelicorum (1760).
240. C. Lanczos, *An iteration method for the solution of the eigenvalue problem of linear differential and integral operators*, J. Research of the National Bureau of Standards **45** (1950), 255–282.

241. L. J. Latecki, R. Lakaemper, and D. Wolter, *Optimal partial shape similarity*, Image and Vision Computing **23** (2005), 227–236.
242. J. C. Lee and E. Milios, *Matching range images of human faces*, Proc. Int'l Conf. Computer Vision (ICCV), 1990, pp. 722–726.
243. S. H. Lee, Y. C. Shin, S. Yang, H. H. Moon, and R. H. Park, *Adaptive motion-compensated interpolation for frame rate up-conversion*, IEEE Trans. Consumer Electronics **48** (2002), no. 3, 444–450.
244. Y. Lee and J. Shim, *Curvature-based human face recognition using depth-weighted Hausdorff distance*, Proc. Int'l Conf. Image Processing (ICIP), 2004, pp. 1429–1432.
245. Y. Lee, H. Song, U. Yang, H. Shin, and K. Sohn, *Local feature based 3D face recognition*, Proc. Audio and Video-based Biometric Person Authentication (AVBPA), Lecture Notes in Computer Science, no. 3546, Springer, 2005, pp. 909–918.
246. R. B. Lehoucq, D. C. Sorensen, and C. Yang, *ARPACK users' guide: Solution of large-scale eigenvalue problems with implicitly restarted Arnoldi methods*, SIAM, 1998.
247. G. Leibon, *Random Delaunay triangulations, the Thruston-Andreev theorem, and metric uniformization*, Ph.D. thesis, UCSD, 1999.
248. G. Leibon and D. Letscher, *Delaunay triangulations and Voronoi diagrams for Riemannian manifolds*, Proc. Symp. Computational Geometry, 2000, pp. 341–349.
249. S. Leopoldseder, H. Pottmann, and H. Zhao, *The d2-tree: A hierarchical representation of the squared distance function*, Tech. report, Institute of Geometry, Vienna University of Technology, 2003.
250. B. Lévy, *Laplace-Beltrami eigenfunctions towards an algorithm that "understands" geometry*, Int'l Conf. Shape Modeling and Applications, 2006.
251. H. Li and R. Hartley, *The 3D-3D registration problem revisited*, Proc. Int'l Conf. Computer Vision (ICCV), 2007.
252. X. Li and H. Zhang, *Adapting geometric attributes for expression-invariant 3D face recognition*, Int'l Conf. Shape Modeling and Applications, 2007, pp. 21–32.
253. N. Linial, *Finite metric spaces: combinatorics, geometry and algorithms*, Proc. ICM, vol. 3, 2002, pp. 573–586.
254. N. Litke, M. Droske, M. Rumpf, and P. Schroder, *An image processing approach to surface matching*.
255. S. P. Lloyd, *Least squares quantization in PCM*, Bell Telephone Laboratories Paper (1957).
256. _____, *Least squares quantization in PCM*, IEEE Trans. Information Theory **28** (1982), no. 2.
257. C. Lombroso, *Antropometria di 400 delinquenti*, 1872.
258. F. Losasso, H. Hoppe, S. Schaefer, and J. Warren, *Smooth geometry images*, Proc. Symp. Geometry Processing (SGP), 2003, pp. 138–145.
259. X. Lu, D. Colbry, and A. K. Jain, *Matching 2.5D scans for face recognition*, Proc. Int'l Conf. Pattern Recognition (ICPR), 2004, pp. 362–366.
260. J. MacQueen, *Some methods for classification and analysis of multivariate observations*, Proc. Berkeley Symp. Mathematical Statistics and Probability, vol. 1, 1967, pp. 281–297.
261. D. Marr and T. Poggio, *Cooperative computation of stereo disparity*, Science **194** (1976), no. 4262, 283.

262. N. Mavridis, F. Tsalakanidou, D. Pantazis, S. Malassiotis, and M. G. Strintzis, *The HISCORE face recognition application: Affordable desktop face recognition based on a novel 3D camera*, Proc. Int'l Conf. Augmented Virtual Environments and 3D Imaging, pp. 157–160.
263. J. Max, *Quantizing for minimum distortion*, IEEE Trans. Information Theory **6** (1960), no. 1, 7–12.
264. S. F. McCormick, *Multigrid methods*, Frontiers in Applied Mathematics, vol. 3, SIAM, 1987.
265. H. McKean and I. Singer, *Curvature and the eigenvalues of the Laplacian*, J. Differential Geometry **1** (1967), 43–69.
266. L. Mealey, R. Bridgstock, and G. C. Townsend, *Symmetry and perceived facial attractiveness: a monozygotic co-twin comparison*, J. Personality and Social Psychology **76** (1999), no. 1, 151–158.
267. G. Medioni and R. Waupotitsch, *Face recognition and modeling in 3D*, Proc. Analysis and Modeling of Faces and Gestures (AMFG), 2003, pp. 232–233.
268. F. Mémoli and G. Sapiro, *Fast computation of weighted distance functions and geodesics on implicit hyper-surfaces*, J. Computational Physics **173** (2001), no. 1, 764–795.
269. _____, *A theoretical and computational framework for isometry invariant recognition of point cloud data*, Foundations of Computational Mathematics **5** (2005), 313–346.
270. M. Mešina, *Convergence acceleration for the iterative solution of the equations $x = ax + f$*, Computer Methods in Applied Mechanics and Engineering (CMAME) **10** (1977), no. 2, 165–173.
271. M. Meyer, M. Desbrun, P. Schroder, and A. H. Barr, *Discrete differential-geometry operators for triangulated 2-manifolds*, Visualization and Mathematics III (2003), 35–57.
272. _____, *Discrete differential-geometry operators for triangulated 2-manifolds*, Visualization and Mathematics III (2003), 35–57.
273. P. W. Michor and D. Mumford, *Riemannian geometries on spaces of plane curves*, Arxiv preprint math.DG/0312384 (2003).
274. K. Miettinen, *Nonlinear multiobjective optimization*, Kluwer, 1999.
275. R. S. Millman and G. D. Parker, *Geometry: A metric approach with models*, Springer, 1990.
276. J. S. B. Mitchell, D. M. Mount, and C. H. Papadimitriou, *The discrete geodesic problem*, SIAM J. Computing **16** (1987), no. 4, 647–668.
277. N. J. Mitra, N. Gelfand, H. Pottmann, and L. Guibas, *Registration of point cloud data from a geometric optimization perspective*, Proc. Symp. Geometry Processing (SGP), 2004, pp. 23–32.
278. N. J. Mitra, L. Guibas, and M. Pauly, *Partial and approximate symmetry detection for 3D geometry*, ACM Trans. Graphics **25** (2006), no. 3, 560–568.
279. A. F. Möbius, *Der baryzentrische Calcul*, 1827.
280. B. Mohar, *The laplacian spectrum of graphs*, Graph Theory, Combinatorics, and Applications (Y. Alavi, G. Chartrand, O. R. Oellermann, and A. J. Schwenk, eds.), vol. 2, Wiley, 1991, pp. 871–898.
281. S. Montiel and A. Ros, *Curves and surfaces*, AMS, 2005.
282. A. B. Moreno, A. Sánchez, J. F. Vélez, and F. J. Díaz, *Face recognition using 3D surface-extracted descriptors*, Proc. Irish Machine Vision and Image Processing Conf., 2003, pp. 233–238.

283. A. Morrison, G. Ross, and M. Chalmers, *Fast multidimensional scaling through sampling, springs, and interpolation*, Information Visualization **2** (2003), no. 1, 68–77.
284. J. M. Morvan and B. Thibert, *On the approximation of a smooth surface with a triangulated mesh*, Computational Geometry: Theory and Applications **23** (2002), no. 3, 337–352.
285. L. Mosheyev and M. Zibulevsky, *Penalty/barrier multiplier algorithm for semidefinite programming*, Optimization Methods and Software **13** (2000), no. 4, 235–261.
286. I. Mpiperis, S. Malassiotis, and M. G. Strintzis, *Expression compensation for face recognition using a polar geodesic representation*, Proc. Int'l Symp. 3D Data Processing, Visualization and Transmission, 2006.
287. I. Mpiperis, S. Malassiotis, and M. G. Strintzis, *3-D face recognition with the geodesic polar representation*, IEEE Trans. Information Forensics and Security **2** (2007), no. 3, 537–547.
288. D. Mumford and J. Shah, *Boundary detection by minimizing functionals*, Image Understanding (1990).
289. T. Nagamine, T. Uemura, and I. Masuda, *3D facial image analysis for human identification*, Proc. Int'l Conf. Pattern Recognition (ICPR), 1992, pp. 324–327.
290. J. Nash, *The imbedding problem for Riemannian manifolds*, Annals of Mathematics **63** (1956), 20–63.
291. S. Nash, *A multigrid approach to discretized optimization problems*, J. Optimization Methods and Software **14** (2000), 99–116.
292. _____, *A survey of truncated-Newton methods*, J. Computational and Applied Mathematics **124** (2000), 45–59.
293. A. Nemirovski, *Efficient methods in convex programming*, Lecture notes, 1995, online, http://iew3.technion.ac.il/Home/Users/Nemirovski.html.
294. _____, *Numerical methods for nonlinear continuous optimization*, Lecture notes, 1999, online, http://iew3.technion.ac.il/Home/Users/Nemirovski.html.
295. _____, *Lectures on modern convex optimization*, Lecture notes, 2003, online, http://iew3.technion.ac.il/Home/Users/Nemirovski.html.
296. Y. Nesterov and A. Nemirovski, *Interior point polynomial methods in convex programming*, SIAM, 1994.
297. I. Newton, *De analysi per æquationes numero terminorum infinitas*, 1699.
298. J. Nocedal and S. Wright, *Numerical optimization*, Springer, New York, 1999.
299. M. Novotni and R. Klein, *Computing geodesic distances on triangular meshes*, Proc. Int'l Conf. in Central Europe on Computer Graphics, Visualization and Computer Vision, 2002.
300. B. O'Neil, *Elementary differential geometry*, Academic Press (1997).
301. K. Onishi and J. Itoh, *Estimation of the necessary number of points in Riemannian Voronoi diagram*, Proc. Canadian Conf. Computational Geometry (CCCG) (2003), 19–24.
302. J. Ortega-Garcia, J. Bigun, D. Reynolds, and J. Gonzalez-Rodriguez, *Authentication gets personal with biometrics*, IEEE Signal Processing Magazine **21** (2004), no. 2, 50–62.
303. A. Osyczka, *Multicriterion optimisation in engineering with FORTRAN programs*, Wiley, 1984.

304. J. D. Owens, D. Luebke, N. Govindaraju, M. Harris, J. Kruger, A. E. Lefohn, and T. J. Purcell, *A survey of general-purpose computation on graphics hardware*, Eurographics 2005, State of the Art Reports (2005), 21–51.
305. G. Pan, S. Han, Z. Wu, and Y. Wang, *3D face recognition using mapped depth images*, IEEE Workshop on Face Recognition Grand Challenge Experiments, 2005.
306. G. Pan, Z. Wu, and Y. Pan, *Automatic 3D face verification from range data*, Proc. Int'l Conf. Acoustics, Speech, and Signal Processing (ICASSP), vol. III, 2003, pp. 193–196.
307. V. Pareto, *Manuale di economia politica*, 1906.
308. K. Pearson, *On lines and planes of closest fit to systems of points in space*, Philosophical Magazine **2** (1901), no. 6, 559–572.
309. C. Pelletan, *Les guerres de la révolution*, Colas, 1884.
310. A. Pentland, *Recognition by parts*, Proc. Int'l Conf. Computer Vision (ICCV), 1987, pp. 612–620.
311. C. Perrault and E. Le Cain, *Cinderella, or the little glass slipper*, Faber and Faber, 1972.
312. G. Peyre and L. Cohen, *Surface segmentation using geodesic centroidal tesselation*, Proc. Int'l Symp. 3D Data Processing Visualiztion Transmission, 2004, pp. 995–1002.
313. G. Peyré and L. Cohen, *Geodesic remeshing using front propagation*, Int'l J. Computer Vision (IJCV) **69** (2006), no. 1, 145–156.
314. J. C. Platt, *FastMap, MetricMap, and Landmark MDS are all Nyström algorithms*, Proc. 10th Int'l Workshop on Artificial Intelligence and Statistics, 2005, pp. 261–268.
315. B. T. Polyak, *Minimization of non-smooth functionals*, USSR Computational Mathematics and Mathematical Physics **9** (1969), 14–29.
316. H. Pottmann and M. Hofer, *Geometry of the squared distance function to curves and surfaces*, Visualization and Mathematics III, 2003, pp. 221–242.
317. H. Pottmann, Q. X. Huang, Y. L. Yang, and S. M. Hu, *Geometry and convergence analysis of algorithms for registration of 3D shapes*, Int'l J. Computer Vision (IJCV) **67** (2006), no. 3, 277–296.
318. H. Pottmann, S. Leopoldseder, and M. Hofer, *Registration without ICP*, Computer Vision and Image Understanding **95** (2004), no. 1, 54–71.
319. M. J. D. Powell, *A method for nonlinear constraints in minimization problems*, Optimization (R. Fletcher, ed.), Academic Press, 1969, pp. 283–298.
320. R. J. Prokop and A. P. Reeves, *A survey of moment-based techniques for unoccluded object representation and recognition*, Computer Vision, Graphics, and Image Processing (CVGIP): Graphical Models and Image Processing **54** (1992), no. 5, 438–460.
321. N. Rajeevan, K. Rajgopal, and G. Krishna, *Vector extrapolated fast maximum likelihood estimation algorithm for emission tomography*, IEEE Trans. Medical Imaging **11** (1992), no. 1, 9–20.
322. D. Raviv, A. M. Bronstein, M. M. Bronstein, and R. Kimmel, *Symmetries of non-rigid shapes*, Proc. Workshop on Non-rigid Registration and Tracking through Learning (NRTL), 2007.
323. M. Reuter, F.-E. Wolter, and N. Peinecke, *LaplaceBeltrami spectra as "shape-DNA" of surfaces and solids*, Computer Aided Design **38** (2006), 342–366.

324. B. Riemann, *Über die Hypothesen, welche der Geometrie zu Grunde liegen*, Abhandlungen der Königlichen Gesellschaft der Wissenschaften zu Göttingen **13** (1867).
325. E. Rivlin, S. Dickenson, and A. Rosenfeld, *Recognition by functional parts*, Proc. Computer Vision and Pattern Recognition (CVPR), 1992, pp. 267–275.
326. G. Rosman, A. M. Bronstein, M. M. Bronstein, A. Sidi, and R. Kimmel, *Fast multidimensional scaling using vector extrapolation*, Tech. Report CIS-2008-01, Dept. of Computer Science, Technion, Israel, 2008.
327. G. Rosman, M. M. Bronstein, A. M. Bronstein, and R. Kimmel, *Topologically constrained isometric embedding*, Workshop on Human Motion Understanding, Modeling, Capture and Animation (D. Metaxas, R. Klette, and B. Rosenhahn, eds.), Lecture Notes in Computer Science, vol. 36, Springer, 2007, pp. 243–262.
328. S. T. Roweis and L. K. Saul, *Nonlinear dimensionality reduction by locally linear embedding*, Science **290** (2000), no. 5500, 2323–2326.
329. S. Rusinkiewicz and M. Levoy, *Efficient variants of the ICP algorithm*, Proc. 3D Digital Imaging and Modeling (2001), 145–152.
330. R. M. Rustamov, *Laplace-Beltrami eigenfunctions for deformation invariant shape representation*, Proc. Symp. Geometry Processing (SGP), 2007, pp. 225–233.
331. O. Sacks, *The man who mistook his wife for a hat*, 1986.
332. H. D. Saffrey, *Ageômetrètos mèdeis eisitô: une inscription légendaire*, Revue des Études grecques **81** (1968), 67–87.
333. M. E. Salukwadze, *Vector-valued optimization problems in control theory*, Academic Press, 1979.
334. T. Samoilov and G. Elber, *Self-intersection elimination in metamorphosis of two-dimensional curves*, Visual Computer **14** (1998), 415–428.
335. P. Sander, Z. Wood, S. Gortler, J. Snyder, and H. Hoppe, *Multi-chart geometry images*, Proc. Symp. Geometry Processing (SGP), 2003, pp. 146–155.
336. L. K. Saul and S. T. Roweis, *Think globally, fit locally: unsupervised learning of low dimensional manifolds*, J. Machine Learning Research (JMLR) **4** (2003), 119–155.
337. H. Schedel, *Liber chronicarum*, 1493.
338. C. Scheidegger, S. Fleishman, and C. T. Silva, *Triangulating point-set surfaces with bounded error*, Proc. Symp. Geometry Processing (SGP) (M. Desbrun and H. Pottmann, eds.), 2005, pp. 63–72.
339. R. J. Schmidt, *On the numerical solution of linear simultaneous equations by an iterative method*, Philosophical Magazine **7** (1941), no. 32, 369–383.
340. I. J. Schoenberg, *Remarks to Maurice Fréchet's article "sur la définition axiomatique dúne classe déspace distanciés vectoriellement applicable sur léspace de Hilbert"*, Annals of Mathematics **36** (1935), no. 3, 724–732.
341. _____, *On certain metric spaces arising from Euclidean spaces by a change of metric and their imbedding in Hilbert space*, Annals of Mathematics **38** (1937), no. 4, 787–793.
342. J. Schreiner, C. Scheidegger, S. Fleishman, and C. T. Silva, *Direct (re)meshing for efficient surface processing*, Proc. Computer Graphics Forum **25** (2006), no. 3, 527–536.
343. J. Schreiner, C. Scheidegger, and C. T. Silva, *High quality extraction of isosurfaces from regular and irregular grids*, IEEE Trans. Visualization and Computer Graphics **12** (2006), no. 5, 1205–1212.

344. P. Schröder and W. Sweldens, *Digital geometry processing*, ACM SIGGRAPH Course Notes (2001).
345. E. L. Schwartz, A. Shaw, and E. Wolfson, *A numerical solution to the generalized mapmaker's problem: flattening nonconvex polyhedral surfaces*, IEEE Trans. Pattern Analysis and Machine Intelligence (PAMI) **11** (1989), 1005–1008.
346. H. A. Schwarz, *Sur une définition erronée de laire dune surface courbe*, Ges. Math. Abhandl. **2** (1890), 309–311 and 369–370.
347. H. Schweitzer, *Template matching approach to content based image indexing by low dimensional euclidean embedding*, Proc. Int'l Conf. Computer Vision (ICCV), 2001.
348. J. A. Sethian, *A fast marching level set method for monotonically advancing fronts*, Proc. National Academy of Sciences (PNAS) **93** (1996), no. 4, 1591–1595.
349. _____, *Level set methods and fast marching methods*, University Press, Cambridge, 1999.
350. J. A. Sethian and A. Vladimirsky, *Fast methods for the eikonal and related Hamilton-Jacobi equations on unstructured meshes*, Proc. National Academy of Science (PNAS) **97** (2000), no. 11, 5699–5703.
351. D. Shanks, *Nonlinear transformations of divergent and slowly convergent sequences*, J. Mathematical Physics **34** (1955), no. 1.
352. D. F. Shanno, *Conditioning of quasi-newton methods for function minimizations*, Mathematics of Computation **24** (1970), 647–657.
353. M. Shapira and A. Rappoport, *Shape blending using the star-skeleton representation*, IEEE Trans. Computer Graphics and Application **15** (1995), no. 2, 44–51.
354. I. Shatz, A. Tal, and G. Leifman, *Paper craft models from meshes*, Visual Computer **22** (2006), no. 9, 825–834.
355. R. N. Shepard, *The analysis of proximities: multidimensional scaling with an unknown distance function*, Psychometrika **27** (1964), 219–246.
356. J. W. Shepherd, G. M. Davies, and H. D. Ellis, *Studies of cue saliency*, Perceiving and remembering faces (G. M. Davies, H. D. Ellis, and J. W. Shepherd, eds.), Academic Press, 1981.
357. K. Siddiqi and B. Kimia, *Parts of visual form: Computational aspects*, IEEE Trans. Pattern Analysis and Machine Intelligence (PAMI) **17** (1995), 239–251.
358. _____, *A shock grammar for recognition*, Proc. Computer Vision and Pattern Recognition (CVPR), 1996, pp. 507–513.
359. A. Sidi, *Efficient implementation of minimal polynomial and reduced rank extrapolation methods*, J. Computational and Applied Mathematics **36** (1991), no. 3, 305–337.
360. _____, *Efficient implementation of minimal polynomial and reduced rank extrapolation methods*, J. Computational and Applied Mathematics **36** (1991), 305–337.
361. _____, *Vector extrapolation methods with applications to solution of large systems of equations and to PageRank computations*, (2007), no. CIS-2007-09.
362. A. Sidi and J. Bridger, *Convergence and stability analyses for some vector extrapolation methods in the presence of defective iteration matrices*, J. Computational and Applied Mathematics **22** (1988), 35–61.
363. A. Sidi and Y. Shapira, *Upper bounds for convergence rates of acceleration methods with initial iterations*, Numerical Algorithms **18** (1998), 113–132.

364. A. Singer, *From graph to manifold Laplacian: the convergence rate*, Applied and Computational Harmonic Analysis **21** (2006), 128–134.
365. L. Sirovich and M. Kirby, *Low-dimensional procedure for the characterization of human faces*, J. Optical Society of America (JOSA) A **2** (1987), 519–524.
366. D. A. Smith, W. F. Ford, and A. Sidi, *Extrapolation methods for vector sequences*, SIAM Review **29** (1987), no. 2, 199–233.
367. A. Spira and R. Kimmel, *An efficient solution to the eikonal equation on parametric manifolds*, Interfaces and Free Boundaries **6** (2004), no. 4, 315–327.
368. T. Surazhsky and G. Elber, *Matching free form surfaces*, Computers and Graphics **26** (2001), no. 1.
369. V. Surazhsky and C. Gotsman, *Controllable morphing of compatible planar triangulations*, ACM Trans. Graphics **20** (2001), no. 4, 203–231.
370. _____, *Intrinsic morphing of compatible triangulations*, Int'l J. Shape Modeling **9** (2003), no. 2, 191–201.
371. V. Surazhsky, T. Surazhsky, D. Kirsanov, S. Gortler, and H. Hoppe, *Fast exact and approximate geodesics on meshes*, Proc. SIGGRAPH, 2005, pp. 553–560.
372. H. T. Tanaka, M. Ikeda, and H. Chiaki, *Curvature-based face surface recognition using spherical correlation principal directions for curved object recognition*, Proc. Int'l Conf. Automated Face and Gesture Recognition, 1998, pp. 372–377.
373. J. W. H. Tangelder and R. C. Veltkamp, *A survey of content based 3D shape retrieval methods*, Proc. Shape Modeling and Applications, 2004, pp. 145–156.
374. G. Taubin, *A signal processing approach to fair surface design*, ACM, 1995.
375. G. Taubin, *Geometric signal processing on polygonal meshes*, Eurographics State of the Art Report (2000).
376. M. R. Teague, *Image analysis via the general theory of moments*, J. Optical Society of America (JOSA) **70** (1979), 920–930.
377. J. B. Tenenbaum, V. de Silva, and J. Langford, *A global geometric framework for nonlinear dimensionality reduction*, Science **290** (2000), no. 5500, 2319–2323.
378. A. H. Thiessen, *Precipitation averages for large areas*, Monthly Weather Review **39**, no. 7, 1082–1089.
379. W. S. Torgerson, *Multidimensional scaling I - theory and methods*, Psychometrika **17** (1952), 401–419.
380. M. Trosset and R. Mathar, *On the existence of nonglobal minimizers of the stress criterion for metric multidimensional scaling*, Proc. Statistical Computing Section, American Statistical Association, 1997, pp. 158–162.
381. U. Trottenberg, C. Oosterlee, and A. Schüller, *Multigrid*, Academic Press, London, 2001.
382. Y. R. Tsai, L. T. Cheng, S. Osher, and H. Zhao, *Fast sweeping algorithms for a class of Hamilton-Jacobi equations*, SIAM J. Numerical Analysis (SINUM) **41** (2003), no. 2, 673–694.
383. J. N. Tsitsiklis, *Efficient algorithms for globally optimal trajectories*, IEEE Trans. Automatic Control **40** (1995), no. 9, 1528–1538.
384. M. Turk and A. Pentland, *Face recognition using eigenfaces*, Proc. Computer Vision and Pattern Recognition (CVPR), 1991, pp. 586–591.
385. A. Tversky, *Feature of similarity*, Psychological Review **84** (1977), no. 4, 327–350.

386. R. Varadhan and C. Roland, *Squared extrapolation methods (SQUAREM): A new class of simple and efficient numerical schemes for accelerating the convergence of the EM algorithm*, Tech. Report 63, Department of Biostatistics Working Paper, Johns Hopkins University, 2004.
387. R. C. Veltkamp, *Shape matching: similarity measures and algorithms*, Int'l Conf. Shape Modeling and Applications, 2001, pp. 188–197.
388. R. C. Veltkamp and M. Hagedoorn, *Shape similarity measures, properties, and constructions*, (2001).
389. M. F. Vignéras, *Variétés riemanniennes isospectrales et non isométriques*, Annals of Mathematics **112** (1980), no. 2, 21–32.
390. G. F. Voronoi, *Nouvelles applications des paramétres continus á la théorie des formes quadratiques.*, J. für die Reine und Angewandte Mathematik **138** (1908), 198–287.
391. J. T.-L. Wang, X. Wang, K.-I. Lin, D. Shasha, B. A. Shapiro, and K. Zhang, *Evaluating a class of distance-mapping algorithms for data mining and clustering*, Proc. Int'l Conf. Knowledge Discovery and Data Mining (KDD), 1999, pp. 307–311.
392. M. Wardetzky, S. Mathur, F. Kälberer, and E. Grinspun, *Discrete Laplace operators: no free lunch*, Proc. Symp. Geometry Processing (SGP), 2007, pp. 33–37.
393. P. Wesseling, *An introduction to multigrid methods*, John Wiley, Chichester, 1992.
394. H. Weyl, *Über die Asymptotische Verteilung der Eigenwerte*, Nachrichten von der Königlichen Gesellschaft der Wissenschaften zu Göttingen (1911), 110–117.
395. H. Weyl, *Symmetry*, Princeton University Press, 1983.
396. C. Wheatstone, *Contributions to the physiology of vision. Part the first: On some remarkable, and hitherto unobserved, phenomena of binocular vision*, Philosophical Trans. Royal Society of London **128** (1838), 371–394.
397. C. K. I. Williams, *Advances in neural information processing systems*, vol. 13, ch. On a connection between kernel PCA and metric multidimensional scaling, MIT press, 2001.
398. M. Williams and T. Munzner, *Steerable, progressive multidimensional scaling*, Proc. INFOVIS, 2004, pp. 57–64.
399. L. Wiskott, *Labeled graphs and dynamic link matching for face recognition and scene analysis*, Reihe Physik **53** (1995).
400. F. E Wolter, *Cut locus and medial axis in global shape interrogation and representation*, MIT Ocean Engineering Design Laboratory, Memorandum 92-2, 1992.
401. J. D. Wolter, T. C. Woo, and R. A. Volz, *Optimal algorithms for symmetry detection in two and three dimensions*, The Visual Computer **1** (1985), 37–48.
402. G. Xu, *Convergence of discrete Laplace-Beltrami operators over surfaces*, Computers and Mathematics with Applications **48** (2004), no. 3-4, 347–360.
403. _____, *Discrete Laplace-Beltrami operators and their convergence*, Computer Aided Geometric Design **21** (2004), no. 8, 767–784.
404. L. Yatziv, A. Bartesaghi, and G. Sapiro, *O(N) implementation of the fast marching algorithm*, J. Computational Physics **212** (2006), no. 2, 393–399.
405. F. W. Young and R. M. Hamer, *Theory and applications of multidimensional scaling*, Eribaum Associates, 1994.
406. G. Young and A. S. Householder, *Discussion of a set of point in terms of their mutual distances*, Psychometrika **3** (1938), 19–22.

407. L. A. Zadeh, *Fuzzy sets*, Information and Control **8** (1965), 338–353.
408. _____, *Fuzzy sets as a basis for a theory of possibility*, Fuzzy Sets and Systems **1** (1978), no. 1, 3–28.
409. D. Zhang and G. Lu, *Review of shape representation and description techniques*, Pattern Recognition **37** (2004), 1–19.
410. Z. Zhang and H. Zha, *Principal manifolds and nonlinear dimension reduction via local tangent space alignment*, Technical Report CSE-02-019, Department of Computer Science and Engineering, Pennsylvania State University, 2002.
411. H. Zhao, *Fast sweeping method for eikonal equations*, Mathematics of Computation **74** (2005), 603–627.
412. W. Zhao, R. Chellappa, and A. Krishnaswamy, *Subspace linear discriminant analysis for face recognition*, Proc. Int. Conf. Automatic Face and Gesture Recognition, 1998, pp. 336–341.
413. W. Zhao, R. Chellappa, A. Rosenfeld, and P. J. Phillips, *Face recognition: A literature survey*, ACM Computing Surveys (2003), 399–458.
414. M. Zibulevsky, *Penalty/barrier multiplier methods for large-scale nonlinear and semidefinite programming*, Ph.D. thesis, Technion, 1996.
415. G. Zigelman, R. Kimmel, and N. Kiryati, *Texture mapping using surface flattening via multi-dimensional scaling*, IEEE Trans. Visualization and Computer Graphics (TVCG) **9** (2002), no. 2, 198–207.
416. H. J. Zimmermann, *Fuzzy set theory and its applications*, Kluwer, 2001.

Subject Index

acceleration, 27, 106, 160, 163–166
adjacency, 52, 53
albedo, 273, 274
aliasing, 259
angle, 54, 63, 79, 82, 134, 192, 273
    acute, 77, 100, 105
    obtuse, 78, 79, 282
anthropometry, 267, 271
applicable, 31, 32, 107, 125, 154, 264
arclength, 27, 53, 204, 240
area, 23, 25, 36, 42, 43, 49, 61, 62, 64, 65, 97, 141, 142, 151, 177, 180, 183, 222, 224, 225, 233, 234, 242, 283
Armijo rule, *see* backtracking
Arnoldi algorithm, 172, 176
artificial variable, 150, 210
atlas, 20–22
authentication, *see* one-to-one

backtracking, 88, 89, 99
ball
    circumscribed, 54
    closed, 13, 54
    metric, 43, 46, 54, 156, 176
    open, 13
barrier, 113, 114, 118
barycentric coordinates, 55, 195, 197, 200, 254, 255
basin of attraction, 109

Bellman principle, 69
bending, 31–34, 255
BFGS algorithm, 108
bijection, 14, 15, 20, 39, 240, 241
bilinear, 29, 30
biometric, 261, 266, 267, 275
biorthonormality, 135
block matrix, 146
body painting, 248, 250
boundedness, 13–15, 63, 85, 124, 144, 157, 254, 291
brightness, 273, 274

canonical form, 139, 142, 150, 154, 157, 165, 169–176, 192–194, 208, 241, 269, 270, 272–274
cartography, 20, 139, 140
Cauchy-Green deformation tensor, 244
centaur, 217, 218, 223–225, 232, 238
center of gravity, 123
characteristic, 5, 36, 37, 75, 83, 84, 88–90, 124, 236, 261, 267, 273, 290
chart, 20, 24, 91
classic scaling, 184
compactness, 13, 14, 18, 30, 36, 43, 59, 63, 64, 178
compatible meshing, 252

327

Subject Index

completeness, 18, 19, 30, 55, 124, 125, 165, 275
condition number, 101, 104
congruence, 28, 29, 31, 32, 119, 125, 126, 241
connectedness, 13, 14, 18, 19, 30, 33, 36, 53, 71, 82, 89, 165, 211, 234
connectivity, *see* adjacency
connectivity pattern, 57, 82, 83
constraint
    equality, 110, 111, 114
    inequality, 110, 113, 114
continuity
    bi-Lipschitz, 15, 16
    Lipschitz, 14, 15, 37, 107
contraction, 14
convergence, 13, 91, 99, 100, 102–110, 129, 131, 133, 147, 149, 153, 154, 160, 161, 164, 167, 171, 182, 184, 203, 229
convergence rate, 102–104
convexity, 19, 33, 34, 40, 47, 48, 54, 55, 63, 64, 94, 96–98, 102, 106, 109–111, 116, 117, 132, 134, 144, 147, 167, 190, 195, 200, 204, 220, 238, 254–258, 282, 283, 289
correction, 156–159
correspondence
    intrinsic, 6, 245, 258
    non-rigid, 257
    rigid, 133, 239, 240
cost, *see* objective
covering, 22, 43–46, 48, 63, 64, 116, 154, 178, 280
curvature
    Gaussian, 28, 34–36, 38, 39, 60, 139, 180, 242, 244
    geodesic, 27, 38
    mean, 28, 280
    normal, 27, 28
    principal, 27, 34, 35, 60, 127, 128

decimation, 156, 158, 159
delta function, 236
descent
    gradient, 89, 101–106, 144–146
    normalized steepest, 101
    projected, 191
    steepest, 101, 104–106, 108, 118, 200
determinant, 25, 34, 38
diffeomorphism, 40
differentiability, 14, 19, 21, 91, 94, 105, 107, 149, 204
differential, 24–26, 35, 37, 40, 49, 63, 74, 75, 88, 89, 91, 158, 177, 204, 242, 243
diffusion map, 175
Dijkstra algorithm, 185
dilation, 14–16, 151
Dirichlet energy, 244
disjoint union, 208
disparity, 242
distance
    embedding, 207–209, 231
    fuzzy Pareto, 228, 236
    Gromov-Hausdorff, 17, 205, 208, 209, 211, 212, 214, 215, 217, 224, 231, 232
    Hausdorff, 61, 62, 126, 133, 137, 208, 211, 219, 229, 267
    Manhattan, 72
    Pareto, 223, 224
    point-to-plane, 77, 126–129, 134
    point-to-surface, 127, 128, 134
    Pottmann-Hofer, 128
    Salukwadze, 226
    surface-to-surface, 125, 126, 132
domain of dominance, 47
dynamic programming, 69, 70, 75, 89

eccentricity, 123
eigendecomposition, 118, 122, 147, 171, 172, 174–177, 184, 185, 289
eigenface, 264
eigenfunction, 178–180, 182
eigenmap, 174–176

Subject Index   329

eikonal equation, 75, 78, 80, 82–84, 89–91
elasticity, 244
embedding
    Euclidean, 139, 187, 274
    isometric, 20, 139–141, 169
    non-Euclidean, 187, 193, 274
    spherical, 191, 203, 204, 274
epigraph, 3, 9, 97, 117
equal error rate, 264
equidistant, 46, 47, 54, 59, 60, 81, 271
equivalence class, 139, 206
Euler characteristic, 36, 180

facial expression, 6, 265, 267, 268, 270, 273–275
false acceptance, 264
false rejection, 264
fast marching, 63, 73–76, 81–86, 88–90, 147, 150, 197, 269, 271
fast sweeping, 84
feasibility, 68, 110, 111, 114, 248, 272
FMM, *see* fast marching
Fourier transform, 275
frame rate conversion, 257
functional, 18, 30, 130, 204, 236, 242–244
fuzzy set, 226–228, 231, 236

gallery, 263, 264, 269, 270
Gauss-Bonnet theorem, 36, 40, 235
general position, 46
genus, 36, 234, 235
geodesic mask, 271, 272
geometric moment, 123, 124, 133–135, 283
    high-order, 125, 284
    second-order, 122
gradient, 75, 80, 83, 88, 89, 95, 96, 98, 100, 101, 106, 107, 110, 111, 115, 144, 147–150, 157, 167, 177, 200, 201, 204, 243, 283, 285, 287
    extrinsic, 74, 82, 90, 236
    intrinsic, 74, 75, 82, 90, 236, 243

Gram matrix, 170, 172, 184, 286
graph, 18, 38, 52–55, 57, 65, 67, 68, 71–73, 76, 86, 88, 91, 94, 97, 116, 173, 174, 180, 184
    undirected, 18, 52, 68
graphics processing unit, 90, 91
great circle, 188, 190, 204
grid
    Cartesian, 53, 56, 57, 71, 195, 241
group, 15, 16, 33, 188, 192, 213, 247
    isometry, 15, 16, 33, 37, 119, 188, 192, 193, 204, 241
    orthogonal, 204
Guttman transform, 146

half-plane, 47
heap, 83, 91
    Fibonacci, 71
Hessian, 96, 102–108, 117, 146, 147, 167, 175, 177
    frozen, 106
holistic, 265
Hopf-Rinow theorem, 30

ICP, *see* iterative closest point algorithm
identification, *see* one-to-many
ill-conditioned, 104
illumination, 265, 266, 273
image
    canonical, 274
    depth, 57
    geometry, 56–58, 195, 196, 269
inequality
    Cauchy-Schwartz, 177
    triangle, 12, 206, 207, 217, 238, 289, 290
initialization, 70, 76, 83, 98, 103, 109, 113, 133, 153–156, 158, 159, 161, 162, 203
injection, 15, 19, 37, 40, 279
interpolation, 154–156, 158, 159, 197, 198, 202–204, 255, 257
invariance
    deformation-, 5–7

330 Subject Index

expression-, 6, 7, 268, 269, 274, 275
isometry-, 138, 178, 192, 194, 205, 206, 240, 242, 244, 269, 274, 275
isometry
   almost-, 16, 17, 206
   arcwise, 20
   rigid, 128, 131, 134, 192, 193, 204
   self-, 15, 245
isospectrality, 178, 180, 184
iterative closest point algorithm, 125, 126, 128–134, 137, 139, 192, 203, 204, 217, 229–231, 239, 267

Jacobian, 23, 75, 243

Klein bottle, 23

Lagrange multiplier, 111, 114, 115, 117, 225, 283, 288
Lagrangian, 111, 112, 114, 115
   augmented, 114–116
Lambert cosine law, 273
Lanczos algorithm, 172
landmark, 121, 154, 264
Laplace operator, *see* Laplacian
Laplace-Beltrami operator, 178, 180–182, 184, 186
Laplacian, 174–178, 180–183, 185
latitude, 22, 23, 188
least action principle, *see* Fermat principle
least squares, 142, 162, 172, 174, 241, 244
   iteratively reweighted, 152
Legendre moment, 124, 135
limit, 13, 14, 19, 39, 62, 125, 141, 190, 229
line search, 99–102, 105, 108, 144, 147, 158, 201
LLE, *see* locally linear embedding
local match error, 242, 244
locally linear embedding, 173, 175, 184

Hessian, 175
longitude, 22, 188

Möbius stripe, 23
majorization, 109, 110, 144, 145
manifold, 9, 20–23, 29, 30, 36, 40, 56–59, 118, 180, 181, 184, 185, 187, 188, 204, 234, 236, 240, 255, 257
   Riemannian, 30, 31, 40, 48, 51, 52, 63, 177, 180, 181
MDS, *see* multidimensional scaling
measure
   area, 215
   fuzzy, 227
medial axis, 59, 60, 63, 134, 135
membership function, 226, 227, 230
mesh overlay, 252
metric
   geodesic, 137
   intrinsic, 19, 25, 30, 31, 38, 65, 67, 68, 72
   length, 18, 19, 25, 30, 53, 141
   restricted, 19
   Riemannian, 29, 30, 40, 177, 178
min-max, 150, 210
minimal geodesic, 30, 48, 53, 67, 88, 118
minimal polynomial extrapolation, 163
minimum
   global, 93, 97, 110, 117
   local, 94, 96, 97, 108–111, 116, 117, 129, 133, 154
morphing, 7, 249–252, 254, 255, 257, 258
   self-intersection free, 257–259
MPE, *see* minimal polynomial extrapolation
multidimensional scaling, 109, 110, 143–149, 151–157, 159, 160, 165–167, 169, 171, 172, 174, 176, 184, 185, 187, 189, 191, 193, 194, 197, 202–204, 207, 213, 217, 241, 258, 274

generalized, 193, 194
multigrid, 158–161, 163, 165

Nash embedding theorem, 30, 31, 40, 141
neighborhood, 13, 20, 21, 52, 56, 129, 155, 156, 173, 175, 202, 254, 255, 283
net, *see* covering
Newton's algorithm, 105–108, 116–118, 131, 146–149, 163
    quasi-, 107, 108, 116
    truncated, 116
Newton-Raphson method, 118
non-rigid world, 2, 3, 6, 7, 9, 11, 67, 277
norm, 12, 101, 104–106, 124, 141, 147–149, 152, 177, 226, 242, 243, 290
    Frobenius, 171, 175, 243, 285
    Hilbert-Schmidt, 243, 259
normal, 23, 26, 27, 39, 61–63, 90, 127, 135, 242, 249, 259, 273, 279, 283

objective, 93, 94, 98, 106, 108–113, 128–131, 134, 142, 146, 150, 171, 222, 224, 228, 234, 283
octree, 132
one-to-many, 264
one-to-one, 264
optical flow, 242
optimality, 46, 54, 89, 94, 96–98, 105, 116, 146, 222, 223, 226, 236, 288
optimization
    alternating, 51, 229
    constrained, 110, 112, 116, 117, 167, 185, 210, 224, 230, 283
    continuous, 191, 195
    convex, 101
    global, 94, 133
    local, 94, 97, 109
    multicriterion, 222, 226, 228, 234, 236
    non-convex, 133, 142
    unconstrained, 93, 98, 108, 113–115, 225
order relation
    total, 224, 240
orientability, 23, 36
outlier, 126, 150

parameterization
    arclength, 240
    canonical, 240
    global, 22, 56, 195, 201, 245
    local, 22, 195, 196, 202
Pareto frontier, 223, 224, 226, 228, 234, 238
Pareto optimum, 222–224, 228, 236, 238, 290
partiality, 222–228, 230, 234
    fuzzy, 227
path, 18–20, 27, 29, 30, 38, 39, 68–71, 73–76, 88, 93, 138, 188, 196, 204
path search, 201
penalty, 111–115, 118, 157, 167, 173
point cloud, 41, 44, 52, 55, 56, 165
polylinear, 201
positive-definiteness, 29, 30, 96, 104, 106, 108, 117
preconditioning, 105, 117
preimage, 14, 39
probe, 263, 264, 269, 270
product
    inner, 23, 29, 94, 95, 177, 242, 279, 286, 288
    outer, 25
projection, 2, 27, 64, 101, 111, 121, 122, 184, 191, 204
prolongation, *see* interpolation

radius
    convexity, 48
    sampling, 43, 156
raster scan algorithm, *see* Danielsson algorithm

Subject Index

reduced rank extrapolation, 162–164, 167
reflectance, 273
reflection, 16, 172, 246, 274
reflexivity, 205, 206
regularity, 22, 23, 38, 53, 71, 74, 83–85, 91, 104, 111, 195, 234, 235, 241, 243, 283
regularization, 236, 243
relaxation, 16, 159–161, 226
remarkable theorem, *see* theorema egregium
residual, 156, 162
restriction, *see* decimation
ROC, *see* receiver operating characteristic
Rock, paper, scissors, 4
rotation, 27, 119–122, 130, 131, 133, 172, 192, 229, 274, 284, 285
RRE, *see* reduced rank extrapolation

safeguard, 100, 105, 161
Salukwadze optimum, 226, 236, 290
sampling
    centroidal Voronoi, 50
    farthest point, 44, 50, 51, 63, 67, 154
    uniform, 44
sellion, 271
separated set, 43–45, 63
shock, 89
shortest path problem, 68–70, 116
similarity
    extrinsic, 61, 129, 137, 165, 229
    full, 219, 220, 223, 224, 228, 229
    intrinsic, 5, 6, 137, 164, 165, 224, 229–232, 245
    partial, 125, 217, 219–226, 229, 230, 232, 233, 236, 238
    rigid, 137, 139, 230
skeleton, *see* medial axis
SMACOF algorithm, 145–149, 152, 153, 158, 160, 161, 163–166, 171, 269

smoothness, 14, 17, 19, 21, 22, 24, 26, 27, 29, 30, 33, 36, 40, 55, 59, 64, 105, 113, 133, 149, 181, 182, 252, 257, 259, 261
source, 69–71, 74, 76, 77, 81, 88, 91, 250–253, 255, 281
space
    embedding, 139–141, 157, 172, 173, 187, 189, 190, 192–195, 202, 209, 273–275
    length, 18–21, 30, 39
    metric, 11–13, 15–18, 25, 30, 37, 38, 43, 44, 63, 137, 139, 140, 187, 193, 208, 213
    quotient, 206
    tangent, 22, 25, 29, 39, 75, 82, 177, 236, 243, 256
    topological, 56
    vector, 12, 93, 97, 283
sparsity, 43, 53, 150, 155, 156, 174, 176
spherical harmonic, 124, 274, 275
splitting, 79, 132
SSE, 86, 91
stereoscopic imaging, 242
Stokes theorem, 177
stopping criterion, 44, 98, 113, 147–149, 160
strain, 171, 172, 184, 185
stress
    generalized, 195, 197, 203
    modified, 157
    relative, 151
    weighted, 151, 152, 231
subgradient, 149
suboptimality, 102–104, 106
successive approximation, 70, 75
super-resolution, 259
surface
    embedded, 21, 25, 26, 29–31, 188
    flat, 32, 34, 35
    hyperbolic, 28
surjection, 16, 20, 207, 289
Swiss roll, 147, 149, 160, 161, 163, 164

symmetry, 12, 15, 16, 23, 29, 33, 38, 52, 59, 96, 108, 122, 126, 129, 146, 152, 172, 181–183, 197, 205–207, 212, 213, 221, 229, 231, 245, 246, 290

Taylor expansion, 94, 100, 105, 127, 130, 281, 284
tessellation, 47, 48, 50, 51, 54, 60, 64, 65
    Delaunay, 54, 59
    Voronoi, 47, 48, 50, 51, 54, 93
texture mapping, 7, 240, 247, 266
texture transfer, 248, 258, 259
theorema egregium, 35–37, 139, 140
three-dimensional scanner, 257, 269
time of arrival, 74, 281
topological ϵ-algorithm, 163
topology, 13, 14, 19–21, 36, 37, 47, 48, 52, 55, 58, 164, 165, 195, 234, 235, 251, 269, 272
trajectory, 27, 252, 253, 255, 257, 259
transitivity, 206
translation, 16, 119–121, 134, 157, 172, 200, 229, 242, 274, 275, 284

triangular mesh, 2, 55, 57–63, 65, 74, 81, 88, 90, 147, 150, 156, 165, 181, 194–197, 212, 283
triangulation, 58, 60, 62, 65, 156, 194, 195, 255
    Delaunay, 59, 65
twins, 261, 272
two-grid algorithm, 157–160

unitary matrix, 122, 163, 192, 285
utopia, 226

V-cycle, 159, 160
vector extrapolation, 160–163, 165, 166
velocity, 27, 30, 74, 252
virtual connection, 83
virtual makeup, 247–249
viscosity solution, 91
visual agnosia, 219
Voronoi edge, 46, 47, 53, 54, 289
Voronoi region, 46, 47, 289
Voronoi vertex, 46, 54

wavefront, 75–77, 79, 83, 89, 90, 197, 203, 204, 281, 282

Author Index

Achermann, B. 267
Acheroy, M. P. 266
Aggarwal, J. K. 268
Alexa, M. 241, 252, 255, 258
Alliez, P. 51
Alt, H. 213
Amari, S. 118
Amenta, N. 60, 63
Anguelov, D. 9
Arnoldi, W. E. 172
Arthur, D. 133
Arya, S. 132
Ashbourn, J. 275
Atallah, M. J. 213
Audibert, J.-Y. 181
Aurenhammer, F. 54, 63

Baglama, J. 172, 185
Bajcsy, R. 220
Bakhvalov, N. S. 158
Barr, A. H. 63, 182
Bartesaghi, A. 91
Basri, R. 220, 236
Basso, C. 243
Belhumeur, P. N. 265
Belkin, M. 174, 181, 185
Bellman, R. E. 89
Benoit, C. 106

Bentley, J. L. 132
Berger, M. 37, 180, 184
Bern, M. 60, 63
Bertillon, A. 266
Bertsekas, D. P. 99, 116
Besl, P. J. 125, 126
Beumier, C. 266
Biderman, I. 220
Bigun, J. 275
Binford, T. O. 220
Blanz, V. 243, 265
Bledsoe, W. W. 264
Blum, H. 59
Boiman, O. 220
Bonnet, P. O. 36
Borg, I. 143, 165, 167, 184, 185, 287
Bossen, F. 56
Bovik, A. C. 268
Bowyer, K. W. 265, 266, 275
Boyd, S. 99, 101, 116, 118, 236
Boykov, Y. 91
Brady, M. 220
Brand, M. 185
Brandt, A. 158
Brezinski, C. 161, 163, 165, 167
Bridger, J. 165

Bridgstock, R. 213
Briggs, W. L. 165
Bronstein, A. M. 56, 83–87, 89–91, 157, 160, 165, 193, 197, 201, 203, 211, 213, 214, 230, 234, 236, 238, 249, 250, 258, 261, 269, 272–275, 280–282
Bronstein, M. M. 56, 83–87, 89–91, 157, 160, 165, 193, 197, 201, 203, 211, 213, 214, 230, 234, 236, 238, 249, 250, 258, 261, 269, 272–275, 280–282
Brooks, R. 180, 220
Broyden, C. G. 108
Bruce, V. 264
Bruckstein, A. M. 63, 203, 236, 238
Bruner, I. S. 264
Bunke, H. 267
Burago, D. 17, 37, 39, 214, 279, 289
Burago, Y. 17, 37, 39, 214, 279, 289
Buser, P. 180

Author Index

Cabay, S. 163
Calvetti, D. 172, 185
Cartoux, J. Y. 266
Cauchy, A. L. 33
Chalmers, M. 153
Chan, T. F 236
Chang, K. 265, 266, 275
Charpiat, G. 133
Chellappa, R. 265, 275
Chen, Y. 125, 127
Chen, Y. K. 257
Cheng, L. T. 89
Cheng, S. W. 63
Chiaki, H. 267
Chiang, F. P. 259
Chladni, E. F. F. 178
Choi, B. T. 257
Chua, C. 267
Chung, F. R. K. 174, 184
Cohen, L. 51, 63, 75
Cohen-Or, D. 255
Cohen-Steiner, D. 51
Cohn-Vossen, S. E. 34
Coifman, R. R. 175
Colbry, D. 267
Conn, A. R. 116
Connell, J. H. 220
Connelly, R. 33
Cook, W. J. 116
Costa, L. 220, 236
Cox, I. 265
Cox, M. A. A. 165, 203
Cox, T. F. 165, 203
Crandal, M. G. 91
Cunningham, W. H. 116

Danielsson, P. 83
Darwin, C. 268
Dattorro, J. 184
Davies, G. M. 264
D'Azevedo, E. F. 65
de Leeuw, J. 144, 146, 165

de Rooij, S. 238
de Saint-Exupéry, A. 1
de Silva, V. 153, 184, 185, 255
Delaunay, B. 54
Dennis, J. E. 108
Desbrun, M. 51, 63, 182
Deuflhard, P. 159
Devir, Y. S. 84–87, 89–91, 280–282
Dey, T. K. 63
Diamond, P. 236
Díaz, F. J. 267
Dickenson, S. 220
Dijkstra, E. W. 70
Dirichlet, G. L. 64
do Carmo, M. P. 37
Donoho, D. L. 175
Droske, M. 241, 244
Du, Q. 50, 63
Duda, R. O. 50
Dupuis, P. 83

Eckart, C. 171
Eddy, R. P. 162
Edgeworth, F. Y. 238
Efrat, A. 133, 135
Ekman, P. 275
Elad, A. 139, 154, 172, 203
Elad, M. 133, 135
Elber, G. 242, 255
Eldar, Y. 44, 63
Ellis, H. D. 264
Erlebacher, G. 267
Etemad, K. 265
Euler, L. 36
Ezra, E. 133, 135

Faber, V. 50, 63
Faloutsos, C. 153
Faugeras, O. 133
Federer, H. 59, 60, 64
Fedorenko, R. P. 158
Fleishman, S. 63, 293

Fletcher, R. 108
Floater, M. S. 253–255, 258
Flynn, P. 265, 266, 275
Ford, W. F. 165, 167
Forsyth, D. A. 152
Fourer, R. 116
Fu, J. 63
Funkhouser, T. 275

Galton, F. 266
Garland, M. 63, 293
Gauduchon, P. 180
Gauss, C. F. 35
Gdalyahu, Y. 238
Geiger, D. 220, 236
Gelfand, N. 129, 133, 203
Gersho, A. 50
Gheorghiades, A. S. 265
Ghosn, J. 265
Gluck, H. 33
Goldfarb, D. 108
Goldstein, E. 255
Golub, G. H. 106, 133, 135, 163, 174, 184
Gomes, J. 258
Gonzalez-Rodriguez, J. 275
Gonzalez, T. F. 44
Gordon, C. 180
Gordon, E. 56, 275
Gordon, G. 267
Gortler, S. 56, 63, 90, 195
Gotsman, C. 184, 253–255, 258
Gould, N. I. M. 116
Govindaraju, N. 91
Gower, J. C. 170, 171
Gray, R. M. 50
Green, G. 17
Grimes, C. 175
Grinspun, E. 181, 183, 184

Groemer, H. 124
Groenen, P. 143, 145, 165, 167, 184, 185, 287
Gromov, M. 208, 214
Gruber, P. M. 51
Gu, X. 56, 63, 195
Gueziec, A. 56
Guibas, L. 129, 133, 203, 213
Gunzburger, M. 50, 63
Gupta, S. 268
Guskov, I. 249
Guttman, L. 146

Hackbusch, W. 158
Hagedoorn, M. 133
Ham, J. 185
Hamer, R. M. 165, 185
Hampel, F. R. 152
Han, F. 267
Han, S. 267
Harris, M. 91
Hart, P. E. 50
Hartley, R. 133
Hausdorff, F. 61
Heckbert, P. S. 63, 293
Hein, M. 181
Heiser, W. 146
Hel-Or, Y. 220
Hesher, C. 267
Hespanha, J. P. 265
Hestenes, M. R. 114
Hildebrandt, K. 182
Ho, Y. K. 267
Hofer, M. 127, 133
Hoffman, D. 217, 220
Hoppe, H. 56, 63, 90, 195
Hormann, K. 258
Horn, B. K. P. 128, 242
Hotelling, H. 134
Householder, A. S. 171
Hu, S. M. 133
Huang, P. S. 259

Huang, Q. X. 133
Huber, P. J. 152

Ikeda, M. 267
Irani, M. 220
Itoh, J. 48
Ivanov, S. 17, 37, 39, 214, 279, 289

Jackson, L. 163
Jacobs, D. 220, 236, 238
Jain, A. K. 267
Jain, V. 241
Jiang, X. 267

Kac, M. 179
Kälberer, F. 181, 183, 184
Kamvysselis, M. 60, 63
Kanade, T. 264
Kao, C. 89
Karni, Z. 184
Kazhdan, M. 275
Kearsley, A. 146, 167
Kelley, C. T. 116, 294
Kelly, M. D. 264
Kepler, J. 213
Keriven, R. 133
Kilian, M. 255, 258
Kimia, B. 217, 220
Kimmel, R. 9, 56, 59, 63, 74, 75, 79, 81, 83–91, 139, 154, 157, 160, 165, 172, 193, 197, 201, 203, 211, 213, 214, 236, 238, 240, 249, 250, 258, 269, 272–275, 280–282
Kirby, M. 264
Kirsanov, D. 90
Kiryati, N. 63, 240
Klein, R. 54, 63, 90, 91, 281
Klir, G. J. 236

Ko, S. J. 257
Koenderink, J. J. 220
Kohonen, T. 165
Kolmogorov, V. 91
Kriegman, D. J. 265
Krishna, G. 161
Krishnaswamy, A. 265
Kruger, J. 91
Kruskal, J. B. 143
Kuhn, H. W. 111
Kung, S. Y. 257
Kunze, R. 63
Kupeev, K. 220

Lafon, S. 175
Lagendijk, R. L. 257
Lakaemper, R. 220, 222, 236, 238
Lambert, J. H. 273
Lanczos, C. 172
Langford, J. 184, 185, 255
Lapreste, J. T. 266
Latecki, L. J. 220, 222, 236, 238
Le Cain, E. 134
Lee, A. B. 175
Lee, D. D. 185
Lee, J. C. 267
Lee, S. H. 257
Lee, Y. 267
Lefohn, A. E. 91
Lehoucq, R. B. 172
Leibon, G. 48, 54, 60, 63
Leifman, G. 33
Leopoldseder, S. 132, 133
Letscher, D. 48, 54, 60, 63
Levin, D. 255
Levoy, M. 132, 133
Lévy, B. 183, 184
Li, H. 133
Li, X. 273

Lin, K.-I. 153
Lindenbaum, M. 44, 63
Linial, N. 140
Lions, P.-L. 91
Litke, N. 241, 244
Lloyd, S. P. 51, 65
Lombroso, C. 267
Losasso, F. 63, 195
Lu, G. 133
Lu, X. 267
Luebke, D. 91

MacQueen, J. 51
Maggioni, M. 175
Malassiotis, S. 267, 268
Markey, M. K. 268
Marr, D. 242
Masuda, I. 266
Mathar, R. 142
Mathur, S. 181, 183, 184
Mavridis, N. 267
Max, J. 51, 65
Mazet, E. 180
McCormick, S. F. 165
McKay, N. D. 125, 126
McKean, H. 180
Mealey, L. 213
Medioni, G. 125, 127, 267
Mehlhorn, K. 213
Mémoli, F. 89, 208, 214
Mešina, M. 162
Meyer, M. 63, 182
Michor, P. W. 258
Miettinen, K. 236
Mika, S. 185
Milanfar, P. 133, 135
Milios, E. 267
Millman, R. S. 37
Mitchell, J. S. B. 90
Mitra, N. J. 129, 133, 203, 213, 255, 258
Möbius, A. F. 55
Mohar, B. 174, 184

Montiel, S. 37
Moon, H. H. 257
Moré, J. J. 108
Moreno, A. B. 267
Morrison, A. 153
Morvan, J. M. 61–63, 65
Mosheyev, L. 106
Mount, D. M. 90, 132
Mpiperis, I. 268
Mumford, D. 236, 258
Munzner, T. 153

Nadler, B. 175
Nagamine, T. 266
Nagel, R. N. 59
Nash, J. 30
Nash, S. 116, 118, 158, 159
Nemirovski, A. 116
Nesterov, Y. 116
Netanyahu, N. S. 132
Newton, I. 118
Niyogi, P. 174, 181, 185
Nocedal, J. 116
Novotni, M. 90, 91, 281

Oliensis, J. 83
O'Neil, B. 37
Onishi, K. 48
Oosterlee, C. 165
Ortega-Garcia, J. 275
Osher, S. 89
Osyczka, A. 236
Owens, J. D. 91

Pan, G. 267
Pan, Y. 267
Pantazis, D. 267
Papadimitriou, C. H. 90
Pareto, V. 223
Park, R. H. 257
Parker, G. D. 37
Pauly, M. 213
Pearson, K. 134

Peinecke, N. 178
Pelletan, C. 263
Pentland, A. 219, 220, 264
Perrault, C. 134
Peyré, G. 63
Phillips, P. J. 275
Platt, J. C. 167
Poggio, T. 242, 243
Polthier, K. 182
Polyak, B. T. 149
Ponce, J. 152
Porat, M. 44, 63
Pottmann, H. 127, 129, 132, 133, 203, 255, 258
Powell, M. J. D. 114
Prokop, R. J. 133
Pulleyblank, W. R. 116
Purcell, T. J. 91

Rajeevan, N. 161
Rajgopal, K. 161
Ramos, E. A. 63
Rappoport, A. 255
Rausch, T. 63
Raviv, D. 203, 213, 214
Ray, T. 63
Reeves, A. P. 133
Reichel, L. 172, 185
Reuter, M. 178
Reynolds, D. 275
Richards, W. 217, 220
Richetin, M. 266
Riemann, B. 29
Rivlin, E. 220
Roland, C. 161
Ronchetti, E. M. 152
Ros, A. 37
Rosenfeld, A. 220, 275
Rosman, G. 160, 165
Ross, G. 153
Rousseeuw, P. J. 152
Roweis, S. T. 173, 175
Rumpf, M. 241, 244

Rusinkiewicz, S. 132, 133, 275
Rustamov, R. M. 178, 183

Sacks, O. 219
Saffrey, H. D. 11
Salukwadze, M. E. 226, 236
Samoilov, T. 255
Sánchez, A. 267
Sander, P. 63, 195
Sapiro, G. 89, 91, 208, 214
Saul, L. K. 173, 175
Schaefer, S. 63, 195
Schedel, H. 221
Scheidegger, C. 63, 293
Schmidt, R. J. 165
Schoenberg, I. J. 170, 171, 185
Schölkopf, B. 185
Schreiner, J. 63, 293
Schrijver, A. 116
Schröder, P. 182, 249, 258
Schüller, A. 165
Schunck, B. G. 242
Schwartz, E. L. 143, 185, 240
Schwarz, H. A. 61, 65
Schweitzer, H. 143
Sethian, J. A. 63, 73, 74, 79, 88, 89
Sezan, M. I. 257
Shah, J. 236
Shaked, D. 63
Shanks, D. 165
Shanno, D. F. 108
Shapira, M. 255
Shapira, Y. 165
Shapiro, B. A. 153
Sharir, M. 133, 135
Shasha, D. 153
Shatz, I. 33

Shaw, A. 143, 185, 240
Shepard, R. N. 143
Shepherd, J. W. 264
Shim, J. 267
Shin, H. 267
Shin, Y. C. 257
Siddiqi, K. 217, 220
Sidi, A. 160, 161, 163, 165, 167
Silva, C. T. 56, 63, 293
Silverman, R. 132
Simpson, R. B. 65
Singer, A. 181
Singer, I. 180
Singh, M. 217
Sirovich, L. 264
Smith, D. A. 165, 167
Snyder, J. 63, 195
Sohn, K. 267
Solina, F. 220
Song, H. 267
Sorensen, D. C. 172
Spira, A. 81, 83, 89, 275
Srivastava, A. 267
Stahel, W. A. 152
Stoop, I. 144
Stork, D. G. 50
Strintzis, M. G. 267, 268
Sun, H. 257
Surazhsky, T. 90, 242
Surazhsky, V. 90, 254, 258
Sweldens, W. 249, 258

Tagiuri, R. 264
Tal, A. 33
Tanaka, H. T. 267
Tangelder, J. W. H. 133
Tapia, R. 146, 167
Taubin, G. 56, 258
Teague, M. R. 124, 135
Tenenbaum, J. B. 153, 184, 185, 255
Thibert, B. 61–63, 65

Thiessen, A. H. 64
Toint, P. 116
Toint, Ph. L. 116
Torgerson, W. S. 143, 171
Townsend, G. C. 213
Trosset, M. 142, 146, 167
Trottenberg, U. 165
Tsai, Y. 89
Tsai, Y. R. 89
Tsalakanidou, F. 267
Tse, R. 180
Tsitsiklis, J. N. 73, 91
Tucker, A. W. 111
Turk, M. 264
Tversky, A. 238

Uemura, T. 266

van de Velden, M. 145
van Doorn, A. J. 220
van Kaick, O. 241
van Loan, C. F. 106, 163, 174, 184
Vandenberghe, L. 99, 101, 116, 118, 236
Varadhan, R. 161
Vassilvitskii, S. 133
Veltkamp, R. C. 133
Vélez, J. F. 267
Vese, L. A. 236
Vetro, A. 257
Vetter, T. 243, 265
Vidimče, K. 249
Vignéras, M. F. 180
Vitanyi, P. 238
Vladimirsky, A. 89
Volz, R. A. 213
von Luxburg, U. 181
Voronoi, G. F. 46, 64

Wagener, H. 213
Wang, J. T.-L. 153
Wang, X. 153

Wang, Y. 267
Wardetzky, M. 181–184
Warner, F. 175
Warren, J. 63, 195
Waupotitsch, R. 267
Webb, D. L. 180
Weber, O. 84–87, 89–91, 280–282
Weinshall, D. 238
Weiser, M. 159
Welzl, E. 213
Werman, M. 220
Wesseling, P. 165
Weyl, H. 180, 212
Wheatstone, C. 242
Williams, C. K. I. 184
Williams, M. 153
Wiskott, L. 264
Wolfson, E. 143, 185, 240
Wolfson, H. 220
Wolpert, S. 180
Wolter, D. 220, 222, 236, 238
Wolter, F.-E. 178
Wolter, J. D. 213
Woo, T. C. 213
Wood, Z. 63, 195
Wright, S. 116
Wu, A. Y. 132
Wu, Z. 267

Xu, G. 182–184

Yang, C. 172
Yang, S. 257
Yang, U. 267
Yang, Y. L. 133
Yatziv, L. 91
Yavneh, I. 157, 165
Yianilos, P. 265
Young, F. W. 165, 185
Young, G. 171
Yuan, B. 236

Zadeh, L. A. 226
Zeevi, Y. Y. 44, 63
Zha, H. 175
Zhang, C. P. 259
Zhang, D. 133
Zhang, H. 241, 273
Zhang, K. 153
Zhang, Z. 175
Zhao, H. 84, 89, 132
Zhao, W. 265, 275
Zibulevsky, M. 106, 118
Zigelman, G. 240
Zimmermann, H. J. 236
Zucker, S. W. 175

Printed in the United States of America